Introduction to Materials Chemistry

Introduction to Materials Chemistry

Sean Fraser

Larsen & Keller
www.larsen-keller.com

Introduction to Materials Chemistry
Sean Fraser
ISBN: 978-1-64172-489-0 (Hardback)

▤ Larsen & Keller

Published by Larsen and Keller Education,
5 Penn Plaza,
19th Floor,
New York, NY 10001, USA

Cataloging-in-Publication Data

Introduction to materials chemistry / Sean Fraser.
       p. cm.
Includes bibliographical references and index.
ISBN 978-1-64172-489-0
1. Chemical engineering. 2. Materials. 3. Chemistry, Technical. I. Fraser, Sean.
TP155 .I58 2020
660--dc23

For more information regarding Larsen and Keller Education and its products, please visit the publisher's website www.larsen-keller.com

Table of Contents

Preface

The use of chemistry to design and synthesize materials with potentially useful physical attributes is known as materials chemistry. It is a subfield of materials science. It integrates concepts from physics and engineering. This field seeks to develop materials based on magnetic, structural, optical or catalytic properties. The processing, characterization and molecular-level understanding of these substances is also studied under this discipline. There are many applications of materials chemistry such as producing engineering ceramics, composite materials, polymers and metal alloys. This book provides comprehensive insights into the field of materials chemistry. It traces the progress of this field and highlights some of its key concepts. The extensive content of this book provides the readers with a thorough understanding of the subject.

To facilitate a deeper understanding of the contents of this book a short introduction of every chapter is written below:

Chapter 1- The branch of chemistry that deals with the design and synthesis of materials using their magnetic, optical, structural and catalytic properties is referred to as materials chemistry. Organic materials, inorganic materials, biological materials, polymers, etc. are some of the types of materials studies within it. All these types and related aspects of materials chemistry have been carefully analyzed in this chapter.

Chapter 2- Any material whose electrical conductivity value falls between that of an insulator and a conductor is known as a semiconductor. They are mainly categorized into intrinsic and extrinsic semiconductors. Silicon, germanium, gallium arsenide, cadmium selenide, etc. are a few of the examples. This chapter has been carefully written to provide an easy understanding of semiconductors.

Chapter 3- Polymer chemistry studies the synthesis, chemical and physical properties, and structure of polymers and macromolecules. Polymerization mechanisms include addition polymerization, step growth polymerization and catalysis. The topics elaborated in this chapter will help in gaining a better perspective about the concepts and mechanisms associated with polymer chemistry.

Chapter 4- Materials that have a unit size between 1 to 1000 nanometers are referred to as nanomaterials. Nanochemistry combines the principles of chemistry and nanoscience to study the synthesis of size, shape, surface and defect properties. This chapter closely examines nanomaterials and nanochemistry to provide an extensive understanding of the subject.

Chapter 5- Metals are substances which are good conductors of electricity and heat. They possess properties such as malleability, ductility, and lustre. This chapter delves into the aspects of thermal conductivity of metals, their physical and chemical properties, uses of metals and non-metals, and their advantages and disadvantages to provide in-depth knowledge of the subject.

I owe the completion of this book to the never-ending support of my family, who supported me throughout the project.

Sean Fraser

Materials Chemistry: An Introduction

The branch of chemistry that deals with the design and synthesis of materials using their magnetic, optical, structural and catalytic properties is referred to as materials chemistry. Organic materials, inorganic materials, biological materials, polymers, etc. are some of the types of materials studies within it. All these types and related aspects of materials chemistry have been carefully analyzed in this chapter.

"Materials Chemistry" can be defined as the branch of chemistry aimed at the preparation, characterization, and understanding of substances/systems that have some specific useful function (or *potentially* useful function). It involves 4 primary components: preparation /synthesis ("How are materials made?"), structure ("How are they put together"), characterization ("How do they behave?") and applications ("What are they good for?"). It integrates elements from all four classical areas of chemistry, but puts an intellectual focus on the fundamental scientific issues that are unique to "materials".

Materials Chemistry largely involves the study of chemistry of condensed phases (solids, liquids, polymers) and interfaces between different phases. Because many of these materials have direct technological applications, materials chemistry has a strong link between basic science and many existing and newly-emerging technologies. While chemistry-focused, the Materials Chemistry Program also serves as a bridge between chemistry and the engineering and life sciences.

Here are just a few examples of some of the many materials chemistry projects that are ongoing:

- Rational chemical synthesis of nanoscale and nanostructured materials.
- Development of new surface-based methods for detecting and manipulating biological molecules (bio-chips, bio-electronics).
- Intermetallic Compounds.
- New organic or inorganic materials for electronics, photonics, and spintronics.
- Surface/interface chemistry of microelectronic materials (silicon, diamond, gold).
- Organic-inorganic interfaces.
- Nano-Calorimetry to probe properties of extremely stable glasses.
- Theoretical studies of soft condense matter.
- Single-Molecule Techniques to study electronic properties of conjugated polymers.
- Supercooled liquids/diffusion in thin films.
- Dynamics in actively-deformed polymer glasses.
- Ion transport membranes.

- Synthesis and characterization of new polymeric materials.
- Semiconductor-based chemical sensors.
- Synthesis and characterization of organic non-linear optical materials.
- Coherent multidimensional spectroscopy and its application in investigating materials.

Different Types of Materials

Inorganic Materials

These materials include metals, clays, sand rocks, gravels, minerals and ceramics and have mineral origin.

These materials are formed due to natural growth and development of living organisms and are not biological materials.

Rocks are the units which form the crust of the earth.

The three major groups of rocks are:

- Igneous Rocks: These rocks are formed by the consolidation of semi-liquid of liquid material (magma).

  These are called as Plutonic if their consolidation takes place deep within the earth and volcanic if lava or magma solidifies on the earth's surface.

  Basalt is igneous volcanic where as granite is igneous plutonic.

- Sedimentary Rocks:

  When broken down remains of existing rocks are consolidated under pressure, then the rocks are named as sedimentary rocks, e.g., shale and sandstone rocks.

  The required pressure for the formation of sedimentary rocks is supplied by the overlying rocky material.

- Metamorphic Rocks:

  These rocks are basically sedimentary rocks which are changed into new rocks by intense heat and pressure, e.g., marble and slates. The structure of these rocks is in between igneous rocks and sedimentary rocks.

Rock materials are widely used for the construction of buildings, houses, bridges, monuments, arches, tombs, etc. The slate, which has got great hardness is still used as roofing material. Basalt, dolerite and rhyolite are crushed into stones and used as concrete aggregate and road construction material.

Another type of materials, i.e. Pozzolanics, are of particular interest to engineers because they are naturally occurring or synthetic silicious materials which hydrate to form cement. Volcanic ash, blast furnace slag, some shales and fly ash are examples of pozzolanic materials. When the cement contains 10- 20% ground blast furnace slag, then it is called pozzolans-portland cement, which sets more slowly than ordinary portland cement and has greater resistance to sulphate solutions and sea water.

Rocks, stone, wood, copper, silver, gold etc. are the naturally occurring materials exist in nature in the form in which they are to be used. However, naturally occurring materials are not many in number. Nowadays, most of the materials are manufactured as per requirements. Obviously, the study of engineering materials is also related with the manufacturing process by which the materials are produced to acquire the properties as per requirement.

Copper, silver, gold, etc. metals, which occur in nature, in their free state are mostly chemically inert and highly malleable and ductile as well as extremely corrosion resistant. Alloys of these metals are harder than the basic metals. Carbonates, sulphates and sulphide ores are more reactive metals.

Biological Materials

Leather, limestone, bone, horn, wax, wood etc. are biological materials. Wood is fibrous composition of hydrocarbon, cellulose and lignin and is used for many purposes.

Apart from these components a small amount of gum, starch, resins, wax and organic acids are also present in wood. One can classify wood as soft wood and hard wood. Fresh wood contains high percentage of water and to dry out it, seasoning is done. If proper seasoning is not done, defects such as cracks, twist, wrap etc. may occur. Leather is obtained from the skin of animals after cleaning and tanning operations.

Nowadays, it is used for making belts, boxes, shoes, purses etc. To preserve the leather, tanning is used.

Following two tanning techniques are widely used:

- Vegetable Tanning: It consists of soaking the skin in tanning liquor for several days and then dried to optimum conditions of leather.

- Chrome Tanning: This technique involves pickling the skin in acid solution and then revolving in a drum which contains chromium salt solution. After that the leather is dried and rolled. Limestone is an important material which is not organic but has biological origin. It mainly consist of calcium carbonate and limestone. It is widely used to manufacture cement. In Iron and Steel Industries, limestone in pure form is used as flux. In early days bones of animals were used to make tools and weapons. Nowadays bones are used for the manufacture of glue, gelatin etc. Bones are laminate of organic substances and phosphates and carbonates of calcium. These are stronger in compression as compared to tension.

Organic Materials

Organic materials are carbon compounds and their derivatives. They are solids composed of long molecular chains. The study of organic compounds is very important because all biological systems are composed of carbon compounds. There are also some materials of biological origin which do not possess organic composition, e.g., limestone.

These materials are carbon compounds in which carbon is chemically bonded with hydrogen, oxygen and other non-metallic substances. The structure of these compounds is complex. Common organic materials are plastics and synthetic rubbers which are termed as organic polymers.

Other Examples of Organic Materials

They are wood, many types of waxes and petroleum derivatives.

Organic polymers are prepared by polymerisation reactions, in which simple molecules are chemically combined into long chain molecules or three-dimensional structures. Organic polymers are solids composed of long molecular chains. These materials have low specific gravity and good strength.

The two important classes of organic polymers are:

- Thermoplastics: On heating, these materials become soft and hardened again upon cooling, e.g., nylon, polythene, etc.

- Thermosetting plastics: These materials cannot be resoftened after polymerisation, e.g., urea-formaldehyde, phenol formaldehyde, etc. Due to cross-linking, these materials are hard, tough, non-swelling and brittle. These materials are ideal for moulding and casting into components. They have good corrosion resistance.

The excellent resistance to corrosion, ease of fabrication into desired shape and size, fine lusture, light weight, strength, rigidity have established the polymeric materials and these materials are fast replacing many metallic components.

PVC (Polyvinyl Chloride) and polycarbonate polymers are widely used for glazing, roofing and cladding of buildings.

Plastics are also used for reducing weight of mobile objects, e.g., cars, aircrafts and rockets. Polypropylenes and polyethylene are used in pipes and manufacturing of tanks.

Thermo-plastic films are widely used as lining to avoid seepage of water in canals and lagoons. To protect metal structure from corrosion, plastics are used as surface coatings.

Plastics are also used as main ingredients of adhesives. The lower hardness of plastic materials compared with other materials makes them subjective to attack by insects and rodents.

Because of the presence of carbon, plastics are combustible. The maximum service temperature is of the order of 100°C. These materials are used as thermal insulators because of lower thermal conductivity. Plastic materials have low modulus of rigidity, which can be improved by addition of filters, e.g., glass fibres. Natural rubber, which is an organic material of biological origin, is a thermoplastic material. It is prepared from a fluid, provided by the rubber trees. Rubber materials are widely used for tyres of automobiles, insulation of metal components, toys and other rubber products.

Polymers

Polymers include the familiar plastic and rubber materials. Many of them are organic compounds that are chemically based on carbon, hydrogen, and other non-metallic elements (viz. O, N, and Si).

Furthermore, they have very large molecular structures, often chain-like in nature that have a backbone of carbon atoms.

Some of the common and familiar polymers are polyethylene (PE), nylon, poly(vinyl chloride) (PVC), polycarbonate (PC), polystyrene (PS), and silicone rubber.

These materials typically have low densities, whereas their mechanical characteristics are generally dissimilar to the metallic and ceramic materials—they are not as stiff nor as strong as these other material types.

However, on the basis of their low densities, many times their stiffness and strengths on a per mass basis are comparable to the metals and ceramics. In addition, many of the polymers are extremely ductile and pliable (i.e., plastic), which means they are easily formed into complex shapes. In general, they are relatively inert chemically and unreactive in a large number of environments.

One major drawback to the polymers is their tendency to soften and/or decompose at modest temperatures, which, in some instances, limits their use.

Furthermore, they have low electrical conductivities and are nonmagnetic. Fiberglass is sometimes also termed a "glass fiber-reinforced polymer" composite, abbreviated "GFRP". Many of these are organic substances and derivatives of carbon and hydrogen. Polymers include the familiar plastic and rubber materials.

Usually polymers are classified into three categories: thermoplastic polymers, thermosetting polymers and elastomers, better called as rubbers.

Polymers have very large molecular structures. Most plastic polymers are light in weight and are soft in comparison to metals.

Polymer materials have typically low densities and may be extremely flexible and widely used as insulators, both thermal and electrical.

Typical examples of polymers are polyesters, phenolics, polyethylene, nylon and rubber.

The overriding consideration of the selection of a given polymer is whether or not the material can be processed into the required article easily and economically. Crude oil supplies the majority of the raw material for the production of polymers, also called plastics.

Polymers can be divided into 3 categories:

- Thermoplastics: Usually soft and easy to be recycled.

- Thermosetting plastics: Usually stiff and not easy to be recycled.

- Elastomers: Flexible (rubbers).

Ceramic Materials

The word ceramic is derived from the Greek word keramikos. The term covers inorganic non -metallic materials whose formation is due to the action of heat Clays, bricks, cements, glass are the most important ones. These are crystalline compounds between metallic and non-metallic elements. They are most frequently oxides, nitrides and carbides.

Ceramics are compounds between metallic and nonmetallic elements; they are most frequently oxides, nitrides, and carbides.

For example, some of the common ceramic materials include aluminum oxide (or alumina, Al_2O_3), silicon dioxide (or silica, SiO_2), silicon carbide (SiC)& silicon nitride (Si_3N_4).

In addition, some are referred to as the traditional ceramics—those composed of clay minerals (i.e., porcelain), as well as cement, and glass.

With regard to mechanical behavior, ceramic materials are relatively stiff and strong— stiffness and strengths are comparable to those of the metals. In addition, ceramics are typically very hard.

On the other hand, they are extremely brittle (lack ductility), and are highly susceptible to fracture.

These materials are typically insulative to the passage of heat and electricity (i.e., have low electrical conductivities, and are more resistant to high temperatures and harsh environments than metals and polymers.

With regard to optical characteristics, ceramics may be transparent, translucent, or opaque, and some of the oxide ceramics (e.g., Fe O) exhibit magnetic behavior.

Nowadays graphite is also categorized in ceramics. The wide range of materials which falls within this classification include ceramics that are composed of clay minerals, cement and glass.

Glass is grouped with this class because it has similar properties but most glasses are amorphous.

Ceramics are characterised by high hardness, abrasion resistance, brittleness and chemical inertness. Ceramics are typically insulative to the passage of electricity and heat, and are more resistant to high temperatures and harsh environments than metals and polymers.

With regard to mechanical behaviour, these materials are hard but very brittle. These materials are widely categorized into oxide and non-oxide ceramics. Examples: Ceramics Glasses, Glass ceramics, Graphite, Diamond.

Composites

A composite is a composition of two or more materials in the first three categories, e.g. metals, ceramics and polymers, that has properties from its constituents. Large number of composite materials have been engineered. Few typical examples of composite materials are wood, clad metals, fibre glass, reinforced plastics, cemented carbides, etc.

Fibre glass is a most familiar composite material, in which glass fibres are embedded within a polymeric material.

A composite is designed to display a combination of the best characteristics of each of the component materials. Fibre glass acquires strength from the glass and the flexibility from the polymer.

A composite is composed of two (or more) individual materials, which come from the categories discussed above—viz., metals, ceramics, and polymers. The design goal of a composite is to achieve a combination of properties that is not displayed by any single material, and also to incorporate the best characteristics of each of the component materials.

A large number of composite types exist that are represented by different combinations of metals, ceramics, and polymers. Furthermore, some naturally-occurring materials are also considered to be composites—for example, wood and bone. However, most of those we consider in our discussions are synthetic (or man-made) composites.

One of the most common and familiar composites is fiber-glass, in which small glass fibers are embedded within a polymeric material (normally an epoxy or polyester).

The glass fibers are relatively strong and stiff (but also brittle), whereas the polymer is ductile (but also weak and flexible). Thus, the resulting fiberglass is relatively stiff, strong, flexible, and ductile.

In addition, it has a low density. Another of these technologically important materials is the "carbon fiber-reinforced polymer" (or "CFRP") composite—carbon fibers that are embedded within a polymer.

These materials are stiffer and stronger than the glass fiber-reinforced materials, yet they are more expensive.

The CFRP composites are used in some aircraft and aerospace applications, as well as hightech sporting equipment (e.g., bicycles, golf clubs, tennis rackets, and skis/snowboards).

A true composite structure should show matrix material completely surrounding its reinforcing material in which the two phases act together to exhibit desired characteristics. These materials as a class of engineering material provide almost an unlimited potential for higher strength, stiffness, and corrosion resistance over the 'pure' material systems of metals, ceramics and polymers.

Many of the recent developments of materials have involved composite materials. Probably, the composites will be the steels of this century.

Nowadays, the rapidly expanding field of nano composites is generating many exciting new materials with novel properties. The general class of nano composite organic or inorganic material is a fast growing field of research.

Significant efforts are going on to obtain control of nano composite materials depend not only on the properties of their individual parents but also on their morphology and interfacial characteristics.

The lamellar class of intercalated organic/ inorganic nano composites and namely those systems that exhibit electronic properties in at least one of the composites offers the possibility of obtaining well ordered systems some of which may lead to unusual electrical and mechanical properties.

Polymer-based nano composites are also being developed for electronic applications such as thin-film capacitors in integrated circuits and solid polymer electrolytes for batteries. No doubt, the field of nano composites is of broad scientific interest with extremely impressive technological promise.

Other materials coming under this group are: Reinforced plastics; Metal-matrix composites; Ceramic-matrix composites; Sandwich structures; and Concrete.

Advanced Materials

Materials that are utilized in high-technology (or high-tech) applications are sometimes termed advanced materials. By high technology we mean a device or product that operates or functions using relatively intricate and sophisticated principles; examples include electronic equipment (camcorders, CD/DVD players, etc.), computers, fiber-optic systems, spacecraft, aircraft, and military rocketry.

These advanced materials are typically traditional materials whose properties have been enhanced, and, also newly developed, high-performance materials.

Furthermore, they may be of all material types (e.g., metals, ceramics, polymers), and are normally expensive.

Advanced materials include semiconductors, biomaterials, and what we may term "materials of the future" (that is, smart materials and nano engineered materials), which we discuss below. The properties and applications of a number of these advanced materials -for example, materials that are used for lasers, integrated circuits, magnetic information storage, liquid crystal displays (LCDs), and fiber optics.

These are new engineering materials which exhibit high strength, great hardness, and superior thermal, electrical, optical and chemical properties. Advanced materials have dramatically altered communication technologies, reshaped data analysis, restructured medical devices, advanced space travel and transformed industrial production process.

These materials are often synthesized from the biproducts of conventional commodity materials and often possess following characteristics:

- These materials are created for specific purposes,
- These materials are highly processed and possess a high value-to weight ratio,
- These materials are developed and replaced with high frequency,
- These materials are frequently combined into new composites.

Nowadays, there is considerable interest in making advanced materials that are usually graded by chemical composition, density or coefficient of thermal expansion of material or based on micro-structural features, e.g. a particular arrangement of second-phase particles or fibres in a matrix.

Such materials are referred as functionally graded materials.

Instead of having a step function, one may strive to achieve a gradual change. Such gradual change will reduce the chances of mechanical and thermal stresses, generally present otherwise. We may note that the concept of a functionally graded material is applicable to any material metal, polymer or ceramic.

Semiconductors

These materials have electrical properties that are intermediate between electrical conductors and insulators. Moreover, the electrical characteristics of semiconducting materials are extremely sensitive to the presence of minute concentrations of impurity atoms; these concentrations may be controlled over very small spatial regions. Silicon, Germanium and some more compounds form the vast majority of semiconducting crystals.

Semiconductors have electrical properties that are intermediate between the electrical conductors (viz. metals and metal alloys) and insulators (viz. ceramics and polymers).

Furthermore, the electrical characteristics of these materials are extremely sensitive to the presence of minute concentrations of impurity atoms, for which the concentrations may be controlled over very small spatial regions.

Semiconductors have made possible the advent of integrated circuitry that has totally revolutionized the electronics and computer industries (not to mention our lives) over the past three decades. These semiconducting materials are used in a number of solid state devices, e.g. diodes, transistors, photoelectric devices, solar batteries, radiation detectors, thermistors and lasers. The semiconductors have made possible the advent of integrated circuitary that has completely revolutionized the electronics and computer industries.

Biomaterials

Biomaterials are employed in components implanted into the human body for replacement of diseased or damaged body parts. These materials must not produce toxic substances and must be compatible with body tissues (i.e., must not cause adverse biological reactions).

All of the above materials—metals, ceramics, polymers, composites, and semiconductors— may be used as biomaterials.

Some biomaterials that are utilized in artificial hip replacements.

Materials of the Future - Smart Materials

Smart (or intelligent) materials are a group of new and state-of-the-art materials now being developed that will have a significant influence on many of our technologies.

The adjective "smart" implies that these materials are able to sense changes in their environments and then respond to these changes in predetermined manners traits that are also found in living organisms.

In addition, this "smart" concept is being extended to rather sophisticated systems that consist of both smart and traditional materials.

Actuators may be called upon to change shape, position, natural frequency, or mechanical characteristics in response to changes in temperature, electric fields, and/or magnetic fields.

Four types of materials are commonly used for actuators: shape memory alloys, piezoelectric ceramics, magnetostrictive materials, and electrorheological/ magnetorheological fluids.

Shape memory alloys are metals that, after having been deformed, revert back to their original shapes when temperature is changed.

Piezoelectric ceramics expand and contract in response to an applied electric field (or voltage); conversely, they also generate an electric field when their dimensions are altered.

The behavior of magnetostrictive materials is analogous to that of the piezoelectrics, except that they are responsive to magnetic fields. Also, electrorheological and magnetorheological fluids are liquids that experience dramatic changes in viscosity upon the application of electric and magnetic fields, respectively.

Materials/devices employed as sensors include optical fibers, piezoelectric materials (including some polymers), and microelectromechanical devices.

For example, one type of smart system is used in helicopters to reduce aerodynamic cockpit noise that is created by the rotating rotor blades.

Piezoelectric sensors inserted into the blades monitor blade stresses and deformations; feedback signals from these sensors are fed into a computer-controlled adaptive device, which generates noise-canceling antinoise. Smart or intelligent materials form a group of new and state of art materials now being developed that will have a significant influence on many of present-day technologies.

The adjective 'smart' implies that these materials are able to sense changes in their environments and then respond to these changes in predetermined manners—traits that are also found in living organisms. In addition, the concept of smart materials is being extended to rather sophisticated systems that consist of both smart and traditional materials.

The field of smart materials attempts to combine the sensor (that detects an input signal), actuator (that performs a responsive and adaptive function) and the control circuit or as one integrated unit.

Acutators may be called upon to change shape, position, natural frequency, or mechanical characteristics in response to changes in temperature, electric fields, and or magnetic fields.

Usually, four types of materials are commonly used for actuators: Shape memory alloys, piezoelectric ceramics, magnetostrictive materials, and electrorheological/magnetorheological fluids. Shape memory alloys are metals that, after having been deformed, revert back to their original shapes when temperature is changed.

Piezoelectric ceramics expand and contract in response to an applied electric field (or voltage); conversely these materials also generate an electric field when their dimensions are altered. The behaviour of magnetostrictive materials is analogous to that of the piezoelectric ceramic materials, except that they are responsive to magnetic fields.

Also, electrorheological and magnetorheological fluids are liquids that experience dramatic changes in viscocity upon application of electric and magnetic fields, respectively. The combined system of sensor, actuator and control circuit or as one IC unit, emulates a biological system. These are known as smart sensors, Micro System Technology (MST) or Micro Electro Mechanical Systems

(MEMS). Materials/devices employed as sensors include optical fibres, piezoelectric materials (including some polymers) and MEMS.

For example, one type of smart system is used in helicopters to reduce aero-dynamic cockpit noise that is created by the rotating rotor blades.

Piezoelectric sensors inserted into the blades, monitor blade stresses and deformations; feedback signals from these sensors are fed into a computer controlled adaptive device, which generates noise cancelling antidose.

MEMS devices are small in size, light weight, low cost, reliable with large batch fabrication technology. They generally consist of sensors that gather environmental information such as pressure, temperature, acceleration, etc., integrated electronics to process the data collected and actuators to influence and control the environment in the desired manner.

The MEMS technology involves a large number of materials. Silicon forms the backbone of these systems also due to its excellent mechanical properties as well as mature micro-fabrication technology including lithography, etching, and bonding. Other materials having piezoelectric, piezoresistive, ferroelectric and other properties are widely used for sensing and actuating functions in conjunction with silicon.

Nano Materials

Nano-structured (NS) materials are defined as solids having microstructural features in the range of 1–100 nm (= (1–100) × 10–9 m) in at least in one dimension. These materials have outstanding mechanical and physical properties due to their extremely fine grain size and high grain boundary volume fraction.

Usually, the clusters of atoms consisting of typically hundreds to thousands on the nanometer scale are called as nanoclusters. These small group of atoms, in general, go by different names such as nano particles, nanocrystals, quantum dots and quantum boxes. Significant work in being carried out in the domain of nano-structured materials and nano tubes since they were found to have potential for high technology engineering applications.

Nano-structured materials exhibit properties which are quite different from their bulk properties. These materials contain a controlled morphology with atleast one nano scale dimension. Nano crystals, nano wires and nano tubes of a large number of inorganic materials have been synthesized and characterized in the last few years.

Some of the nano materials exhibit properties of potential technological value. This is particularly true for nano-structures of semiconducting materials such as metal chalcogenides and nitrides. The mixing of nano-particles with polymers to form composite materials has been practiced for decades.

For example, the clay reinforced resin known as Bakelite is the first mass-produced polymer-nanoparticle composites and fundamentally transformed the nature of practical household materials. Even before bakelite, nano composites were finding applications in the form of nano particle-toughened automobile tires prepared by blending carbon black, zinc oxide, and/or magnesium sulfate particles with vulcanized rubber.

Despite these early successes, the broad scientific community was not galvanized by nano composites until the early 1990s, when reports revealed that adding mica to nylon produced a five-fold increase in the yield and tensile strength of the material. Subsequent developments have further contributed to the surging interest in polymer nano particle composites.

Nanoengineered Materials

Until very recent times the general procedure utilized by scientists to understand the chemistry and physics of materials has been to begin by studying large and complex structures, and then to investigate the fundamental building blocks of these structures that are smaller and simpler. This approach is sometimes termed "top-down" science.

However, with the advent of scanning probe microscopes, which permit observation of individual atoms and molecules, it has become possible to manipulate and move atoms and molecules to form new structures and, thus, design new materials that are built from simple atomic-level constituents (i.e., "materials by design").

This ability to carefully arrange atoms provides opportunities to develop mechanical, electrical, magnetic, and other properties that are not otherwise possible.

We call this the "bottom-up" approach, and the study of the properties of these materials is termed "nanotechnology"; the "nano" prefix denotes that the dimensions of these structural entities are on the order of a nanometer (10^{-9} m)—as a rule, less than 100 nanometers (equivalent to approximately 500 atom diameters).

Materials produced out of nanoparticles have some special features:

- Very high ductility.

- Very high hardness ~4 to 5 times more than usual conventional materials.

- Transparent ceramics achievable.

- Manipulation of colour.

- Extremely high coercivity magnets.

- Developing conducting inks and polymers.

Material science has expanded from the traditional metallurgy and ceramics into new areas such as electronic polymers, complex fluids, intelligent materials, organic composites, structural composites, biomedical materials (for implants and other applications), biomimetics, artificial tissues, biocompatible materials, "auxetic" materials (which grow fatter when stretched), elastomers, dielectric ceramics (which yield thinner dielectric layers for more compact electronics), ferroelectric films (for non-volatile memories), more efficient photovoltaic converters, ceramic superconductors, improved battery technologies, self-assembling materials, fuel cell materials, optoelectronics, artificial diamonds, improved sensors (based on metal oxides, or conducting polymers), grated light values, ceramic coatings in air (by plasma deposition), electrostrictive polymers, chemical—mechanical polishing, alkali-metal thermoelectric converters, luminescent silicon, planar optical displays without phosphors, MEMS, and super molecular materials.

However, there still remain technological challenges, including the development of even more sophisticated and specialized materials, as well as consideration of the environmental impact of materials production.

Material scientists are interested in green approaches, by entering the field of environmental—biological science, by developing environmentally friendly processing techniques and by inventing more recyclable materials.

The following table shows the properties of materials to be considered for different applications:

Manufacturing processes	Functional requirements	Cost considerations	Operating parameters
Plasticity	Strength	Raw material	Pressure
Malleability	Hardness	Processing	Temperature
Ductility	Rigidity	Storage	Flow
Machinability	Toughness	Manpower	Type of material
Casting properties	Thermal conductivity	Special treatment	Corrosion requirements
Weldability	Fatigue	Inspection	Environment
Heat	Electrical treatment	Packaging properties	Protection from fire
Tooling	Creep	Inventory	Weathering
Surface finish	Aesthetic look	Taxes and custom duty	Biological effects

Materials Science

The combination of physics, chemistry, and the focus on the relationship between the properties of a material and its microstructure is the domain of Materials Science.

Materials Characterization

Information needed about the structure and composition (surface and bulk) of materials: microscopic and nanoscopic domains.

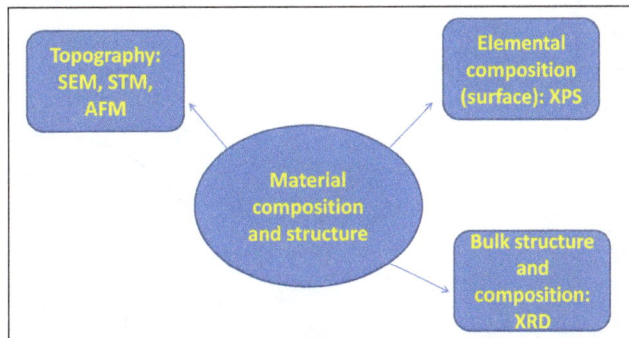

Scanning Electron Microscopy

The scanning electron microscope (SEM) is a type of electron microscope that images the sample surface by scanning it with a high-energy beam of electrons in a raster scan pattern. The electrons interact with the atoms that make up the sample producing signals that contain information about the sample's surface topography, composition and other properties such as electrical conductivity.

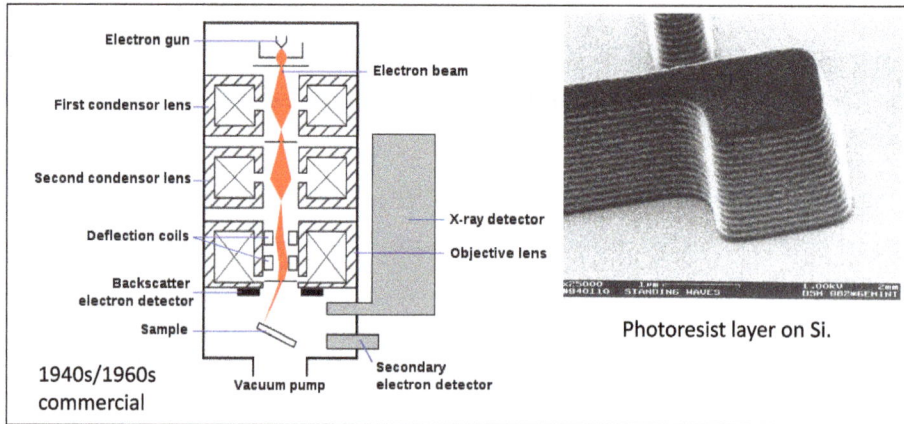

Photoresist layer on Si.

1940s/1960s commercial

Scanning Electron Microscopy

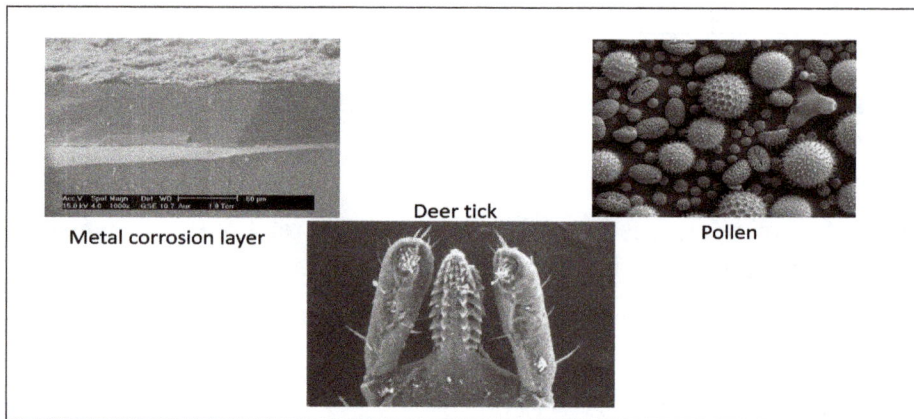

Metal corrosion layer

Deer tick

Pollen

Three different imaging modes: secondary electrons (low energy electrons from the surface of the material), backscattered electrons (more from bulk) and x-rays (from near-surface region).

SEM: X-ray Analysis

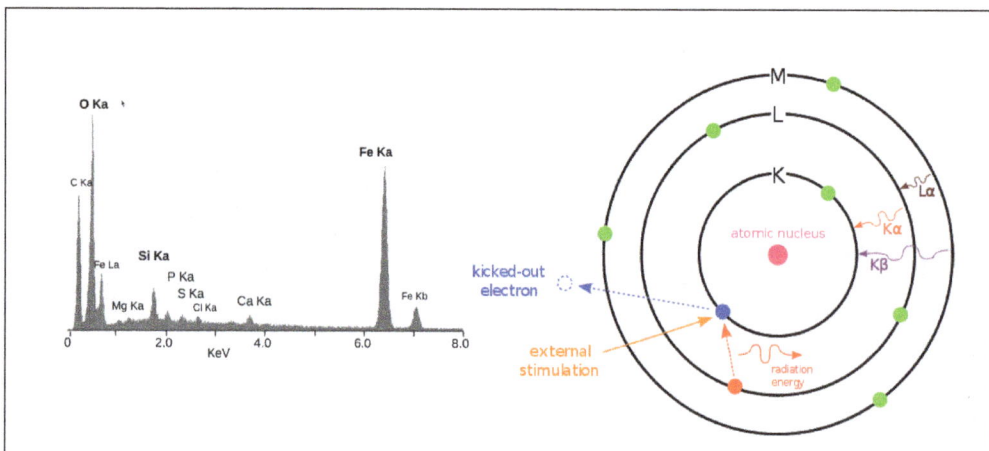

X-ray analysis provides both qualitative (elemental analysis) and quantitative (atomic percent) information about a solid sample. The information content relates to the surface and near-surface region.

Scanning Tunneling Microscopy

A scanning tunneling microscope (STM) is a powerful instrument for imaging surfaces at the atomic level. Its development in 1981 earned its inventors, Gerd Binnig and Heinrich Rohrer (at IBM Zürich), the Nobel Prize in Physics in 1986. For an STM, good resolution is considered to be 0.1 nm lateral resolution and 0.01 nm depth resolution. With this resolution, individual atoms within materials are routinely imaged and manipulated. The STM can be used not only in ultra high vacuum but also in air, water, and various other liquid or gas ambients.

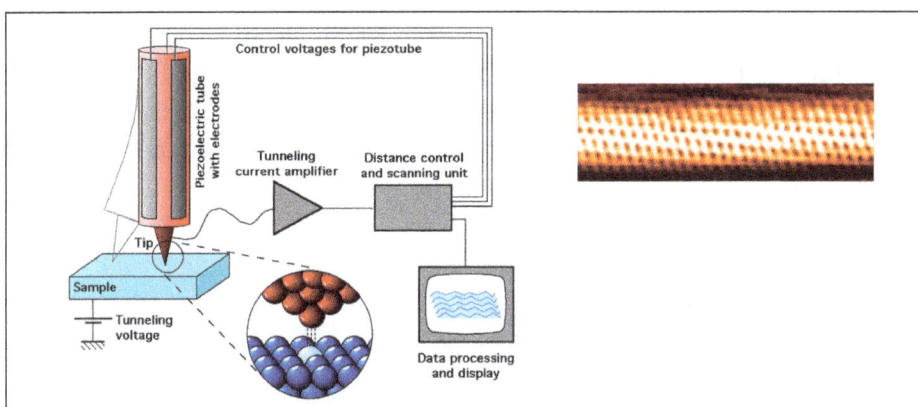

The STM is based on the concept of quantum tunneling. When a conducting tip is brought very near to the surface to be examined, a bias (voltage difference) applied between the two can allow electrons to tunnel through the vacuum between them. The resulting tunneling current is a function of tip position, applied voltage, and the local density of states (LDOS) of the sample. Information is acquired by monitoring the current as the tip's position scans across the surface, and is usually displayed in image form. STM can be a challenging technique, as it requires extremely clean and stable surfaces, sharp tips, excellent vibration control, and sophisticated electronics.

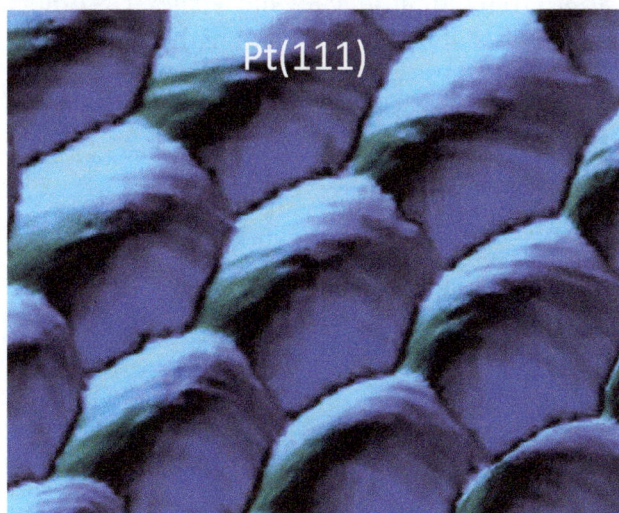

Xe atoms on Ni

STM: Modes of Operation

Constant Height Mode: By using a feedback loop the tip is vertically adjusted in such a way that the current always stays constant. As the current is proportional to the local density of states, the tip follows a contour of a constant density of states during scanning. A kind of a topographic image of the surface is generated by recording the vertical position of the tip.

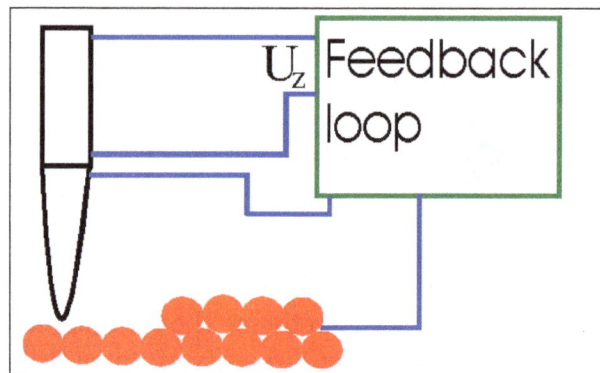

Constant Current Mode: In this mode the vertical position of the tip is not changed, equivalent to a slow or disabled feedback. The current as a function of lateral position represents the surface image. This mode is only appropriate for atomically flat surfaces as otherwise a tip crash would be inevitable. One of its advantages is that it can be used at high scanning frequencies (up to 10 kHz).

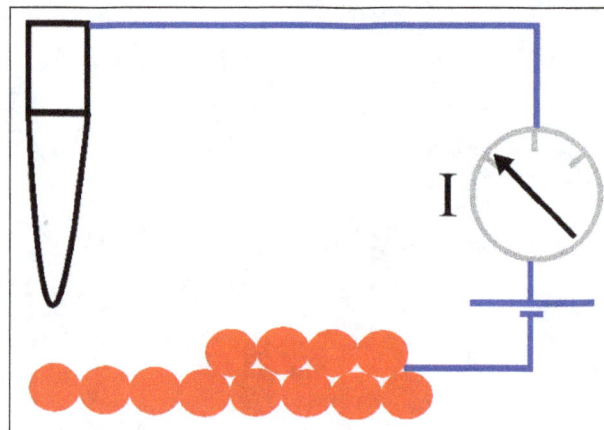

Atomic Image of Au(111)

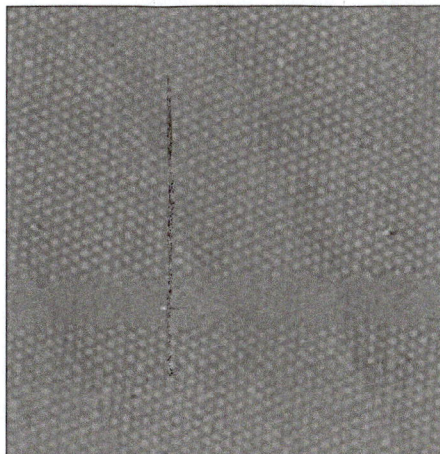

Atomic Force Microscopy

Atomic or near-atomic resolution images of topography of conductors, semiconductors and insulators.

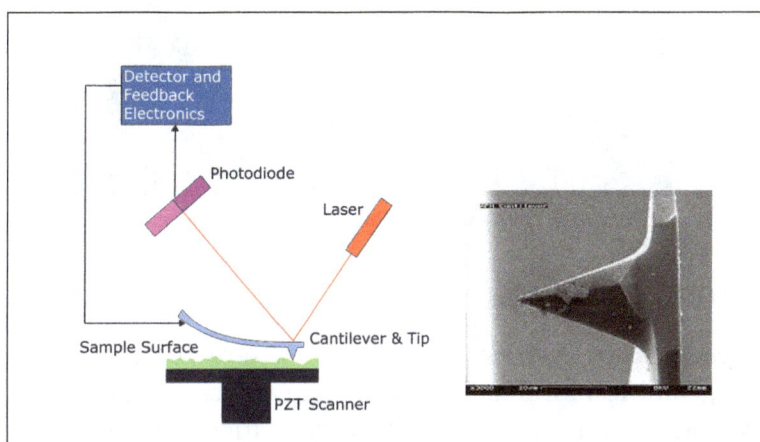

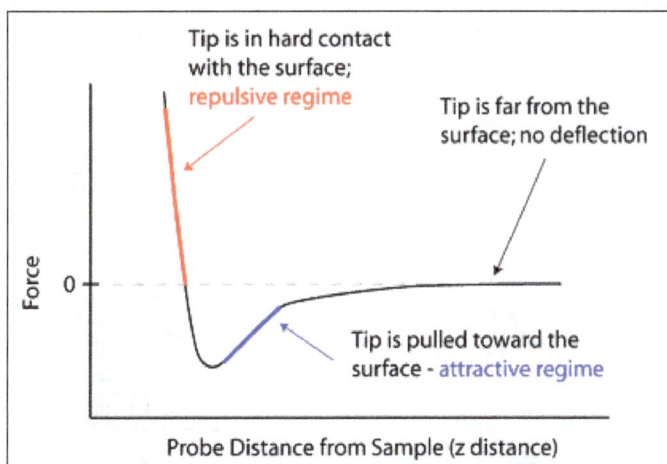

Images generated using contact mode afm.

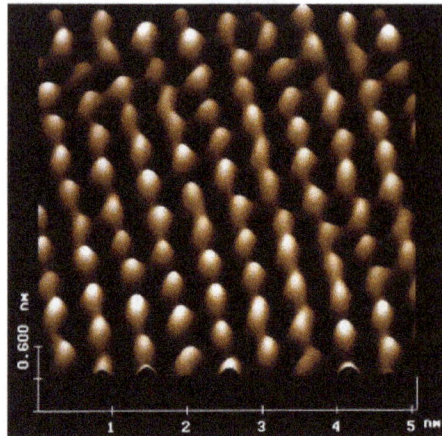

A 5nm scan atomic scale image showing surface atoms on freshly cleaved mica.

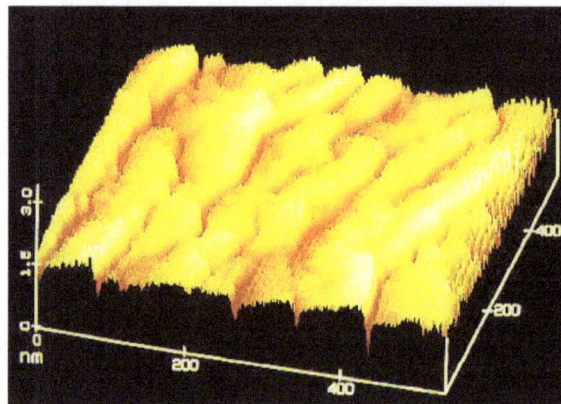

An image of bovine bone obtained in the wet cell.

X-ray Photoelectron Spectroscopy

X-ray photoelectron spectroscopy (XPS) is a quantitative spectroscopic technique that measures the elemental composition, empirical formula, chemical state and electronic state of the elements that exist within a material. XPS spectra are obtained by irradiating a material with a beam of X-rays while simultaneously measuring the kinetic energy and number of electrons that escape from the top 1 to 10 nm of the material being analyzed. XPS requires ultra high vacuum (UHV) conditions.

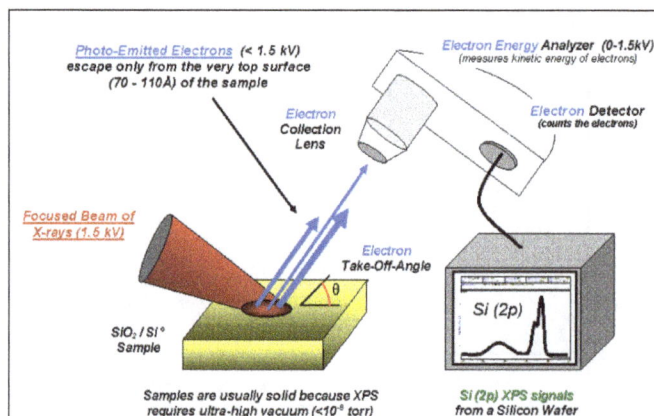

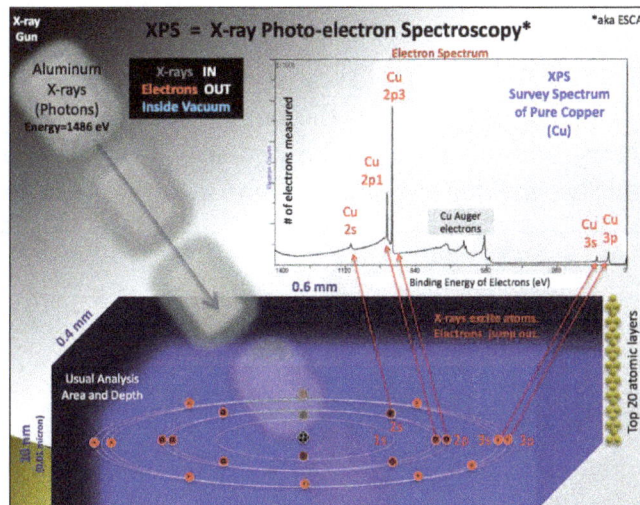

XPS is a surface chemical analysis technique that can be used to analyze the surface chemistry of a material in its "as received" state, or after some treatment.

- XPS detects all elements with an atomic number (Z) of 3 (lithium) and above. It cannot detect hydrogen (Z = 1) or helium (Z = 2) because the diameter of these orbitals is so small, reducing the catch probability to almost zero.

- Detection limits for most of the elements are in the parts per thousand (ppt) (0.1%) range. Detections limits of parts per million (ppm) (0.0001%) are possible, but require special conditions: concentration at top surface or very long collection time (overnight).

- XPS is routinely used to analyze inorganic compounds, metal alloys, semiconductors, polymers, elements, catalysts, glasses, ceramics, paints, papers, inks, woods, plant parts, makeup, teeth, bones, medical implants, biomaterials, viscous oils, glues, ion modified materials and many others.

$$E_{binding} = E_{Photon} - \left(E_{Kinetic} + \phi \right)$$

where $E_{binding}$ is the binding energy (BE) of the electron, E_{Photon} is the energy of the X-ray photons being used, $E_{Kinetic}$ is the kinetic energy of the electron as measured by the instrument and φ is the work function of the spectrometer (not the material).

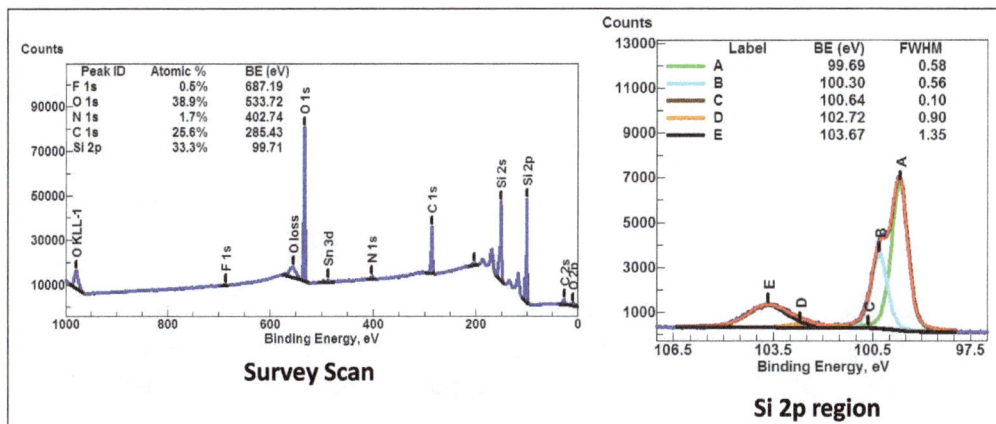

Survey Scan

Si 2p region

X-ray Diffraction

Powder XRD (X-ray Diffraction) is perhaps the most widely used xray diffraction technique for characterizing materials. As the name suggests, the sample is usually in a powdery form, consisting of fine grains of single crystalline material to be studied. The technique is used also widely for studying particles in liquid suspensions or polycrystalline solids (bulk or thin film materials).

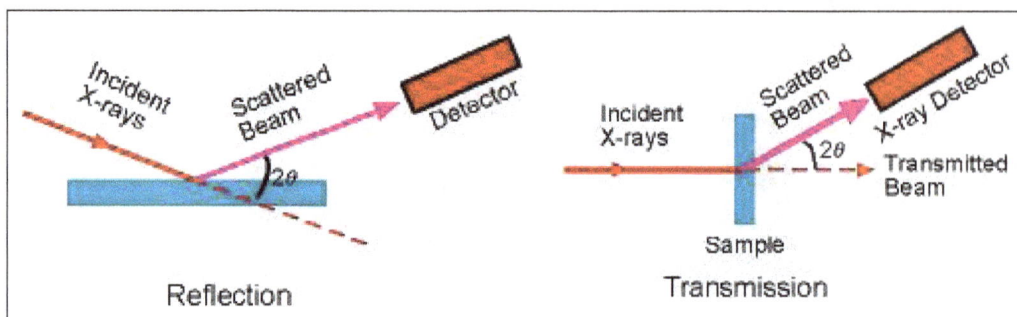

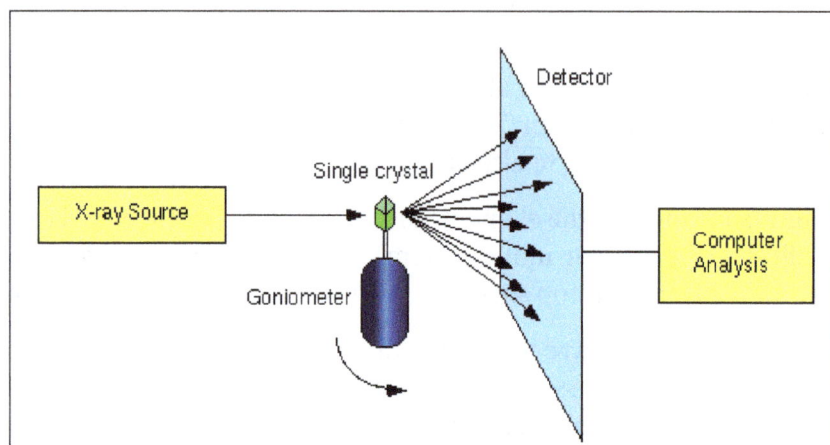

The peaks in a x-ray diffraction pattern are directly related to the atomic distances. Let us consider an incident x-ray beam interacting with the atoms arranged in a periodic manner as shown in 2 dimensions in the following illustrations. The atoms, represented as green spheres in the graph, can be viewed as forming different sets of planes in the crystal (colored lines in graph on left). For a given set of lattice planes with an inter-plane distance of d, the condition for a diffraction (peak) to occur can be simply written as:

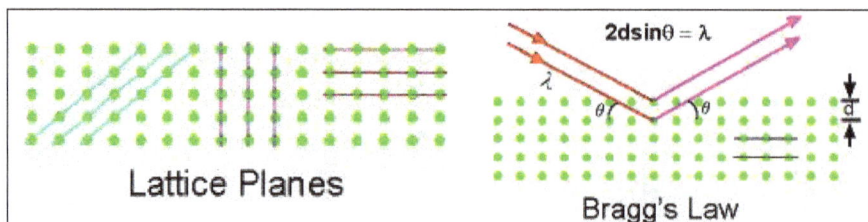

which is known as the Bragg's law, after W.L. Bragg, who first proposed it. In the equation, λ is the wavelength of the x-ray, θ the scattering angle, and n an integer representing the order of the diffraction peak. The Bragg's Law is one of most important laws used for interpreting x-ray diffraction data.

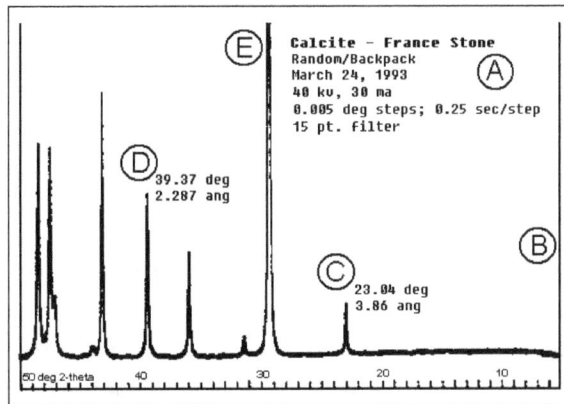

Calcite – France Stone
Random/Backpack
March 24, 1993
40 kv, 30 ma
0.005 deg steps; 0.25 sec/step
15 pt. filter

(D) 39.37 deg
2.287 ang

(C) 23.04 deg
3.86 ang

50 deg 2-theta   40   30   20   10

Bragg's Law refers to the simple equation:

$$n\lambda = 2d\,\sin\Theta$$

derived by the English physicists Sir W.H. Bragg and his son Sir W.L. Bragg in 1913 to explain why the cleavage faces of crystals appear to reflect X-ray beams at certain angles of incidence (Θ, λ). The variable d is the distance between atomic layers in a crystal, and the variable lambda is the wavelength of the incident X-ray beam; n is an integer.

Semiconductors: A Comprehensive Study

Any material whose electrical conductivity value falls between that of an insulator and a conductor is known as a semiconductor. They are mainly categorized into intrinsic and extrinsic semiconductors. Silicon, germanium, gallium arsenide, cadmium selenide, etc. are a few of the examples. This chapter has been carefully written to provide an easy understanding of semiconductors.

The development of semiconductors is clearly among the most significant technological achievements to evolve from the study of solid-state chemistry and physics. Aside from their well-known applications in computers and electronics, semiconductors are also used in a wide variety of optical devices such as lasers, light-emitting diodes, and solar panels. The diversity of applications can be readily understood with only a basic understanding of the theory behind these materials.

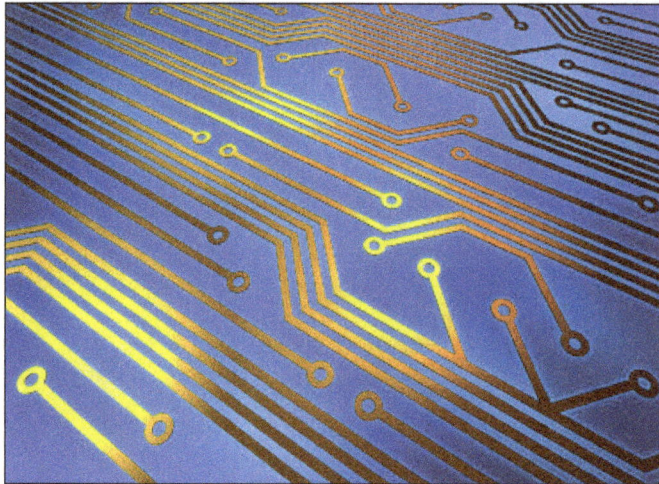

Theory

The operation of semiconductors is best understood using band theory. At its most fundamental level, band theory can be extremely complex, requiring relatively advanced mathematics and physics. When a large number of atoms combine to form a solid, the electrons e^- in the solid are distributed into energy bands among all the atoms in the solid. Each band has a different energy, and the electrons fill these bands from the lowest energy to the highest, similar to the way electrons occupy the orbitals in a single atom. The variation in properties between electrical insulators, conductors (metals), and semiconductors stems from differences in the band structures of these materials. For this discussion, three terms must be defined. The highest energy band that contains electrons is called the valence band, whereas the lowest energy empty band is called the conduction band. The band gap is the difference in energy between the valence and conduction bands. The laws of quantum mechanics forbid electrons from being in the band gap; thus, an electron must always be in one of the bands.

In a metal (e.g., copper or silver), the valence band is only partially filled with electrons. This means that the electrons can access empty areas within the valence band, and move freely across all atoms that make up the solid. A current can therefore be generated when a voltage is applied. In general, for electrons to flow in a solid, they must be in a partially filled band or have access to a nearby empty band. In an electrical insulator, there is no possibility for electron flow, because the valence band is completely filled with electrons, and the conduction band is too far away in energy to be accessed by these electrons (the band gap is too large). A semiconductor is a special case in which the band gap is small enough that electrons in the valence band can jump into the conduction band using thermal energy. That is, heat in the material (even at room temperature) gives some of the electrons enough energy to travel across the band gap. Thus, an important property of semiconductors is that their conductivity increases as they are heated up and more electrons fill the conduction band. The most well-known semiconductor is silicon (Si), although germanium (Ge) and gallium arsenide (GaAs) are also common.

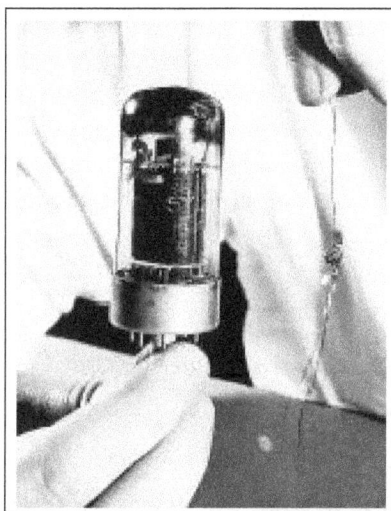

A conventional tube amplifier, at left, and a solid-state memory cell, at right. The size of such semiconductors allows for the manufacturing of smaller devices.

To complete the development of semiconductor theory, the concept of doping must be described. In principle, the idea is to introduce a different kind of atom into a semiconductor in order to modify its electronic structure. Consider, for example, adding a small amount of phosphorus, P, into a silicon host. Phosphorus is one column to the right of silicon in the Periodic Table, so it contains one additional electron. This means that doping P into Si has the effect of introducing additional electrons to the material, such that some e$^-$ must go into the conduction band. Because extra negatively charged electrons are added to the system, phosphorusdoped Si is called an n- type semiconductor, and phosphorus is described as a donor (of electrons). Similarly, a p- type semiconductor can be fabricated by adding an element to the left of Si in the Periodic Table. Boron, B, is a common dopant for a p- type. In this case, the valence band will be missing electrons. These empty locations in a p- type semiconductor are also referred to as holes. Since holes represent the absence of an electron, they carry a positive charge. In p- type semiconductors, boron is referred to as an acceptor (of electrons). From figure, it can be seen that both n- and p- type materials create partially filled bands, allowing for electrical conduction. Dopant concentrations are fairly small, around 10^{16} atoms/cm^3, constituting only about ten-billionths of the total mass of the material.

If p- and n- type materials are layered together, a p-n junction results. Right at the interface, some of the excess electrons from the n- type combine with holes from the p- type. The resulting charge separation creates an energy barrier that impedes any further movement of electrons. In most technological applications, the important properties of semiconductors are the result of the band structure of the p-n junction. A single junction based on the same host material (e.g., one interface of p- and n- doped silicon) is called a homojunction. The homojunction model is used here to describe the properties of many devices that are based on semiconductors. However, it should be noted that real systems are typically composed of multiple p-p, n-n, and p-n junctions, called heterojunctions. Such configurations greatly improve the performance of these materials; in fact, the development of heterojunction devices was critical to the widespread practical application of this technology.

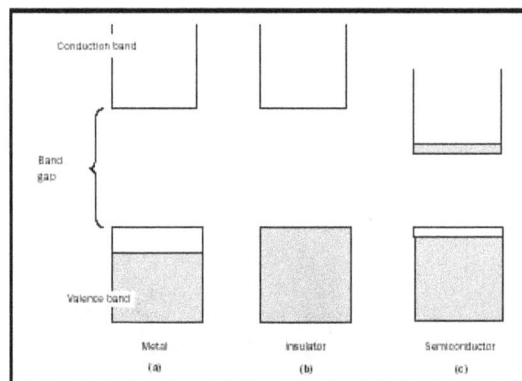

Schematic of the electronic band structures of different types of solids.
Electrons are represented by shaded areas.

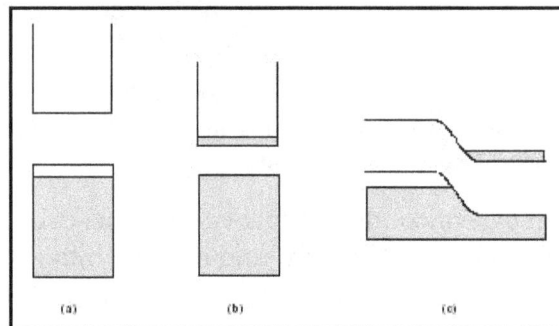

Schematic diagrams of the band structures of (a) p -type semiconductors,
(b) n -type semiconductors, and (c) a p-n junction.

Semiconductors in Electronics

Semiconductors are used extensively in solid-state electronic devices and computers. The majority of materials for these applications are based on doped silicon. An important property of p-n junctions is that they allow electron flow only from the n side to the p side. Such one-way devices are called diodes. Consider the figure again. If a positive voltage (also called a forward bias) is applied that lowers the energy barrier between n and p, then the electrons in the conduction band on the n side can flow across the junction (and holes can flow from p to n). A reverse bias, however, raises the height of the barrier and increases the charge separation at the junction, impeding any flow of electrons from p to n.

Diodes have several important applications in electronics. The power supplied by most electrical utilities is typically alternating current (AC); that is, the direction of current flow switches back and forth with a frequency of sixty cycles per second. However, many electronic devices require a steady flow of current in one direction (direct current or DC). Since a diode only allows current to flow through it in one direction, it can be combined with a capacitor to convert AC input to DC output. For half the AC cycle, the diode passes current and the capacitor is charged up. During the other half of the cycle, the diode blocks any current from the line, but current is provided to the circuit by the capacitor. Diodes applied in this way are referred to as rectifiers.

The by far most important application of semiconductors is as logic gates and transistors in computers. Logic gates, such as OR and AND gates, take advantage of the one-way nature of diodes to compare the presence or absence of current at different locations in a circuit. More complex solid-state transistors are composed of npn or pnp junctions. The device geometry is slightly more complicated than that observed in a diode, but the result is materials that allow for the generation of the zeros and ones required for the binary logic used by computers.

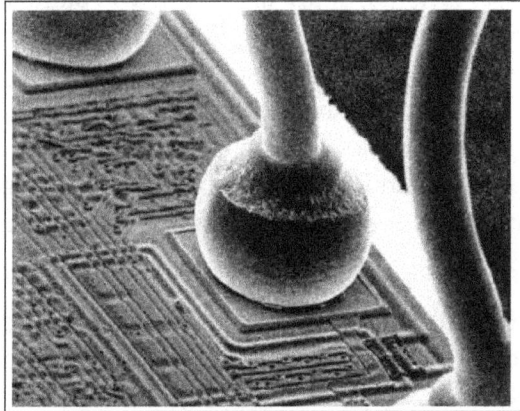

Micro-wires bonded on a silicon chip.

Optoelectronic Devices

Optoelectronic materials are a special class of semiconductors that can either convert electrical energy into light or absorb light and convert it into electrical energy. Light-emitting diodes (LEDs), for example, are commonly used for information display and in automotive interior lighting applications. In an LED, a forward bias applied across the junction moves electrons in the conduction band over holes in the valence band. The electron and hole combine at the junction, and the energy created by this process is conserved via the emission of light. The wavelength of emitted light will depend on the band gap of the material; larger band gaps lead to shorter wavelengths of light. Only certain kinds of semiconductors, called direct gap semiconductors, exhibit this behavior. GaAs is an example of a direct gap semiconductor used in these applications. Silicon is an indirect gap material, and electrons and holes combine with the generation of heat instead of light.

A diode laser operates in essentially the same fashion as an LED. Two additional requirements must be met for a direct gap semiconductor to be an efficient laser. The first is that larger forward bias currents are needed for a laser than for an LED, because lasers require a higher degree of population inversion—a large number of electrons in the conduction band above empty levels in the valence band. Lasers also require an optical cavity; light bounces back and forth within the cavity, building up intensity. In a diode laser, this can be achieved by cleaving and polishing opposite faces of the diode. The smooth faces act like partially reflecting mirrors. This kind of laser is used to read information on compact disks and is also used in laser pointers.

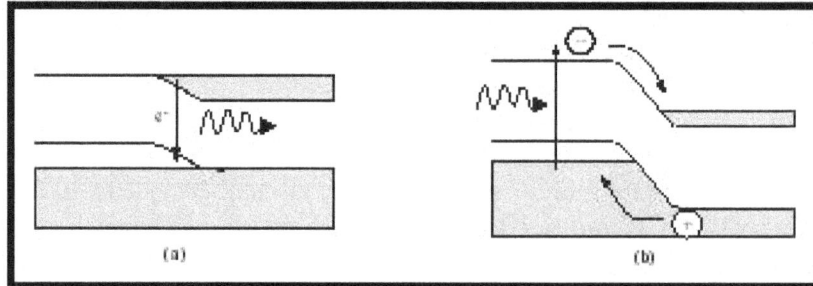

Principle of operation of (a) a light-emitting diode or diode laser and (b) a photodetector or solar cell.

The most common materials for lasers and LEDs are heterojunctions based on GaAs. More complex systems containing Ga, As, P, Al, and N are also used. The band gap of these materials can be tailored to create emission from infrared to yellow. In optical data storage systems, such as compact disks, the amount of information that can be stored is dependent in part on the wavelength of light being used to read the disk—shorter wavelengths allow for denser information storage. Thus, there has been considerable interest in developing larger band gap LEDs and lasers that emit in the blue. This has been achieved in semiconductors based on GaN (gallium nitride). Further refinement of these materials will no doubt lead to significant advances in optoelectronic technology in the coming years.

A final important class of optoelectronic devices based on semiconductors is photovoltaics, such as photodetectors and solar cells. In some respects, these can be regarded as LEDs operating in reverse. Light energy incident on the p-n junction is absorbed by an electron, which then jumps to the conduction band. Once in the conduction band, the electron travels downhill (energetically) to the n side of the junction, with a hole migrating to the p side. This creates a flow of current that is the reverse of what is seen in a forward biased diode. The result is the conversion of light energy to electrical energy. These devices can therefore be used to detect light, as in digital imaging systems or miniature cameras; or the electrical energy can be stored, as in solar cells. Commercial photovoltaics are based on a variety of host materials, including Si, AlGaAs, and InAlAs.

Fabrication

The industrial fabrication of semiconductors can be extremely complex, involving high-purity materials, sophisticated equipment, and hundreds of steps. Most processes begin with the growth of a large single crystal of n- type Si, called a wafer. A dopant (e.g., phosphorus) is added to high-purity molten silicon, and a crystal is then slowly extracted from this melt. The polished wafer is 20 to 30 centimeters (7.9–11.8 inches) in diameter.

The rest of the processing will depend on the nature of the device being produced. A simple p-n junction is usually fabricated via photolithography and etching processes. In this method, a layer of silicon dioxide, SiO_2, is created on the surface of the wafer by heating it in the presence of oxygen. Some of the SiO_2 is then chemically stripped away, or etched, exposing only a portion of the Si wafer. This exposed part of the wafer is made into p- type material by bombarding it with boron ions. As these ions diffuse into the Si wafer, p- type Si is formed. Since the original wafer was n-type, a p-n junction forms where the diffusion of boron stops. Metal contacts can then be added to each side of the junction to create a simple homojunction device.

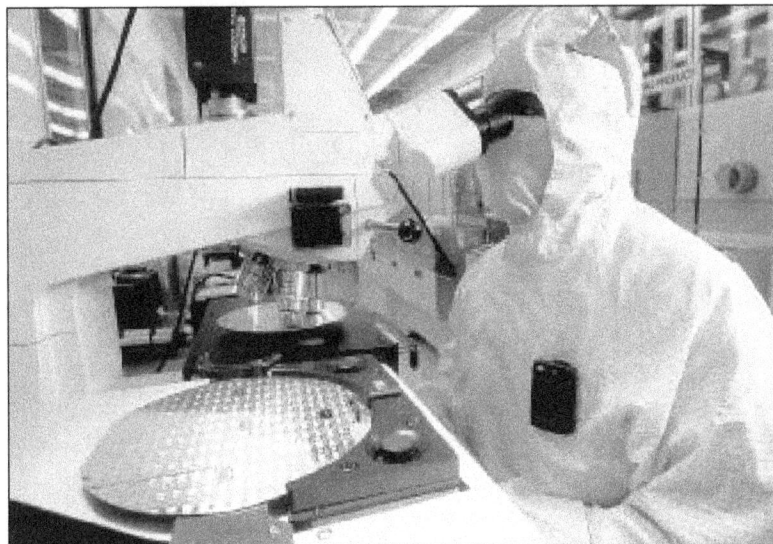

A worker is testing silicon wafers at the Matsushita Semiconductor plant in Puyallup, Washington. Semiconductors are used in many different electronic products, such as computers, lasers, and solar panels.

Fabrication of more complicated devices is achieved via combinations of etching, deposition, and ion implantation steps. In the production of integrated circuits for computers, about 400 chips can be synthesized on a single 30-centimeter (11.8-inch) wafer. Each chip may contain as many as 50 million transistors in a space barely more than 1 centimeter (0.39 inches) on a side—a truly remarkable technological achievement. As faster and faster systems are developed, the demand for smaller and smaller features increases. Such miniaturization is the most significant challenge facing the semiconductor industry today.

Semiconductors are used in a wide variety of electronic and optoelectronic applications. The useful properties of semiconductors arise from the unique behavior of doped materials, the special control of electron flow provided by p-n junctions, and the interaction of light energy with electrons at these junctions. The industry continues to grow, and research in this and related areas (i.e., organic semiconductors and molecular transistors) is occurring at academic institutions around the world.

Materials can be divided into three categories: conductors that allow electrons to flow through them, insulators that prevent the flow of electrons, and semiconductors that only let electrons flow under certain conditions. The difference between them may be best explained by the difference in their band gaps.

A band gap is an energy range in a material where no electron can exist. Conductors have no band gap, so electrons can freely move through them to generate an electric current. Metals including iron, copper, silver, gold, and aluminum are representative conductors. Insulators such as oil, glass, rubber, and ceramics have a large band gap which prevents the flow of electrons. Semiconductors, in contrast, have a small band gap, and the flow of electrons and electron holes can be controlled by adding impurities to the material.

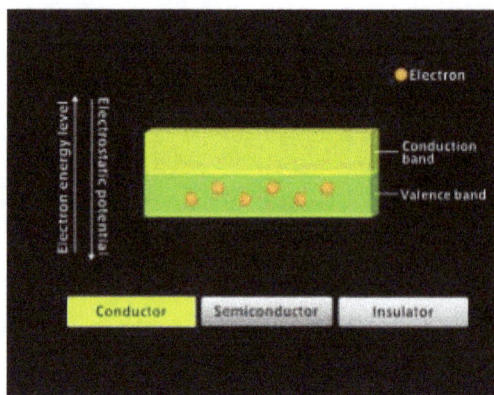

N-type and P-type Semiconductors

Pure silicon and germanium crystals have insulator-like properties, and electricity hardly flows through them even when a voltage is applied. This is because their crystal lattice tightly keeps the electrons in place and hardly lets them move around.

When a very small amount of impurity such as phosphorus is introduced, however, it frees up some electrons and give the crystals conductor-like properties. Semiconductors containing impurities that produce surplus electrons are called n-type semiconductors ("n" stands for negative), and those with impurities such as boron that create a deficiency of electrons are called p-type semiconductors ("p" stands for positive). In a p-type semiconductor, electron holes rather than electrons serve as charge carriers, behaving as if positively charged electrons are flowing.

When p-type and n-type semiconductors are joined, the composite device (called p-n junction diode) produces the rectifier effect in which the flow of electric current is released or stopped depending on the direction of the electric field.

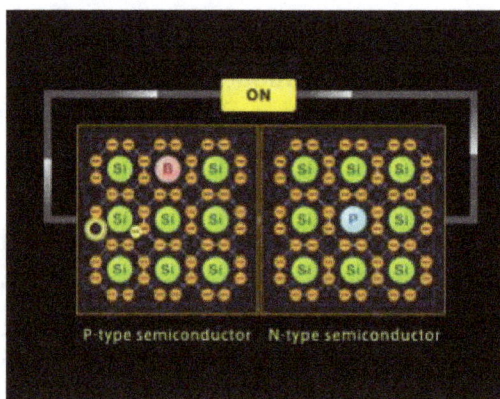

Transistor: A Device for Amplifying Electric Current

A transistor is a semiconductor device used to amplify or switch electrical signals. The name transistor is a combination of the words transfer and resistor. Transistors were developed because, once the rectifier effect had been attained with semiconductors, people needed a semiconductor device for amplifying electrical signals for telegraph and telephone.

The world's first MOS transistor was made in 1960 by Dawon Kahng and M. M. Atalla at Bell Labs. MOS transistors are the most commonly used transistors today. They have two regions of n-type substrates separated by a wall of p-type substrate. When a positive gate voltage is applied, the top of the p-type substrate turns conductive by induction, lowering the barrier and allowing electrons to flow between the two n-type terminals. In effect, slight alterations in the gate voltage amplify changes in the output current.

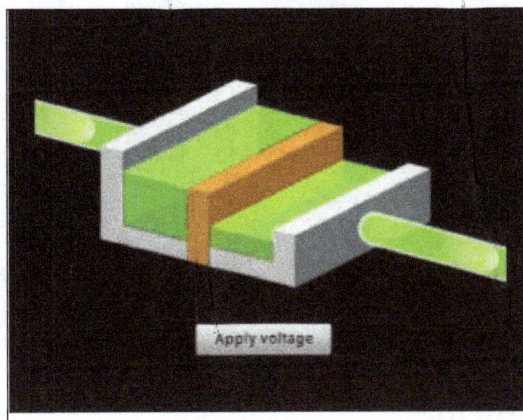

The Expanding Domain of Semiconductors

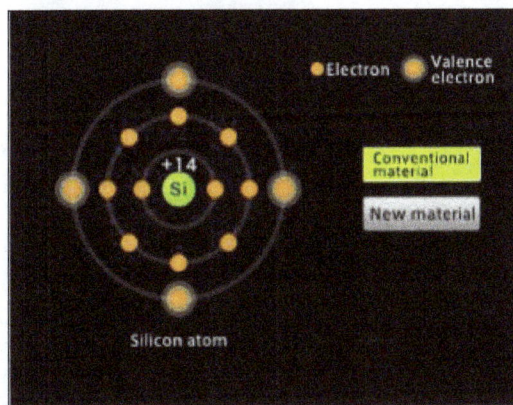

A semiconductor is broadly defined today as a material with electrical conductivity that can be freely controlled by one means or another. In another words, whatever material that can be used as a transistor is a semiconductor.

There was a time when germanium and silicon were exclusively used as semiconductors, and only the group 14 elements in the periodic table were deemed to be semiconductors. As studies on compound semiconductors and organic semiconductors progressed, however, the definition of a semiconductor also changed to include all kinds of semiconducting materials, rather than just a specific group of elements.

Fairly recent additions to the category of semiconductors include carbon nanotubes discovered by Dr. Sumio Iijima, and conductive polymers discovered by Dr. Hideki Shiarakawa and others who won the Nobel Prize in Chemistry. Applications of these semiconductor materials are being studied by researchers around the world.

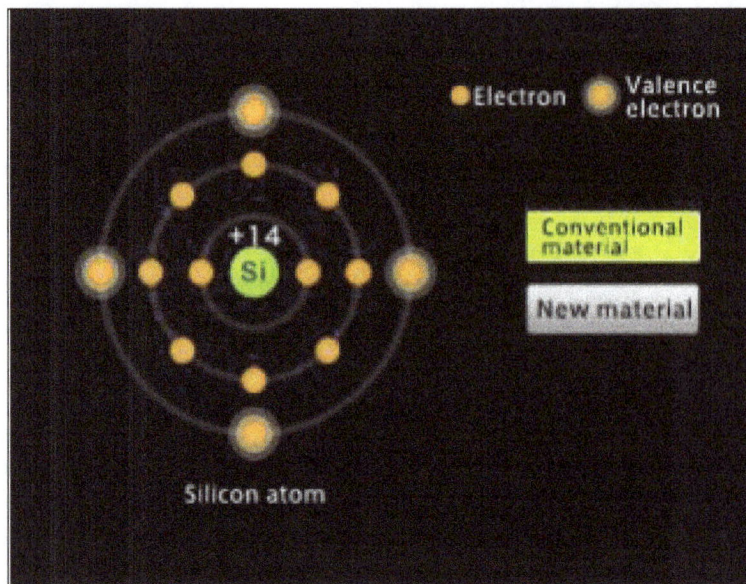

Silicon atom

Properties of Semiconductors

The difference between metals, semiconductors and electrical insulators, we have to define the following terms from solid-state physics:

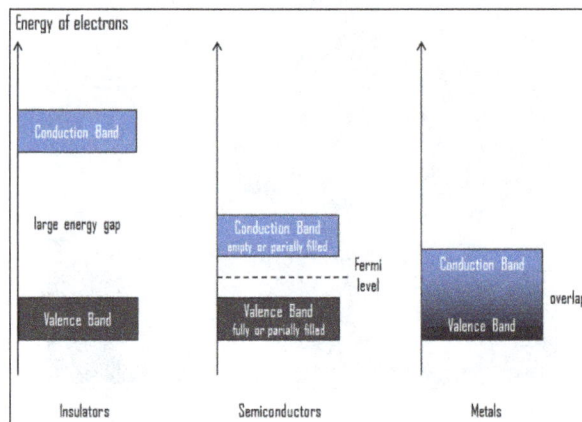

- Valence Band: In solid-state physics, the valence band and conduction band are the bands closest to the Fermi level and thus determine the electrical conductivity of the solid. In electrical insulators and semiconductors, the valence band is the highest range of electron energies in which electrons are normally present at absolute zero temperature. For example, a silicon atom has fourteen electrons. In the ground state, they are arranged in the electron configuration $[Ne]3s^23p^2$. Of these, four are valence electrons, occupying the 3s orbital and two of the 3p orbitals. The distinction between the valence and conduction bands is meaningless in metals, because conduction occurs in one or more partially filled bands that take on the properties of both the valence and conduction bands.

- Conduction Band: In solid-state physics, the valence band and conduction band are the bands closest to the Fermi level and thus determine the electrical conductivity of the solid. In electrical insulators and semiconductors, the conduction band is the lowest range of vacant electronic states. On a graph of the electronic band structure of a material, the valence band is located below the Fermi level, while the conduction band is located above it. In semiconductors, electrons may reach the conduction band, when they are excited, for example, by ionizing radiation (i.e. they must obtain energy higher than E_{gap}). For example, diamond is a wide-band gap semiconductor (E_{gap} = 5.47 eV) with high potential as an electronic device material in many devices. On the other side, germanium has a small band gap energy (E_{gap} = 0.67 eV), which requires to operate the detector at cryogenic temperatures. The distinction between the valence and conduction bands is meaningless in metals, because conduction occurs in one or more partially filled bands that take on the properties of both the valence and conduction bands.

- Band Gap: In solid-state physics, the energy gap or the band gap is an energy range between valence band and conduction band where electron states are forbidden. In contrast to conductors, electrons in a semiconductor must obtain energy (e.g. from ionizing radiation) to cross the band gap and to reach the conduction band. Band gaps are naturally different for different materials. For example, diamond is a wide-band gap semiconductor (E_{gap} = 5.47 eV) with high potential as an electronic device material in many devices. On the other side, germanium has a small band gap energy (E_{gap} = 0.67 eV), which requires to operate the detector at cryogenic temperatures.

- Fermi Level: The term "Fermi level" comes from Fermi-Dirac statistics, which describes a distribution of particles over energy states in systems consisting of fermions (electrons) that obey the Pauli exclusion principle. Since they cannot exist in identical energy states, Fermi level is the term used to describe the top of the collection of electron energy levels at absolute zero temperature. The Fermi level is the surface of Fermi sea at absolute zero where no electrons will have enough energy to rise above the surface. In metals, the Fermi level lies in the hypothetical conduction band giving rise to free conduction electrons. In semiconductors the position of the Fermi level is within the band gap, approximately in the middle of the band gap.

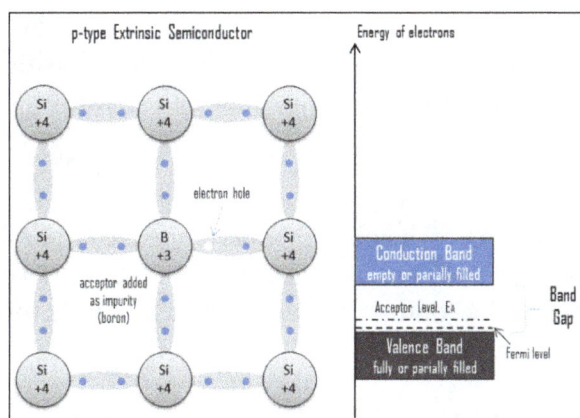

p-type Extrinsic Semiconductor

Energy of electrons

Si +4   Si +4   Si +4

electron hole

Si +4   B +3   Si +4

acceptor added as impurity (boron)

Si +4   Si +4   Si +4

Conduction Band
empty or partially filled

Acceptor Level. Ea

Band Gap

Valence Band
fully or partially filled

Fermi level

- Electron-hole Pair: In the semiconductor, free charge carriers are electrons and electron holes (electron-hole pairs). Electrons and holes are created by excitation of electron from valence band to the conduction band. An electron hole (often simply called a hole) is the lack of an electron at a position where one could exist in an atom or atomic lattice. It is one of the two types of charge carriers that are responsible for creating electric current in semiconducting materials. Since in a normal atom or crystal lattice the negative charge of the electrons is balanced by the positive charge of the atomic nuclei, the absence of an electron leaves a net positive charge at the hole's location. Positively charged holes can move from atom to atom in semiconducting materials as electrons leave their positions. When an electron meets with a hole, they recombine and these free carriers effectively vanish. The recombination means an electron which has been excited from the valence band to the conduction band falls back to the empty state in the valence band, known as the holes.

The conductivity of a semiconductor can be modeled in terms of the band theory of solids. The band model of a semiconductor suggests that at ordinary temperatures there is a finite possibility that electrons can reach the conduction band and contribute to electrical conduction. In the semiconductor, free charge carriers (electron-hole pairs) are created by excitation of electron from valence band to the conduction band. This excitation left a hole in the valence band which behaves as positive charge and an electron-hole pair is created. Holes can sometimes be confusing as they are not physical particles in the way that electrons are, rather they are the absence of an electron in an atom. Holes can move from atom to atom in semiconducting materials as electrons leave their positions.

Electron Excitation in Semiconductors

Energy for the excitation can be obtained by different ways.

Thermal Excitation

Electron-hole pairs are constantly generated from thermal energy as well, in the absence of any external energy source. Thermal excitation does not require any other form of starting impulse. This phenomenon occurs also at room temperature. It is caused by impurities, irregularity in structure lattice or by dopant. It strongly depends on the E_{gap} (a distance between valence and conduction band), so that for lower E_{gap} a number of thermally excited charge carriers increases. Since thermal excitation results in the detector noise, active cooling is required for some types of semiconductors (e.g. germanium). Detectors based on silicon have sufficiently low noise even by room temperature. This is caused by the large band gap of silicon (Egap= 1.12 eV), which allows us to operate the detector at room temperature, but cooling is prefered to reduce noise.

Optical Excitation

Note that, energy of a single photon of visible light spectrum is comparable with these band gaps. Photons of wave lengths 700 nm – 400 nm have energies of 1.77 eV 3.10 eV. As a result, also visible light is able to excite electrons to the conduction band. Actually, this is the principle of photovoltaic panels that generate electric current.

Excitation by Ionizing Radiation

Electrons may reach the conduction band, when they are excited by ionizing radiation (i.e. they must obtain energy higher than Egap). In general, heavy charged particles transfer energy mostly by:

- Excitation: The charged particle can transfer energy to the atom, raising electrons to a higher energy levels.

- Ionization: Ionization can occur, when the charged particle have enough energy to remove an electron. This results in a creation of ion pairs in surrounding matter.

A convenient variable that describes the ionization properties of surrounding medium is the stopping power. The classical expression that describes the specific energy loss is known as the Bethe formula. For alpha particles and heavier particles the stopping power of most materials is very high for heavy charged particles and these particles have very short ranges.

In addition to these interactions, beta particles also lose energy by radiative process known as the bremsstrahlung. From classical theory, when a charged particle is accelerated or decelerated, it must radiate energy and the deceleration radiation is known as the bremsstrahlung ("braking radiation").

Photons (gamma rays and X-rays) can ionize atoms directly (despite they are electrically neutral) through the Photoelectric effect and the Compton effect, but secondary (indirect) ionization is much more significant. Although a large number of possible interactions are known, there are three key interaction mechanisms with matter.

- Photoelectric effect.

- Compton scattering.

- Pair production.

In all cases, a particle of ionizing radiation deposits a portion of its energy along its path. Particle passing through the detector ionizes the atoms of semiconductor, producing the electron-hole pairs. For example, typical thickness of silicon detector is about 300 μm so the number of generated electron-hole pairs by minimum ionizing particle (MIP) passing perpendicular through the detector is about 3.2×10^4. This value is minor in comparison the total number of free carriers in intrinsic semiconductor of a surface of 1 cm^2 and the same thickness. Note that, a sample of pure germanium at 20 °C contains about 1.26×10^{21} atoms, but also contains 7.5×10^{11} free electrons and 7.5×10^{11} holes constantly generated from thermal energy. As can be seen, the signal to noise ratio (S/N) would be minimal. The addition of 0.001% of arsenic (an impurity) donates an extra 10^{15} free electrons in the same volume and the electrical conductivity is increased by a factor of 10,000. In doped material the signal to noise ratio (S/N) would be even smaller. Cooling of the semiconductor is one way to lower this ratio.

Improvement can be reached by use of a reverse-bias voltage to the P-N junction to deplete the detector of free carriers, which is the principle of the most silicon radiation detectors. In this case, negative voltage is applied to the p-side and positive to the second one. Holes in the p-region are attracted from the junction towards the p contact and similarly for electrons and the n contact.

Types of Semiconductors

Semiconductors are mainly classified into two categories:

- Intrinsic Semiconductor.

- Extrinsic Semiconductor.

Intrinsic Semiconductor

An intrinsic semiconductor material is chemically very pure and possesses poor conductivity. It has equal numbers of negative carriers (electrons) and positive carriers (holes).

Extrinsic Semiconductor

Where as an extrinsic semiconductor is an improved intrinsic semiconductor with a small amount of impurities added by a process, known as doping, which alters the electrical properties of the semiconductor and improves its conductivity. Depending on whether the added impurities have "extra" electrons or "missing" electrons determines how the bonding in the crystal lattice is affected as shown in figure, and therefore how the material's electrical properties change.

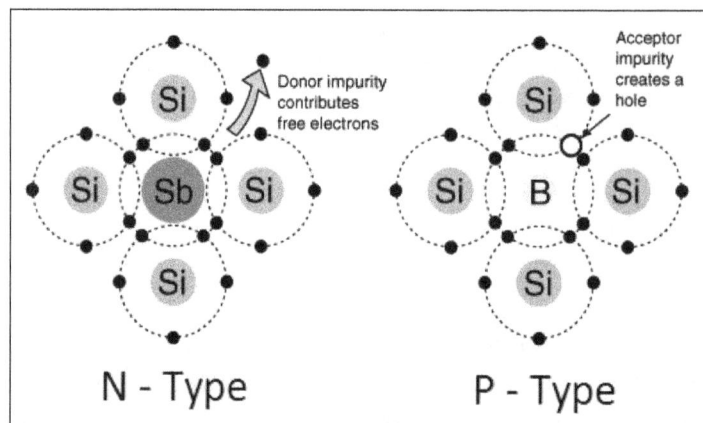

The Doping of Semiconductors

The addition of a small percentage of impurity atoms in the intrinsic semiconductor (pure silicon or pure germanium) produces dramatic changes in their electrical properties. Depending on the type of impurity added, the extrinsic semiconductors can be divided in to two classes:

- N-type Semiconductors.

- P-type Semiconductors.

N-type Semiconductor

Group V dopants are the atoms with an "extra" electron, in other words a valence shell with only one electron. When a semiconductor is doped with a Group V impurity it is called an n-type material, because the addition of these pentavalent impurities such as antimony, arsenic or phosphorous contributes free electrons, greatly increasing the conductivity of the intrinsic semiconductor. In an n-type semiconductor, the majority carrier, or the more abundant charge carrier, is the electron, and the minority carrier, or the less abundant charge carrier, is the hole.

The effect of this doping process on the relative conductivity can be explained by energy band diagram shown in figure. When donor impurities are added to an intrinsic semiconductor, allowable energy levels are introduced at a very small gap below the conduction band, as illustrated in figure. These new allowable levels are essentially a discrete level because the added impurity atoms are far apart in the crystal structure and hence their interaction is small. In the case of Silicon, the gap of the new discrete allowable energy level is only 0.05 eV (0.01 eV for germanium) below the conduction band, and therefore at room temperature almost all of the "fifth" electrons of the donor impurity are raised into the conduction band and the conductivity of the material increases considerable.

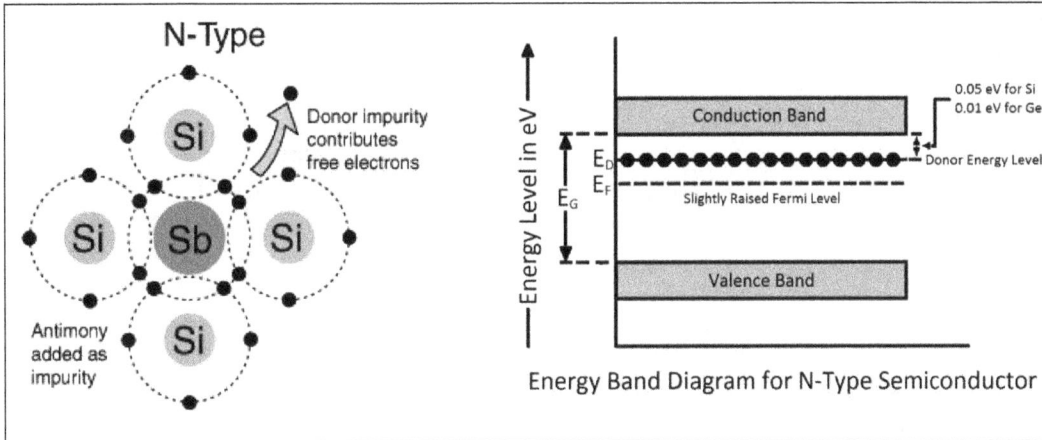

Energy Band Diagram for N-Type Semiconductor

P-type Semiconductor

Group III dopants are the atoms with a hole in their valence shell (only "missing" one electron), When a semiconductor is doped with a Group III impurity it is called a p-type material, The addition of these trivalent impurities such as boron, aluminum or gallium to an intrinsic semiconductor creates deficiencies of valence electrons, called "holes". In an p-type semiconductor, the majority carrier, or the more abundant charge carrier, is the hole, and the minority carrier, or the less abundant charge carrier, is the electron.

The effect of this doping process on the relative conductivity can be explained by energy band diagram shown in figure. When accepter impurities or P type impurities are added to the intrinsic semiconductor, they produce an allowable discrete energy levels which is just above the valance band, as shown in figure. Since a very small amount of energy (0.08 eV in case of Silicon and 0.01 eV in case of Germanium) is required for an electron to leave the valence band and occupy the accepter energy level, holes are created in the valence band by these electrons.

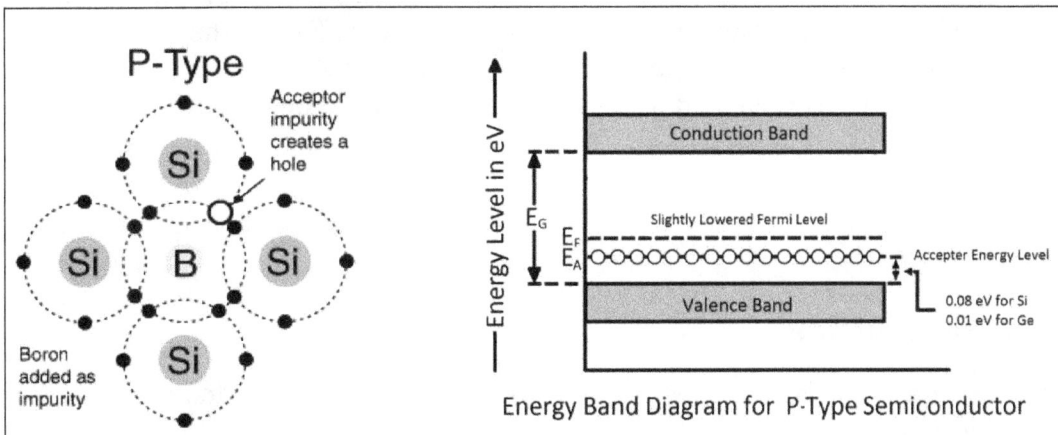

Energy Band Diagram for P-Type Semiconductor

Difference between Intrinsic and Extrinsic Semiconductor

	Intrinsic Semiconductors	Extrinsic Semiconductors
1	It is pure semi-conducting material and no impurity atoms are added to it.	It is prepared by doping a small quantity of impurity atoms to the pure semi-conducting material.
2	Examples: crystalline forms of pure silicon and germanium.	Examples: silicon "Si" and germanium "Ge" crystals with impurity atoms of As, Sb, P etc. or In B, Aℓ etc.
3	The number of free electrons in the conduction band and the no. of holes in valence band is exactly equal and very small indeed.	The number of free electrons and holes is never equal. There is excess of electrons in n type semi-conductors and excess of holes in p-type semi-conductors.
4	Its electrical conductivity is low.	Its electrical conductivity is high.
5	Its electrical conductivity is a function of temperature alone.	Its electrical conductivity depends upon the temperature as well as on the quantity of impurity atoms doped the structure.

Extrinsic Semiconductors

After some experiments, scientists observed an increase in the conductivity of a Semiconductor when a small amount of impurity was added to it. These materials are Extrinsic Semiconductors or impurity Semiconductors. Another term for these materials is 'Doped Semiconductor'. The impurities are dopants and the process – Doping.

An important condition to doping is that the amount of impurity added should not change the lattice structure of the Semiconductor. To achieve this the size of the dopant and Semiconductor atoms should be the same.

Types of Dopants in Extrinsic Semiconductors

Crystals of Silicon and Germanium are doped using two types of dopants:

- Pentavalent (valency 5); like Arsenic (As), Antimony (Sb), Phosphorous (P), etc.
- Trivalent (valency 3); like Indium (In), Boron (B), Aluminium (Al), etc.

The reason behind using these dopants is to have similarly sized atoms as the pure semiconductor. Both Si and Ge belong to the fourth group in the periodic table. Hence, the choice of dopants is from the third and fifth group. This ensures that size of the atoms is not much different from the fourth group. Hence, the trivalent and pentavalent choices. These dopants give rise to two types of semiconductors:

- N-type.
- P-type.

N-type Semiconductor

An n-type semiconductor is created when pure semiconductors, like Si and Ge, are doped with pentavalent elements.

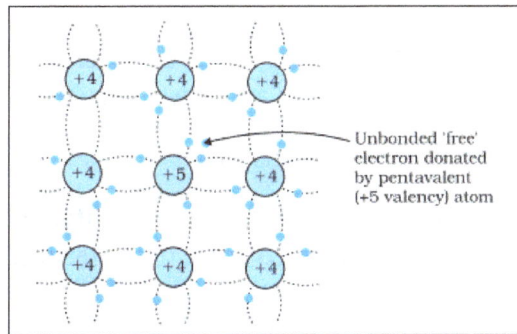

As can be seen in the image above, when a pentavalent atom takes the place of a Si atom, four of its electrons bond with four neighbouring Si atoms. However, the fifth electron remains loosely bound to the parent atom. Hence, the ionization energy required to set this electron free is very small. Thereby, this electron can move in the lattice even at room temperature.

To give you a better perspective, the ionization energy required for silicon at room temperature is around 1.1 eV. On the other hand, by adding a pentavalent impurity, this energy drops to around 0.05 eV.

It is important to remember that the number of electrons made available by the dopant atoms is independent of the ambient temperature and primarily depends on the doping level. Also, as the temperature rises, the Si atoms free some electrons and generate some holes. But, the number of these holes is very small. Hence, at any given point in time, the number of free electrons is much higher than the number of holes. Also, due to recombination, the number of holes reduce further.

when a semiconductor is doped with a pentavalent atom, electrons are the majority charge carriers. On the other hand, the holes are the minority charge carriers. Therefore, such extrinsic semiconductors are called n-type semiconductors. In an n-type semiconductor,

Number of free electrons (n_e) >> Number of holes (n_h)

P-type Semiconductor

A p-type semiconductor is created when trivalent elements are used to dope pure semiconductors, like Si and Ge. As can be seen in the image above, when a trivalent atom takes the place of a Si atom, three of its electrons bond with three neighbouring Si atoms. However, there is no electron to bond with the fourth Si atom.

This leads to a hole or a vacancy between the trivalent and the fourth silicon atom. This hole initiates a jump of an electron from the outer orbit of the atom in the neighbourhood to fill the vacancy. This creates a hole at the site from where the electron jumps. In simple words, a hole is now available for conduction.

It is important to remember that the number of holes made available by the dopant atoms is independent of the ambient temperature and primarily depends on the doping level. Also, as the temperature rises, the Si atoms free some electrons and generate some holes. But, the number of these electrons is very small. Hence, at any given point in time, the number of holes is much higher than the number of free electrons. Also, due to recombination, the number of free electrons reduce further.

When a semiconductor is doped with a trivalent atom, holes are the majority charge carriers. On the other hand, the free electrons are the minority charge carriers. Therefore, such extrinsic semiconductors are called p-type semiconductors. In a p-type semiconductor,

Number of holes (n_h) >> Number of free electrons (n_e)

Important note: The crystal maintains an overall charge neutrality. The charge of additional charge carries is equal and opposite to that of the ionized cores in the lattice.

Energy Bands of Extrinsic Semiconductors

In extrinsic semiconductors, a change in the ambient temperature leads to the production of minority charge carriers. Also, the dopant atoms produce the majority carriers. During recombination, the majority carriers destroy most of these minority carriers. This leads to a decrease in the concentration of the minority carriers.

Therefore, this affects the energy band structure of the semiconductor. In such semiconductors, additional energy states exist:

- Energy state due to donor impurity (E_D).

- Energy state due to acceptor impurity (E_A).

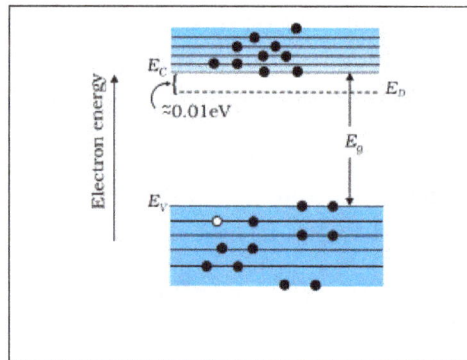

The above energy band diagram is of n-type Si semiconductor. The energy level of the donor (E_D) is lower than that of the conduction band (E_C). Hence, electrons can move into the conduction band with minimal energy (~0.01 eV). Also, at room temperature, most donor atoms and very few Si atoms get ionized. Hence, the conduction band has most electrons from the donor impurities.

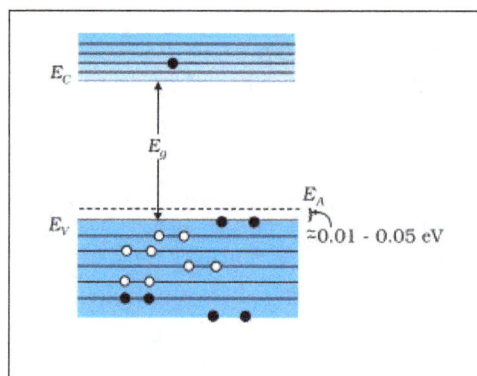

The above energy band diagram is of p-type Si semiconductor. The energy level of the acceptor (E_A) is higher than that of the valence band (E_V). Hence, electrons can move from the valence band to the level Ea, with minimal energy. Also, at room temperature, most acceptor atoms are ionized.

This leaves holes in the valence band. Hence, the valence band has most holes from the impurities. The electron and hole concentration in a semiconductor in thermal equilibrium is:

$$n_e \times n_h = n_i^2$$

Superconductors and Semiconductors

Electronic Properties of Materials: Superconductors and Semiconductors

Modern integrated circuits contain billions of nanoscale transistors and diodes that are essential for logic and memory functions. Both kinds of devices rely on junctions between crystalline silicon regions that contain a few parts per million of boron or phosphorus impurities.

It developed an energy band picture for metals, starting from atomic orbitals and building up the molecular orbitals of the solid metallic crystal. This treatment gave us a useful picture of how electrons behave in metals, moving at very fast speed between scattering events, and migrating in an electric field at a slow drift velocity.

While the band picture works well for most crystalline materials, it does not tell us the whole story of conduction in solids. That is because the band model (like MO theory) is based on a *one-electron* model. Moving from atoms to molecules, we made linear combinations to generate one-electron molecular orbitals (and, in solids, one-electron energy bands). But as in multi-electron atoms, life is not so simple for real molecules and solids that contain many electrons. Electrons repel each other and so their movement in molecules and in solids is *correlated*. While this effect is weak in a "good" metal such as sodium where the wavefunctions are highly delocalized it can be quite important in other materials such as transition metal oxides. Correlated electron effects give rise to metal-insulator transitions that are driven by small changes in temperature, pressure, or composition, as well as to superconductivity the passage of current with zero resistance at low temperatures.

Metal-insulator Transitions

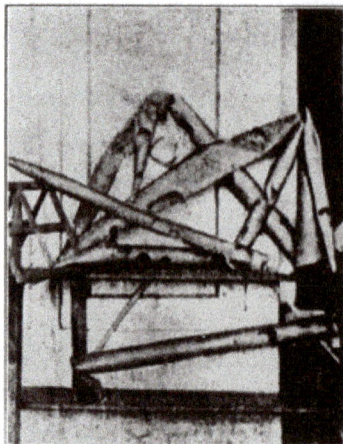

Organ pipes destroyed by "tin blight".

Metals and insulators not only have different electrical properties but also have very different crystal structures. Metals tend to have high coordination numbers (typically 8 or 12) whereas insulators have low coordination numbers that can be rationalized as "octet" bonding arrangements. For example, in crystalline Si or Ge (diamond structure), each atom has four nearest neighbors. There are two electrons per bond, and thus each atom has eight electrons in its valence shell. Sn, the element below Ge, exists in two different forms, one (gray tin) with the diamond structure that is a brittle narrow-gap semiconductor, and the other (white tin) with a body-centered tetragonal structure that is a malleable metal. These two forms are very close in energy, and in fact metallic white tin transforms to the brittle semiconducting gray form at low temperature. Extremely cold weather in 18th century Europe caused many tin organ pipes to break and eventually turn to dust. This transformation has been called tin blight, tin disease, tin pest or tin leprosy. The dust is actually grey tin, which lacks the malleability of its metallic cousin white tin.

Under experimentally accessible temperatures and pressures, Si and Ge are always semiconducting (i.e., insulating), and Pb is always metallic. Why is Sn different? The reason has to do with orbital overlap. Theory tells us in fact that any (and all) insulators should become metallic at high enough pressure, or more to the point, at high enough density. For most insulators, however, the pressures required are far beyond those that we can achieve in the laboratory.

How can we rationalize the transition of insulators to the metallic state? Indeed, how can we understand the existence of insulators at all?

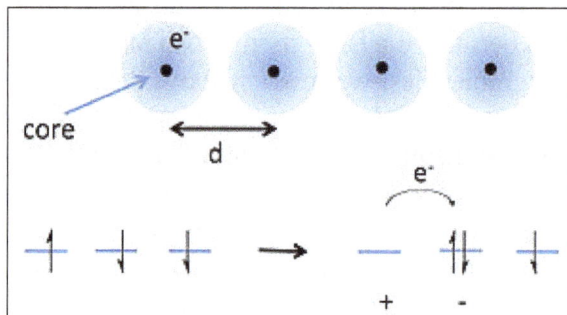

Electron hopping in a 1-D chain of atoms.

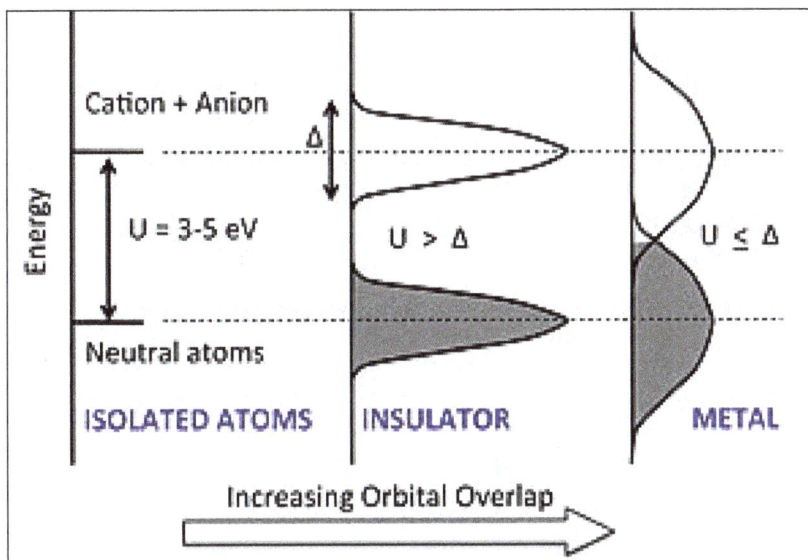

Energy vs. DOS for the chain of atoms as the density and degree of orbital overlap between atoms increases. Increasing overlap broadens the neutral atom and anion-cation states into bands, each of which has a bandwidth Δ. A transition to the metallic state occurs abruptly when Δ exceeds the Hubbard gap U.

The Hubbard Model: Let's consider a chain of a large number (N) of atoms as, because N atoms combine to make N orbitals, and the N valence electrons only fill the band of orbitals halfway. But this conclusion doesn't depend on the density, which creates a paradox. If atoms in the chain are very far apart, we suspect that the electrons should localize on the atoms.

A solution to this problem was proposed by J. Hubbard in 1963. Hubbard considered the energy required to transfer an electron from an atom to its nearest neighbor, as shown in the picture at the right. Because each atom already has one electron (with random spin), moving an electron over by one atom requires overcoming the energy of electron-electron repulsion to make a cation-anion pair. For well-separated atoms this energy (U) is given by:

$$U = IP - EA - e^2 / 4\pi\varepsilon_0 d$$

where IP and EA are the ionization energy and electron affinity, ε_0 is the permittivity of free space, and the last term in the equation represents the coulombic attraction between the cation and the anion. For atoms such as alkali metals, U is on the order of 3–5 eV, which is much larger than the thermal energy kT. Thus we expect there to be very few anion-cation pairs at room temperature, and the chain of atoms should be insulating.

What happens when we squeeze the atoms together? In the Hubbard model, as the distance between atoms decreases, the energies of both the neutral atom states and the anion-cation states broaden into bands, each of which has a band width Δ. The lower band can accommodate exactly N electrons (not $2N$ as in the MO picture we developed earlier) because each orbital can only take one electron without spin-pairing. Thus for small Δ the lower band is full and the upper band is empty. However, as we continue to compress the chain, the orbital overlap becomes so strong that $\Delta \approx U$. At this point, the bands overlap and some of the electrons fill the anion-cation states. The chain then becomes conducting and the material is metallic.

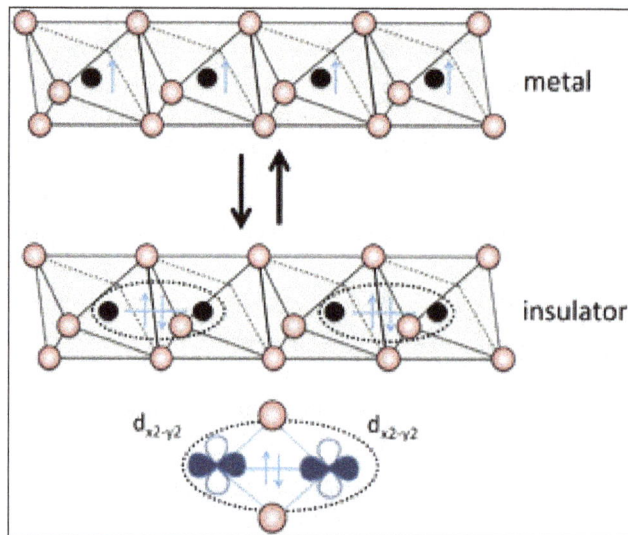

Vanadium dioxide has the rutile structure, and in its undistorted form it is metallic, with one valence electron per V atom. Distortion of the lattice makes pairs of V atoms, resulting in an electronically insulating state. The metal-insulator transition can be driven reversibly by changing the temperature, pressure, or orbital occupancy. Electrical switching of this transition in VO_2 is being studied for applications in high performance thin film transistors.

Some materials, such as Sn and VO_2, happen to have just the right degree of orbital overlap to make the Hubbard transition occur by changing the temperature or pressure. Such materials can be very useful for electrical switching, as illustrated at the right for rutile structure VO_2. Most materials are far away from the transition, either on the metallic or insulating side. An interesting periodic trend that illustrates this concept can be seen among the transition metal monoxides, MO (M = Ti, V, Cr, Mn, Fe, Co, Ni), all of which have the NaCl structure. TiO and VO are metallic, because the 3d orbitals have significant overlap in the structure. However, CrO, MnO, FeO, CoO, and NiO are all insulators, because the 3d orbitals contract (and therefore $\Delta < U$) going across the transition metal series. In contrast, the analogous sulfides (TiS, VS,....NiS) are all metallic. The sulfides have the NiAs structure, in which all the metal atoms are eclipsed along the stacking axis (the hexagonal c-axis). The short metal-metal distances along that axis result in strong orbital overlap, making $\Delta > U$.

The Mott Model: A simpler, less atomistic model of the metal-insulator transition was formulated by Neville Mott. The Mott model considers the behavior of an electron in a material as a function of the density of all the other valence electrons. We know for a one-electron hydrogen-like atom (H, Na, Cs, etc.) the Schrodinger equation contains a potential energy term:

$$V(r) = -(e^2 / 4\pi\varepsilon_0 r)$$

This potential energy function gives rise to familiar ladder of allowed energy levels in the hydrogen atom. However, in a metal, this Coulomb potential must be modified to include the screening of nuclear charge by the other electrons in the solid. In this case there is a screened Coulomb potential:

$$V(r) = -(e^2 / 4\pi\varepsilon_0 r)\exp(-qr)$$

where q, which is the inverse of the screening length, is given by:

$$q^2 = 4m_e^{\,2}\left(3n/\pi\right)^{1/3}\left(2\pi/h\right)$$

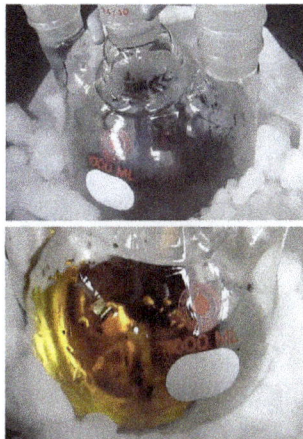

Solutions of lithium metal in liquid ammonia at low (top, ionic conductor)
and higher (bottom, metal) Li concentration.

Here n is the density of atoms (or valence electrons), m_e is the electron mass, and h is Planck's constant. At distances much larger than the screening length q^{-1}, the electron no longer "feels" the charge on the nucleus. Mott showed that there is a critical density of electrons n_c above which the valence electrons are no longer bound by individual nuclei and are free to roam the crystal. This critical density marks the transition to the metallic state, and is given by the Mott criterion:

$$n_c^{\,1/3}a_H \approx 0.26$$

In this equation, a_H is an effective Bohr radius for the valence electrons in the low-density limit, e.g. the average orbital radius of electrons in the 6s shell of a Cs atom when computing the value for Cs metal.

The important concept from the Mott model is that the metal-insulator transition depends very strongly on the density of valence electrons. This is consistent with the orbital overlap model of Hubbard, but also more general in the sense that it does not depend on a periodic structure of atoms. The Mott model is thus applicable to such diverse systems as metal atoms dissolved in liquid ammonia, metal atoms trapped in frozen gas matrices, and dopants in semiconductors. In some systems, it is possible to continuously tune the density of valence electrons with rather striking results. For example, dissolving alkali metal (Li, Na,) in liquid ammonia (bp -33 °C) produces a blue liquid. The solvated alkali cations and negatively charged electrons impart ionic conductivity (as in a salt solution) but not electronic conductivity to the blue liquid ammonia solution. But as the concentration of electrons increases, a reflective, bronze-colored liquid phase forms that floats over the blue phase. This bronze phase is metallic and highly conducting. Eventually, with enough alkali metal added, the entire liquid is converted to the electronically conducting bronze phase.

The electrical switching of VO_2 between insulating and metallic phases can also be rationalized in terms of the Mott transition. Adding more electron density (by chemical or electric field doping) increases the concentration of valence electrons, driving the phase transition to the metallic state.

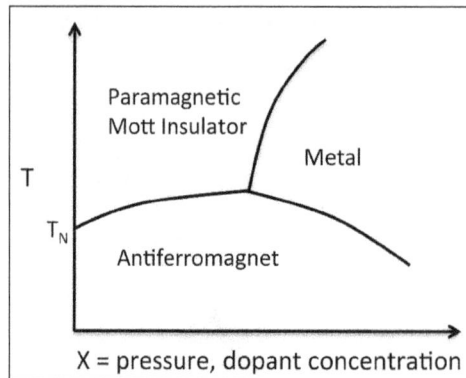

Thermodynamics and phase transitions: Thermodynamically, the metal-insulator transition is a first-order phase transition. In such a transition, the structure and properties change abruptly (think of the breakfast-to-lunch transition at McDonalds there is just no way to get pancakes after, or hamburgers before 10:30 AM!). Thus in the case of Sn metal, the changes in structure (from four- to eight-coordination) and in electronic conductivity (insulator to metal) occur simultaneously. As in other first order phase transitions such as ice to water to steam, there is a latent heat associated with the transition and a discontinuity in derivative properties such as the heat capacity.

A typical phase diagram for a metal-insulator transition is shown at the right for V_2O_3 The octahedrally coordinated V^{3+} ion has a d^2 electron count, so there are two unpaired spins per atom, and at low temperature the spins in the lattice order antiferromagnetically. The Néel temperature an antiferromagnet becomes a paramagnet, which is also a Mott insulator. Increasing the pressure, or doping with electrons (e.g., by substituting some d^3 Cr^{3+} for V^{3+}) pushes the electron density over the Mott transition, the spins pair, and the solid becomes metallic.

Superconductors

A magnet levitating above a high-temperature superconductor, cooled with liquid nitrogen. Persistent electric current flows on the surface of the superconductor, excluding the magnetic field of the magnet. This current effectively forms an electromagnet that repels the magnet.

Superconductivity refers to the flow of electrical current in a material with zero resistance. Such materials are very important for use in electromagnets, e.g., in magnetic resonance imaging (MRI) and nuclear magnetic resonance (NMR) machines, because once the current starts flowing in the coils of these magnets it doesn't stop. Magnetic levitation using superconductors which, below a critical field strength, are perfect diamagnets that are not penetrated by magnetic flux lines is also potentially relevant to future technologies such as magnetically levitated trains.

The phenomenon of superconductivity, first discovered in Hg metal in 1911 by Onnes, continues to be only partially understood. It is of great interest to physicists as a macroscopic quantum phenomenon, and to chemists and materials scientists who try to make better superconductors (especially those that superconduct at higher temperatures) and devices derived from them, such as superconducting quantum interference devices (SQUIDs), which are extremely sensitive magnetometers.

Spin pairing and zero resistance: The transition from the metallic to the superconducting state is related to the quantum phenomena of Bose-Einstein condensation and superfluidity. Individual electrons have spin = 1/2, and as such are fermions (particles with half-integer spin). Because of the Pauli exclusion principle, no more than two fermions can occupy the same quantum state (such as an orbital in a molecule or a solid). The familiar consequence of this rule is the aufbau filling of orbitals with spin-paired electrons in each energy level. In contrast, particles with integer spins which are called bosons do not have this restriction, and any number of bosons can occupy the same quantized energy level.

Superconductivity occurs when electrons spin pair into so-called Cooper pairs, which can travel through the lattice together. The electrons in a Cooper pair, although spin-paired, have a long-distance relationship: the spatial extent of a Cooper pair is a few nanometers in cuprate superconductors, and up to one micron in low T_c superconductors such as aluminum. Because its overall spin angular momentum is zero, a Cooper pair is a boson. When the temperature is low enough, the Cooper pairs "condense" into the lowest energy level. The second lowest energy level which is typically a few meV above the ground state is not accessible to them as long as the energy gap is larger than the thermal energy, kT. The scattering of electrons by the lattice then becomes forbidden by energy conservation because scattering dissipates energy, and the Cooper pairs cannot change their energy state. Thus the resistance (which arises from scattering, as we learned in Ch. 6) drops abruptly to zero below T_c. However, the Cooper pairs can be broken apart when they move fast, and thus superconductors turn back into normal metals (even below T_c) above some critical current density j_c. This phenomenon is also related to the critical magnetic field, H_c, that quenches superconductivity.

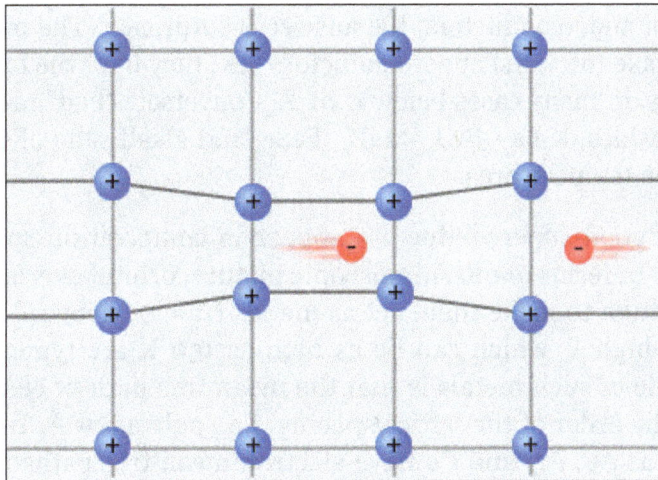

A trampoline for electrons: What causes electrons, which repel each other because of their negative charge, to pair up and travel together in superconductors? The mechanism which must involve some kind of attractive interaction between electrons is well understood for "conventional" super-conductors which have relatively low transition temperatures, but is not yet known with certainty for high temperature oxide superconductors. In conventional, or BCS superconductors, the spin pairing is mediated by the lattice as shown in the figure. A strong electron-lattice interaction causes a distortion in the lattice as an electron moves through. This elastic deformation is felt as an attractive force by a second electron moving in the opposite direction. This can be thought of as analogous to the interaction of two people jumping on a trampoline. The weight of the first person on the trampoline creates a "well" that attracts the second one, and they tend to move together (even if they don't like each other). Strange as this interaction seems, it is supported experimentally by isotope effects on T_c and by quantitative predictions of T_c values in conventional superconductors.

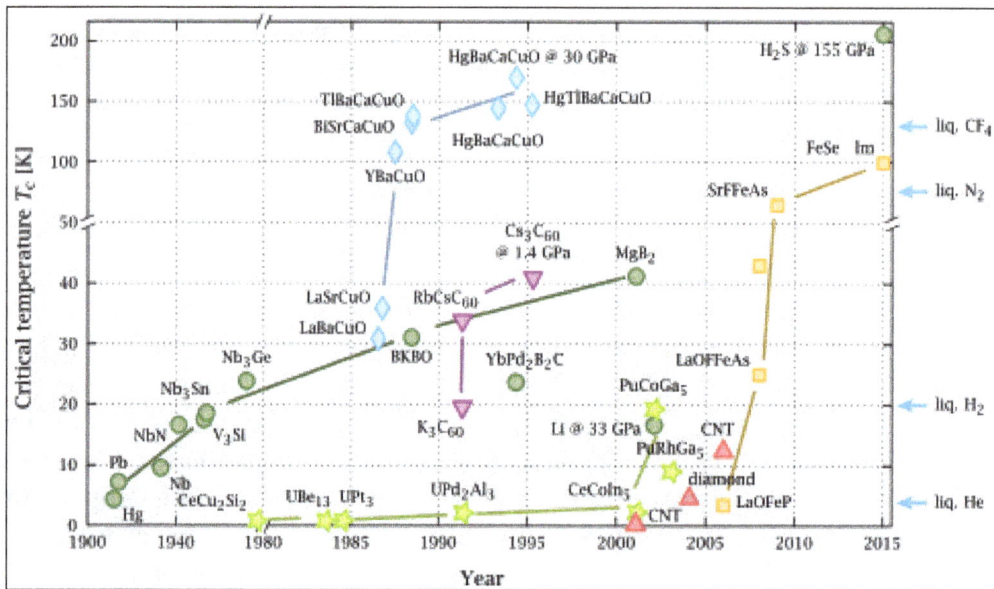

Timeline of superconducting materials, showing T_c vs. year of discovery.

Bad metals make good superconductors: All superconductors are "normal" metals with finite electrical resistance above their critical transition temperature, T_c. If you ask where in the periodic table one might look for superconductors, the answer is surprising. The most conductive metals (Ag, Au, Cu, Cs, etc.) make the worst superconductors, i.e., they have the lowest superconducting transition temperatures, in many cases below 0.01 K. Conversely "bad" metals, such as niobium alloys, certain copper oxides, $K_xBa_{1-x}BiO_3$, MgB_2, FeSe, and alkali salts of C_{60}^{n-} anions, can have relatively high transition temperatures.

We observe that most good superconductors appear in composition space very near a metal-insulator transition. In terms of our microscopic picture, orbital overlap in superconductors is poor, just barely enough to make them act as metals ($\Delta \approx U$) above T_c. In the normal state, superconductors with high T_c which can be as high as 150 K are typically "bad" metals. An important characteristic of such metals is that the mean free path of electrons (in the normal state, above T_c) is on the order of the lattice spacing, i.e., only a few Å. In contrast, we learned that good metals such as Au, Ag, and Cu have electron mean free paths that are two orders of magnitude longer (ca. 40 nm). In a bad metal, the electron "feels" the lattice rather strongly, whereas in a good metal, the electrons are insensitive to small changes in the distance between metals atoms.

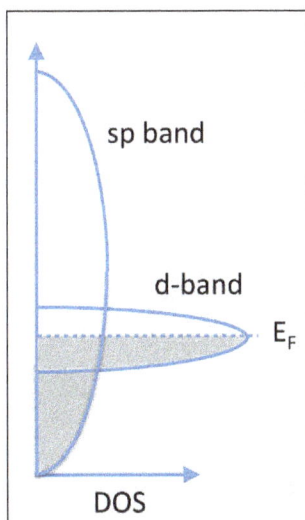

Generic E vs. DOS for a bad metal.

What does the band picture look like for a bad metal? The key point is that because orbital overlap is poor, the metal has a high density of states at the Fermi level. This is a universal property of high temperature superconductors and provides a clue of where to look for new and improved superconducting materials. Recall that transition elements in the middle of the 3d series (Cr, Fe, Co, Ni) were magnetic because of poor orbital overlap and weak d-d bonding. The elements below these especially Nb, Ta, and W - have just barely enough d-d orbital overlap to be on the metallic side of the metal-insulator transition and to be "bad" metals. Carbides and nitrides of these elements are typically superconducting, with the carbon and nitrogen atoms serving to adjust the valence electron density, as illustrated in the table below.

Compound	NbC	Mo_2N	TaC	VN	NbN	TaN	Nb_3Ge
T_c (K)	11.1	5.0	9.7	7.5	15.2	17.8	22.3

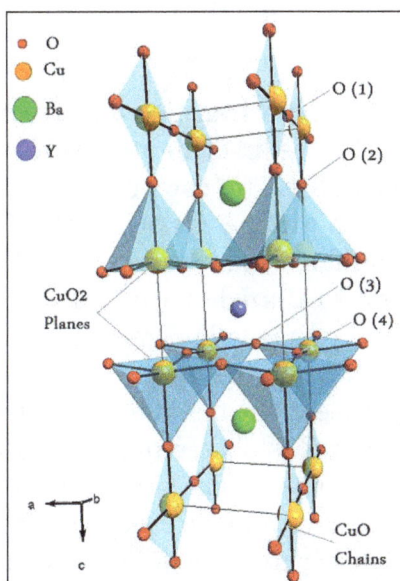

Crystal structure of $YBa_2Cu_3O_{7-\delta}$ (YBCO), the first superconductor with T_c above the boiling point of liquid nitrogen.

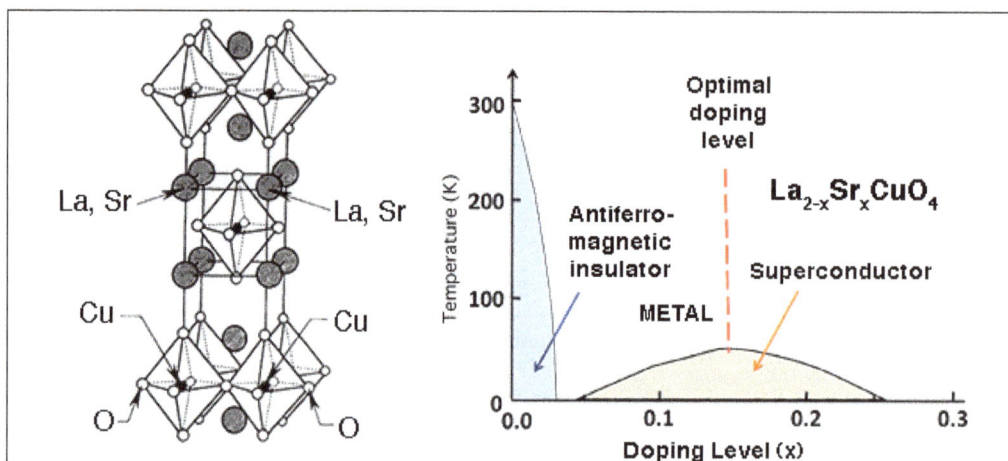

Crystal structure and phase diagram of the cuprate superconductor $La_{2-x}Sr_xCuO_4$. (LSCO).

High T_c superconductors: In addition to having weak orbital overlap in the metallic state which results in a high DOS at E_F, high temperature superconductors also typically contain elements in mixed oxidation states (for example, $Cu^{2+/3+}$ or $Bi^{3+/5+}$) that are close in energy to the O^{2-}/O^- couple in the lattice. At ambient pressure, cuprate superconductors have the highest known T_c values, ranging between about 35 and 150 K. The crystal structures of these materials are almost all variants of the perovskite lattice, as shown at the right for the 1-2-3 superconductor $YBa_2Cu_3O_{7-\delta}$. An ideal perovskite lattice would have formula $ABO_3 = A_3B_3O_9$. In $YBa_2Cu_3O_{7-\delta}$, Y and Ba occupy the A cation sites, Cu occupies the B sites, and two of the nine O atoms are missing.

The $YBa_2Cu_3O_{7-\delta}$ lattice consists of mixed-valent copper(II/III) oxide sheets capped by oxygen atoms to form CuO_5 square pyramids. These sheets encapsulate the Y^{3+} cations. Copper(II) oxide ribbons that share the apical oxygen atoms of the square pyramids run in one direction through the structure. In $YBa_2Cu_3O_{7-\delta}$ and related materials, one component of the structure (here the Cu-O ribbons) acts as a charge reservoir to control the doping of the planar CuO_2 sheets, which are the elements of the structure that carry the supercurrent. Cuprate superconductors with Bi, Tl, or Hg-containing charge reservoir layers and multiple, eclipsed CuO_2 sheets in the unit cell tend to have the highest T_c values.

The connection between the metal-insulator transtion and superconductivity is nicely illustrated in the phase diagram of $La_{2-x}Sr_xCuO_4$, the first cuprate superconductor, which was discovered in 1986 by Georg Bednorz and K. Alex Müller. This compound has a rather simple structure in which rocksalt La(Sr)O layers are intergrown with perovskite $La(Sr)CuO_3$ layers. Undoped La_2CuO_4 contains only Cu^{2+} ions and is an antiferromagnetic insulator. As a small amount of Sr^{2+} is substituted for La^{3+}, some of the Cu^{2+} is oxidized to Cu^{3+}, and the lattice is doped with holes. As the doping level increases, the antiferromagnetic phase undergoes a first-order phase transition to a "bad" metal, and at slightly higher doping density the superconducting phase appears. The proximity of the superconducting phase to the metal-insulator transition is a hallmark of cuprate superconductors. A maximum T_c of 35K is observed at x = 0.15. Doping at higher levels moves the Fermi level beyond the point of highest DOS in the d-band of Cu and the superconducting phase then gradually disappears. It is interesting to compare this phase diagram with that of V_2O_3 (above), which also undergoes an antiferromagnetic insulator to "bad" metal transition as it is doped.

Periodic Trends: Metals, Semiconductors and Insulators

As we consider periodic trends in the electronic properties of materials:

- Going down the periodic table, atoms in solids tend to adopt structures with higher coordination numbers.

- The second row of the periodic table is special, with strong s-p hybridization and π-bonding between atoms.

- Electrons in higher quantum shells are less strongly bound, so the energy difference between bonding and antibonding orbitals becomes smaller for heavier atoms.

We also know that most of the elements in the periodic table are metals, but the elements in the top right corner are insulating under ordinary conditions (1 atm. pressure) and tend to obey the octet rule in their compounds.

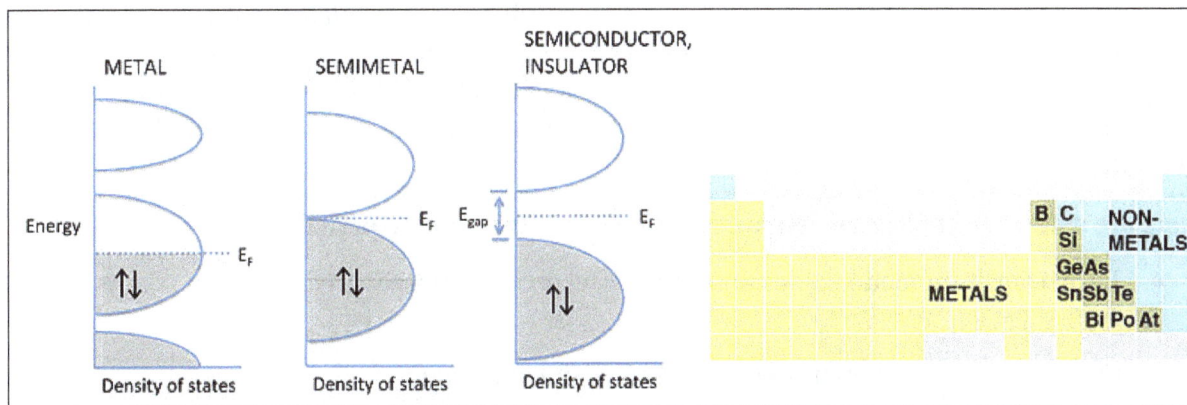

At the transition between metals and non-metals in the periodic table we encounter a crossover in electronic properties, as well as in other properties such as the acidity of the oxides. The group of elements at the border is loosely referred to as the metalloids. Several of these elements (such as C, Sn, and As) can exist as different allotropes that can be metals, insulators, or something in between.

A more rigorous delineation of the electronic properties of these elements (and of many compounds) can be made by considering their band structures and the temperature dependence of the electronic conductivity. Metals have partially filled energy bands, meaning that the Fermi level intersects a partially filled band. With increasing temperature, metals become poorer conductors because lattice vibrations (which are called phonons in the physics literature) scatter the mobile valence electrons. In contrast, semiconductors and insulators, which have filled and empty bands, become better conductors at higher temperature, since some electrons are thermally excited to the lowest empty band. The distinction between insulators and semiconductors is arbitrary, and from the point of view of metal-insulator transitions, all semiconductors are insulators. We typically call an insulator a semiconductor if its band gap (E_{gap}) is less than about 3 eV. A semimetal is a material that has a band gap near zero, examples being single sheets of sp^2-bonded carbon (graphene) and elemental Bi. Like a narrow gap semiconductor, a semimetal has higher conductivity at higher temperature.

Semiconductors: Band Gaps, Colors, Conductivity and Doping

A 2" wafer cut from a GaAs single crystal. GaAs, like many p-block semiconductors, has the zincblende structure.

Semiconductors, as we noted, are somewhat arbitrarily defined as insulators with band gap energy < 3.0 eV (~290 kJ/mol). This cutoff is chosen because, the conductivity of undoped semiconductors drops off exponentially with the band gap energy and at 3.0 eV it is very low. Also, materials with wider band gaps (e.g. $SrTiO_3$, E_{gap} = 3.2 eV) do not absorb light in the visible part of the spectrum.

There are a number of places where we find semiconductors in the periodic table:

- Early transition metal oxides and nitrides, especially those with d^0 electron counts such as TiO_2, TaON, and WO_3.

- Oxides of later 3d elements such as Fe_2O_3, NiO, and Cu_2O.

- Layered transition metal chalcogenides with d^0, d^2 and d^6 electron counts including TiS_2, ZrS_2, MoS_2, WSe_2, and PtS_2.

- d^{10} copper and sliver halides, e.g., CuI, AgBr, and AgI.

- Zincblende- and wurtzite-structure compounds of the p-block elements, especially those that are isoelectronic with Si or Ge, such as GaAs and CdTe. While these are most common, there are other p-block semiconductors that are not isoelectronic and have different structures, including GaS, PbS, and Se.

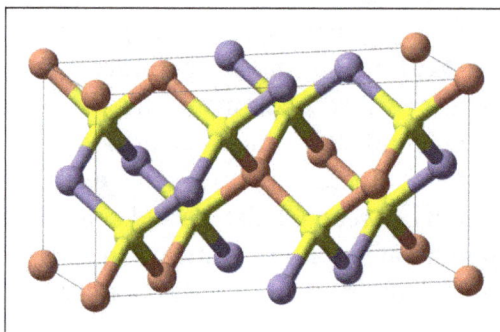

The chalcopyrite structure is adopted by ABX_2 octet semiconductors such as $Cu^IIn^{III}Se_2$ and $Cd^{II}Sn^{IV}P_2$. The unit cell is doubled relative to the parent zincblende structure because of the ordered arrangement of cations. Each anion (yellow) is coordinated by two cations of each type (blue and red).

Zincblende- and wurtzite-structure semiconductors have 8 valence electrons per 2 atoms. These combinations include 4-4 (Si, Ge, SiC), 3-5 (GaAs, AlSb, InP), 2-6 (CdSe, HgTe, ZnO), and 1-7 (AgCl, CuBr) semiconductors. Other variations that add up to an octet configuration are also possible, such as $Cu^I In^{III} Se_2$, which has the chalcopyrite structure, shown in the figure.

How does the band gap energy vary with composition? There are two important trends

- Going down a group in the periodic table, the gap decreases:

$$C \, (\text{diamond}) \;>\; Si \;>\; Ge \;>\; \alpha - Sn$$

$$\mathbf{E_{gap}(eV)}: \qquad 5.4 \qquad\quad 1.1 \quad\; 0.7 \qquad 0.0$$

This trend can be understood by recalling that E_{gap} is related to the energy splitting between bonding and antibonding orbitals. This difference decreases (and bonds become weaker) as the principal quantum number increases.

- For isoelectronic compounds, increasing ionicity results in a larger band gap.

$$Ge \;<\; GaAs \;<\; ZnSe$$
$$0.7 \qquad 1.4 \qquad 2.8 \text{ eV}$$
$$Sn \;<\; InSb \;<\; CdTe \;<\; AgI$$
$$0.0 \qquad 0.2 \qquad 1.6 \qquad\; 2.8\,eV$$

This trend can also be understood from a simple MO picture. As the electronegativity difference $\Delta\chi$ increases, so does the energy difference between bonding and antibonding orbitals.

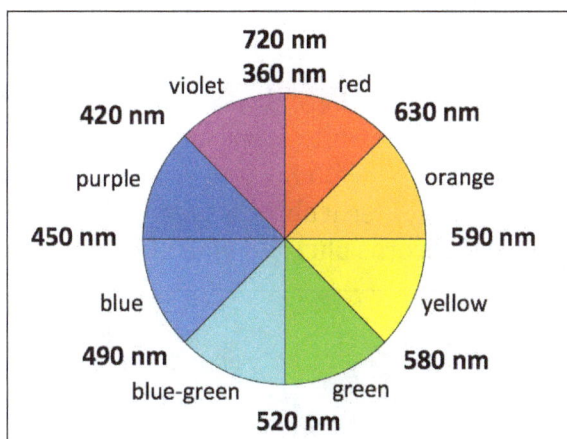

Color wheel showing the colors and wavelengths of emitted light.

The band gap is a very important property of a semiconductor because it determines its color and conductivity. Many of the applications of semiconductors are related to band gaps:

- Narrow gap materials $\left(Hg_x Cd_{1-x} Te, \; VO_2, \; InSb, \; Bi_2 Te_3\right)$ are used as infrared photodetectors and thermoelectrics (which convert heat to electricity).

- Wider gap materials $\left(Si, \; GaAs, \; GaP, \; GaN, \; CdTe, \; CuIn_x Ga_{1-x} Se_2\right)$ are used in electronics, light-emitting diodes, and solar cells.

Fe$_2$O$_3$ powder is reddish orange because of its 2.2 eV band gap.

Semiconductor solid solutions such as GaAs$_{1-x}$P$_x$ have band gaps that are intermediate between the end member compounds, in this case GaAs and GaP (both zincblende structure). Often, there is a linear relation between composition and band gap, which is referred to as Vegard's Law. This "law" is often violated in real materials, but nevertheless offers useful guidance for designing materials with specific band gaps. For example, red and orange light-emitting diodes (LED's) are made from solid solutions with compositions of GaP$_{0.40}$As$_{0.60}$ and GaP$_{0.65}$As$_{0.35}$, respectively. Increasing the mole fraction of the lighter element (P) results in a larger band gap, and thus a higher energy of emitted photons.

Colors of semiconductors: The color of absorbed and emitted light both depend on the band gap of the semiconductor. Visible light covers the range of approximately 390-700 nm, or 1.8-3.1 eV. The color of emitted light from an LED or semiconductor laser corresponds to the band gap energy and can be read off the color wheel shown at the right.

The color of absorbed light includes the band gap energy, but also all colors of higher energy (shorter wavelength), because electrons can be excited from the valence band to a range of energies in the conduction band. Thus semiconductors with band gaps in the infrared (e.g., Si, 1.1 eV and GaAs, 1.4 eV) appear black because they absorb all colors of visible light. Wide band gap semiconductors such as TiO$_2$ (3.0 eV) are white because they absorb only in the UV. Fe$_2$O$_3$ has a band gap of 2.2 eV and thus absorbs light with $\lambda < 560$ nm. It thus appears reddish-orange (the colors of light reflected from Fe$_2$O$_3$) because it absorbs green, blue, and violet light. Similarly, CdS $\left(E_{gap} = 2.6 \text{ eV} \right)$ is yellow because it absorbs blue and violet light.

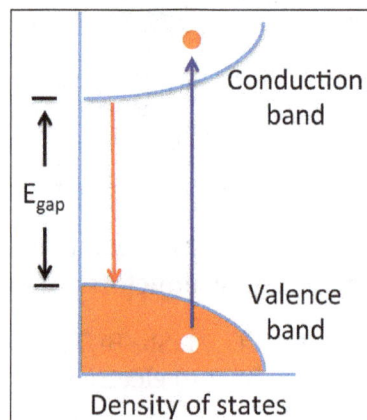

An electron-hole pair is created by adding heat or light energy E > E$_{gap}$ to a semiconductor (blue arrow). The electron-hole pair recombines to release energy equal to E$_{gap}$ (red arrow).

Electrons and holes in semiconductors: Pure (undoped) semiconductors can conduct electricity when electrons are promoted, either by heat or light, from the valence band to the conduction band. The promotion of an electron (e^-) leaves behind a hole (h^+) in the valence band. The hole, which is the absence of an electron in a bonding orbital, is also a mobile charge carrier, but with a positive charge. The motion of holes in the lattice can be pictured as analogous to the movement of an empty seat in a crowded theater. An empty seat in the middle of a row can move to the end of the row (to accommodate a person arriving late to the movie) if everyone moves over by one seat. Because the movement of the hole is in the opposite direction of electron movement, it acts as a positive charge carrier in an electric field.

The opposite process of excitation, which creates an electron-hole pair, is their recombination. When a conduction band electron drops down to recombine with a valence band hole, both are annihilated and energy is released. This release of energy is responsible for the emission of light in LEDs.

At equilibrium, the creation and annihilation of electron-hole pairs proceed at equal rates. This dynamic equilibrium is analogous to the dissociation-association equilibrium of H^+ and OH^- ions in water. We can write a mass action expression:

$$n \times p = K_{eq} = n_i^2 ,$$

where, n and p represent the number density of electrons and holes, respectively, in units of cm⁻³. The intrinsic carrier concentration, n_i, is equal to the number density of electrons or holes in an undoped semiconductor, where $n = p = n_i$.

Note the similarity to the equation for water autodissociation:

$$[H^+][OH^-] = K_w$$

By analogy, n (e.g., by doping), p will decrease, and vice-versa, but their product will remain constant at a given temperature.

Temperature dependence of the carrier concentration: Using the equations $K_{eq} = \exp(-\Delta G^\circ / RT)$ and $\Delta G^\circ = \Delta H^\circ - T\Delta S^\circ$, we can write:

$$n \times p = n_i^2 = \exp(\Delta S^\circ / R)\exp(-\Delta H^\circ / RT)$$

The entropy change for creating electron hole pairs is given by:

$$\Delta S^\circ = R \ln(N_C) + R \ln(N_V) = R \ln(N_C N_V) ,$$

where, N_V and N_C are the effective density of states in the valence and conduction bands, respectively.

and thus we obtain,

$$n_i^2 = N_C N_V \exp(-\Delta H^\circ / RT)$$

Since the volume change is negligible, $\Delta H^\circ \approx \Delta E^\circ$, and therefore $\Delta H^\circ / R \approx E_{gap} / k$, from which we obtain:

$$n_i^2 = N_C N_V \, \exp(-E_{gap} / kT) \, ,$$

and finally,

$$\mathbf{n = p = n_i = \left(N_C N_V \right)^{1/2} \exp\left(-E_{gap} / 2kT \right)}$$

For pure Si $\left(E_{gap} = 1.1 \text{ eV} \right)$ with $N \approx 10^{22}/\text{cm}^3$, we can calculate from this equation a carrier density n_i of approximately $10^{10} / \text{cm}^3$ at 300 K. This is about 12 orders of magnitude lower than the valence electron density of Al, the element just to the left of Si in the periodic table. Thus we expect the conductivity of pure semiconductors to be many orders of magnitude lower than those of metals.

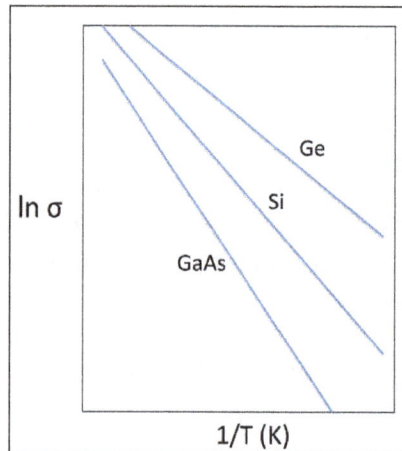

Plots of $\ln(\sigma)$ vs. inverse temperature for intrinsic semiconductors $\text{Ge}\left(E_{gap} = 0.7 \text{ eV} \right)$, Si (1.1 eV) and GaAs (1.4 eV). The slope of the line is - $E_{gap} / 2k$.

Conductivity of intrinsic semiconductors: The conductivity (σ) is the product of the number density of carriers (n or p), their charge (e), and their mobility (μ). μ is the ratio of the carrier drift velocity to the electric field and has units of cm²/Volt-second. Typically electrons and holes have somewhat different mobilities (μ_e and μ_h , respectively) so the conductivity is given by:

$$\sigma = ne\mu_e + pe\mu_h$$

For either type of charge carrier, the mobility μ is given by:

$$\mu = v_{drift} / E = e\tau / m,$$

where e is the fundamental unit of charge, τ is the scattering time, and m is the effective mass of the charge carrier.

Taking an average of the electron and hole mobilities, and using n = p, we obtain,

$$\sigma = \sigma_o \exp\left(-E_{gap} / 2kT \right), \text{ where } \sigma_o = 2(N_C N_V)^{1/2} e\mu.$$

By measuring the conductivity as a function of temperature, it is possible to obtain the activation energy for conduction, which is $E_{gap}/2$. This kind of plot, which resembles an Arrhenius plot, is shown at the right for three different undoped semiconductors. The slope of the line in each case is $E_{gap}/2k$.

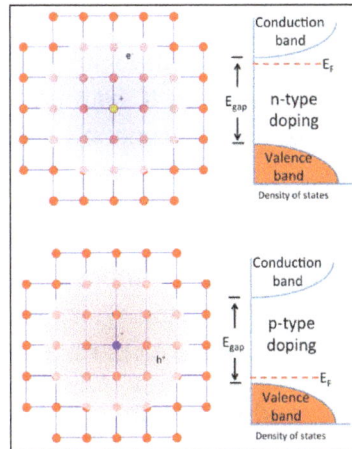

n- and p-type doping of semiconductors involves substitution of electron donor atoms (light orange) or acceptor atoms (blue) into the lattice. These substitutions introduce extra electrons or holes, respectively, which are easily ionized by thermal energy to become free carriers. The Fermi level of a doped semiconductor is a few tens of mV below the conduction band (n-type) or above the valence band (p-type).

Doping of semiconductors: Almost all applications of semiconductors involve controlled doping, which is the substitution of impurity atoms, into the lattice. Very small amounts of dopants (in the parts-per-million range) dramatically affect the conductivity of semiconductors. For this reason, very pure semiconductor materials that are carefully doped both in terms of the concentration and spatial distribution of impurity atoms are needed.

n- and p-type doping: In crystalline Si, each atom has four valence electrons and makes four bonds to its neighbors. This is exactly the right number of electrons to completely fill the valence band of the semiconductor. Introducing a phosphorus atom into the lattice (the positively charged atom in the figure at the right) adds an extra electron, because P has five valence electrons and only needs four to make bonds to its neighbors. The extra electron, at low temperature, is bound to the phosphorus atom in a hydrogen-like molecular orbital that is much larger than the 3s orbital of an isolated P atom because of the high dielectric constant of the semiconductor. In silicon, this "expanded" Bohr radius is about 42 Å, i.e., 80 times larger than in the hydrogen atom. The energy needed to ionize this electron – to allow it to move freely in the lattice is only about 40−50 meV, which is not much larger the thermal energy (26 meV) at room temperature. Therefore the Fermi level lies just below the conduction band edge, and a large fraction of these extra electrons are promoted to the conduction band at room temperature, leaving behind fixed positive charges on the P atom sites. The crystal is n-doped, meaning that the majority carrier (electron) is negatively charged.

Alternatively, boron can be substituted for silicon in the lattice, resulting in p-type doping, in which the majority carrier (hole) is positively charged. Boron has only three valence electrons, and "borrows" one from the Si lattice, creating a positively charged hole that exists in a large hydrogen-like orbital around the B atom. This hole can become delocalized by promoting an electron from the valence band to fill the localized hole state. Again, this process requires only 40–50 meV, and so at room temperature a large fraction of the holes introduced by boron doping exist in delocalized valence band states. The Fermi level (the electron energy level that has a 50% probability of occupancy at zero temperature) lies just above the valence band edge in a p-type semiconductor.

As noted above, the doping of semiconductors dramatically changes their conductivity. For example, the intrinsic carrier concentration in Si at 300 K is about 10^{10} cm^{-3}. The mass action equilibrium for electrons and holes also applies to doped semiconductors, so we can write:

$$n \times p = n_i^{2} = 10^{20} \text{ cm}^{-6} \text{ at 300K}$$

If we substitute P for Si at the level of one part-per-million, the concentration of electrons is about 10^{16} cm^{-3}, since there are approximately 10^{22} Si atoms/cm^3 in the crystal. According to the mass action equation, if $n = 10^{16}$, then $p = 10^4$ cm^{-3}. There are three consequences of this calculation:

- The density of carriers in the doped semiconductor (10^{16} cm^{-3}) is much higher than in the undoped material ($\sim 10^{10}$ cm^{-3}), so the conductivity is also many orders of magnitude higher.

- The activation energy for conduction is only 40–50 meV, so the conductivity does not change much with temperature (unlike in the intrinsic semiconductor).

- The minority carriers (in this case holes) do not contribute to the conductivity, because their concentration is so much lower than that of the majority carrier (electrons).

Similarly, for p-type materials, the conductivity is dominated by holes, and is also much higher than that of the intrinsic semiconductor.

Chemistry of semiconductor doping: Sometimes it is not immediately obvious what kind of doping (n- or p-type) is induced by "messing up" a semiconductor crystal lattice. In addition to substitution of impurity atoms on normal lattice sites (the examples given for Si), it is also possible to dope with vacancies missing atoms and with interstitials extra atoms on sites that are not ordinarily occupied. Some simple rules are as follows:

- For substitutions, adding an atom to the right in the periodic table results in n-type doping, and an atom to the left in p-type doping.

  For example, when TiO_2 is doped with Nb on some of the Ti sites, or with F on O sites, the result is n-type doping. In both cases, the impurity atom has one more valence electron than the atom for which it was substituted. Similarly, substituting a small amount of Zn for Ga in GaAs, or a small amount of Li for Ni in NiO, results in p-type doping.

- Anion vacancies result in n-type doping, and cation vacancies in p-type doping.

  Examples are anion vacancies in CdS_{1-x} and WO_{3-x}, both of which give n-type semiconductors, and copper vacancies in $Cu_{1-x}O$, which gives a p-type semiconductor.

- Interstitial cations (e.g. Li) donate electrons to the lattice resulting in n-type doping. Interstitial anions are rather rare but would result in p-type doping.

Sometimes, there can be both p- and n-type dopants in the same crystal, for example B and P impurities in a Si lattice, or cation and anion vacancies in a metal oxide lattice. In this case, the two kinds of doping compensate each other, and the doping type is determined by the one that is in higher concentration. A dopant can also be present on more than one site. For example, Si can occupy both the Ga and As sites in GaAs, and the two substitutions compensate each other. Si has a slight preference for the Ga site, however, resulting in n-type doping.

Semiconductor p-n Junctions

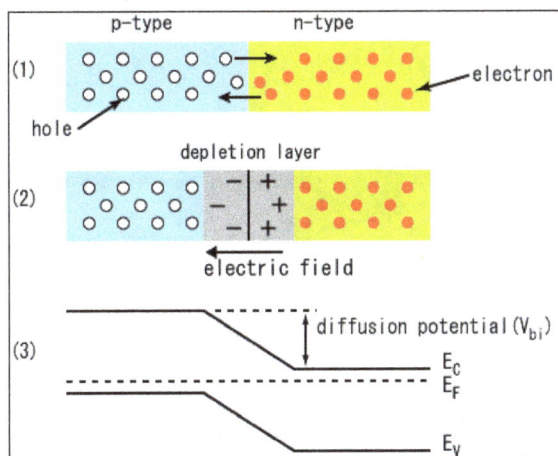

When p-type and n-type semiconductors are joined, electrons and holes are annihilated at the interface, leaving a depletion region that contains positively and negatively charged donor and acceptor atoms, respectively. At equilibrium, the Fermi level $\left(E_F\right)$ is uniform throughout the junction. E_F lies just above the valence band on the p-type side of the junction and just below the conduction band on the n-type side.

Semiconductor p-n junctions are important in many kinds of electronic devices, including diodes, transistors, light-emitting diodes, and photovoltaic cells. To understand the operation of these devices, we first need to look at what happens to electrons and holes when we bring p-type and n-type semiconductors together. At the junction between the two materials, mobile electrons and holes annihilate each other, leaving behind the fixed + and charges of the electron donor and electron acceptor dopants, respectively. For example, on the n-side of a silicon p-n junction, the positively charged dopants are P^+ ions and on the p-side the negatively charged dopants are B^-. The presence of these uncompensated electrical charges creates an electric field, the built-in field of the p-n junction. The region that contains these charges (and a very low density of mobile electrons or holes) is called the depletion region.

The electric field, which is created in the depletion region by electron-hole recombination, repels both the electrons (on the n-side) and holes (on the p-side) away from the junction. The concentration gradient of electrons and holes, however, tends to move them in the opposite direction by diffusion. At equilibrium, the flux of mobile carriers is zero because the field-driven migration flux is equal and opposite to the concentration-driven diffusion flux.

The width of the depletion layer depends on the screening length in the semiconductor, which in turn depends on the dopant density. At high doping levels, the depletion layer is narrow (tens of nanometers across), whereas at low doping density it can be as thick as 1 µm. The depletion region is the only place where the electric field is nonzero, and the only place where the bands bend. Elsewhere in the semiconductor the field is zero and the bands are flat.

In the middle of p-n junction, the Fermi level energy, E_F, is halfway between the valence band, VB, and the conduction band, CB, and the semiconductor is intrinsic $(n = p = n_i)$.

Diodes, LEDs and Solar Cells

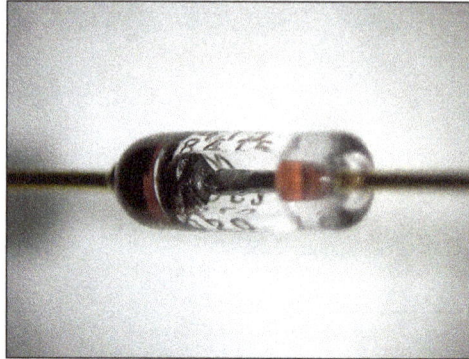

Closeup of a diode, showing the square-shaped semiconductor crystal (*black object on left*).

Diodes are semiconductor devices that allow current to flow in only one direction. Diodes act as rectifiers in electronic circuits, and also as efficient light emitters (in LEDs) and solar cells (in photovoltaics). The basic structure of a diode is a junction between a p-type and an n-type semiconductor, called a p-n junction. Typically, diodes are made from a single semiconductor crystal into which p- and n- dopants are introduced.

If the n-side of a diode is biased at positive potential and the p-side is biased negative, electrons are drawn to the n-side and holes to the p-side. This reinforces the built in potential of the p-n junction, the width of the depletion layer increases, and very little current flows. This polarization direction is referred to as "back bias". If the diode is biased the other way, carriers are driven into the junction where they recombine. The electric field is diminished, the bands are flattened, and current flows easily since the applied bias lowers the built-in potential. This is called "forward bias".

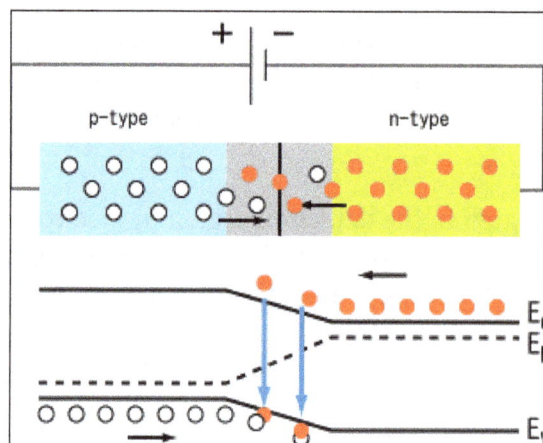

Electrons (red) and holes (white) in a forward-biased diode.

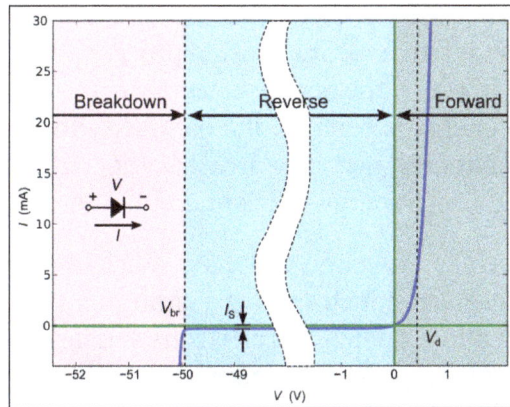

Diode i-V curve.

The figure illustrates a forward-biased diode, through which current flows easily. As electrons and holes are driven into the junction (black arrows in lower figure), they recombine (downward blue arrows), producing light and/or heat. The Fermi level in the diode is indicated as the dotted line. There is a drop in the Fermi level (equal to the applied bias) across the depletion layer. The corresponding diode i-V curve is shown on the right. The current rises exponentially with applied voltage in the forward bias direction, and there is very little leakage current under reverse bias. At very high reverse bias (typically tens of volts) diodes undergo avalanche breakdown and a large reverse current flows.

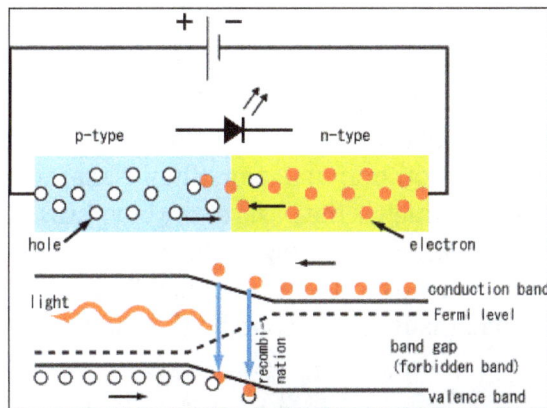

Light-emitting diode (LED).

Prof. Shuji Nakamura holding a blue LED.

A light-emitting diode or LED is a kind of diode that converts some of the energy of electron-hole recombination into light. This radiative recombination process always occurs in competition with non-radiative recombination, in which the energy is simply converted to heat. When light is emitted from an LED, the photon energy is equal to the bandgap energy. Because of this, LED lights have pure colors and narrow emission spectra relative to other light sources, such as incandescent and fluorescent lights. LED lights are energy-efficient and thus are typically cool to the touch.

Direct-gap semiconductors such as GaAs and GaP have efficient luminescence and are also good light absorbers. In direct gap semiconductors, there is no momentum change involved in electron-hole creation or recombination. That is, the electrons and holes originate at the same value of the momentum wavevector k, k is related to the momentum (also a vector quantity) by $p = hk/2\pi$. In a direct-gap semiconductor, the top of the valence band and the bottom of the conduction band most typically both occur at $k = 0$. Since the momentum of the photon is close to zero, photon absorption and emission are strongly allowed (and thus kinetically fast). Polar semiconductors such as GaAs, GaN, and CdSe are typically direct-gap materials. Indirect-gap semiconductors such as Si and Ge absorb and emit light very weakly because the valence band maximum and conduction band minimum do not occur at the same point in k-space. This means that a lattice vibration (a phonon) must also be created or annihilated in order to conserve momentum. Since this "three body" (electron, hole, phonon) process has low probability, the radiative recombination of electrons and holes is slow relative to non-radiative decay the thermalization of electron-hole energy as lattice vibrations in indirect-gap semiconductors. The momentum selection rule thus prevents light absorption/emission and there are no pure Si LEDs or Si-based lasers.

While red, orange, yellow, and green LEDs can be fabricated relatively easily from AlP-GaAs solid solutions, it was initially very difficult to fabricate blue LEDs because the best direct gap semiconductor with a bandgap in the right energy range is a nitride, GaN, which is difficult to make and to dope p-type. Working at Nichia Corporation in Japan, Shuji Nakamura succeeded in developing a manufacturable process for p-GaN, which is the basis of the blue LED. Because of the importance of this work in the development of information storage (Blu-Ray technology) and full-spectrum, energy-efficient LED lighting, Nakamura shared the 2014 Nobel Prize in Physics with Isamu Asaki and Hiroshi Amano, both of whom had made earlier contributions to the development of GaN diodes.

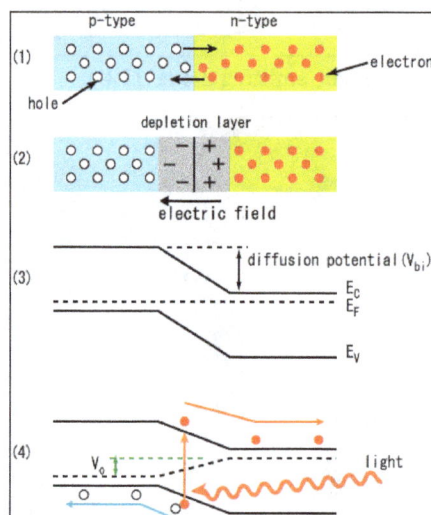

Photovoltaic effect in a semiconductor p-n junction.

A Solar cell, or photovoltaic cell, converts light absorbed in a p-n junction directly to electricity by the photovoltaic effect. Photovoltaics is the field of technology and research related to the development of solar cells for conversion of solar energy to electricity. Sometimes the term solar cell is reserved for devices intended specifically to capture energy from sunlight, whereas the term photovoltaic cell is used when the light source is unspecified.

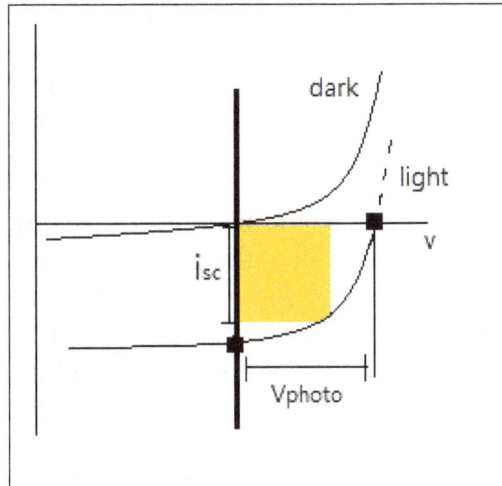

Current-voltage characteristic of a solar cell in the dark and under illumination with band gap light. The short-circuit photocurrent is indicated as i_{sc}, and the open-circuit photovoltage is V_{photo}. The maximum power generated by the solar cell is determined by the area of the orange box.

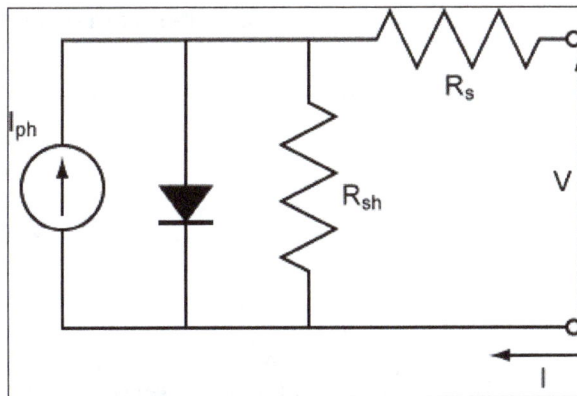

The equivalent circuit of a p-n junction solar cell, which results in the "light" i-V curve shown in the figure above. The solar cell is effectively a diode with a reverse-bias current source provided by light-generated electrons and holes. The shunt resistance (R_{sh}) in the equivalent circuit represents parasitic electron-hole recombination. A high shunt resistance (low recombination rate) and low series resistance (R_s) are needed for high solar cell efficiency.

Photocurrent in p-n junction solar cells flows in the diode reverse bias direction. In the dark, the solar cell simply acts as a diode. In the light, the photocurrent can be thought of as a constant current source, which is added to the i-V characteristic of the diode. The relationship between the dark and light current in a photovoltaic cell is shown in the diagram.

The built-in electric field of the p-n junction separates e⁻ h⁺ pairs that are formed by absorption of bandgap light in the depletion region. The electrons flow downhill, towards the n-type side of the junction, the holes flow uphill towards the p-side. If $h\nu \geq E_{gap}$, light can be absorbed by promoting an electron from the valence band to the conduction band. Any excess energy is rapidly thermalized. Light with $h\nu > E_g$ thus can store only E_g worth of energy in an e^- h^+ pair. If light is absorbed outside of depletion region, i.e., on the n- or p-side of the junction where there is no electric field, minority carriers must diffuse into the junction in order to be collected. This process occurs in competition with electron-hole recombination. Because impurity atoms and lattice defects make efficient recombination centers, semiconductors used in solar cells (especially indirect-gap materials such as Si, which must be relatively thick in order to absorb most of the solar spectrum) must be very pure. Most of the cost of silicon solar cells is associated with the process of purifying elemental silicon and growing large single crystals from the melt.

In the photodiode i-V curve above, V_{photo} is typically only about 70% of the bandgap energy E_{gap}. The photocurrent is limited by the photon flux, the recombination rate, and the re-emission of absorbed light. The area of the orange rectangle indicates the power generated by the solar cell, which can be calculated as $P = i \times V$. In good single crystal or polycrystalline solar cells made of Si, GaAs, CdTe, $CuIn_xGa_{1-x}Se_2$, or $(CH_3NH_3)PbI_3$ the quantum yield (the ratio of short circuit photocurrent to photon flux) is close to unity.

Solar cells have many current applications. Individual cells are used for powering small devices such as electronic calculators. Photovoltaic arrays generate a form of renewable electricity, particularly useful in situations where electrical power from the grid is unavailable such as in remote area power systems, Earth-orbiting satellites and space probes, remote radio-telephones and water pumping applications. Photovoltaic electricity is also increasingly deployed in grid-tied electrical systems.

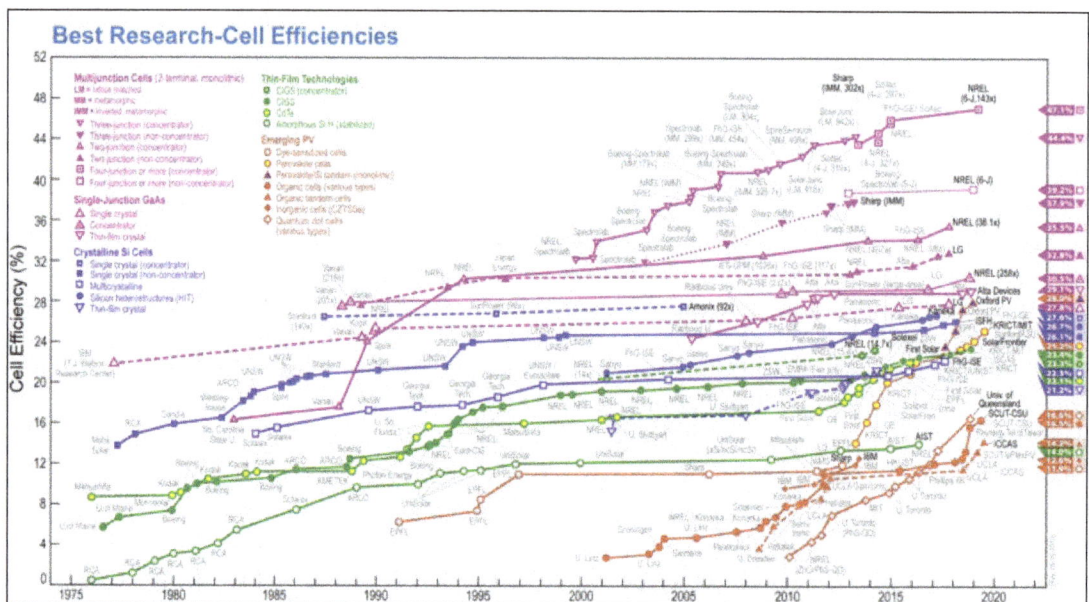

Reported timeline of solar cell energy conversion efficiencies since 1976 (National Renewable Energy Laboratory).

The cost of installed photovoltaics (calculated on a per-watt basis) has dropped over the past decade at a rate of about 13% per year, and has already reached grid parity in Germany and a number of other countries. Photovoltaic grid parity is anticipated in U.S. power markets in the 2020 timeframe. A major driver in the progressively lower cost of photovoltaic power is the steadily increasing efficiency of solar cells, which is shown in the graphic at the right. Higher efficiency solar cells require less area to deliver the same amount of power, and this lowers the "balance of system" costs such as wiring, roof mounting, etc., which scale as the area of the solar panels. Progress towards higher efficiency reflects improved processes for making photovoltaic materials such as silicon and gallium arsenide, as well as the discovery of new materials. Silicon solar cells are a mature technology, so they are now in the flat part of the learning curve and are approaching their maximum theoretical efficiencies. Newer technologies such as organic photovoltaics, quantum dot solar cells, and lead halide perovskite cells are still in the rising part of the learning curve.

A field effect transistor (FET) is a transistor that uses an electric field to control the width of a conducting channel and thus the current in a semiconductor material. It is classified as unipolar transistor, in contrast to bipolar transistors.

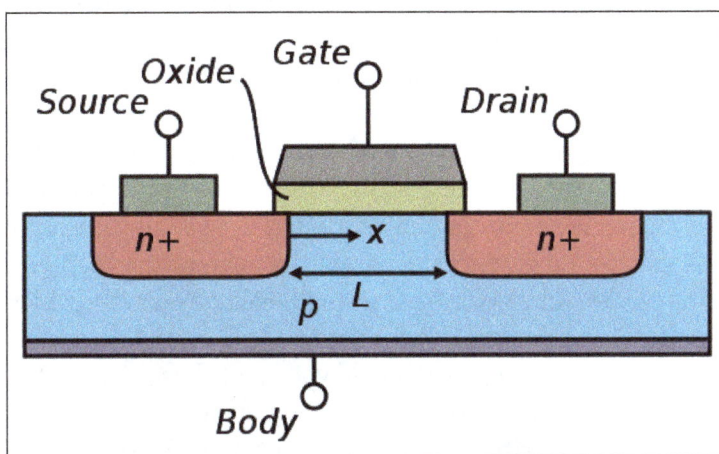

Cross section of an n-type MOSFET.

Field effect transistors function as current amplifiers. The typical structure of Si-based FETs is one in which two n-type regions (the source and the drain) are separated by a p-type region. An oxide insulator over the p-type region separates a metal gate lead from the semiconductor. This structure is called a metal-oxide-semiconductor FET (or MOSFET). When voltage is applied between source and drain, current cannot flow because either the n-p or the p-n junction is back-biased. When a positive potential is applied to the gate, however, electrons are driven towards the gate, and locally the semiconductor is "inverted" to n-type. Then the current flows easily between the n-type source and drain through the n-channel. The current flow between the source and drain is many times larger than the current through the gate, and thus the FET can act as an amplifier. Current flow can also represent a logical "1," so FETs are also used in digital logic.

In electronic devices such as microprocessors, field-effect transistors are kept in the off-state most of the time in order to minimize background current and power consumption. The FET shown, which has n-type source and drain regions, is called an NMOS transistor. In a PMOS transistor, the source and drain regions are p-type and the gate is n-type. In CMOS (complementary metal-oxide semiconductor) integrated circuits, both NMOS and PMOS transistors are used. CMOS circuits are constructed in such a way that all PMOS transistors must have either an input from the voltage source or from another PMOS transistor. Similarly, all NMOS transistors must have either an input from ground or from another NMOS transistor. This arrangement results in low static power consumption.

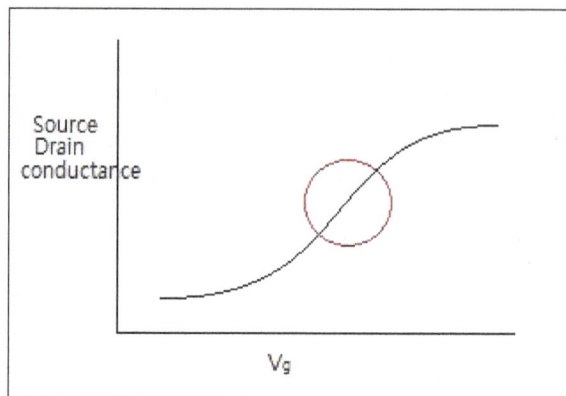

Transistors are most useful in the range of gate voltage (indicated by the red circle in the figure at the left) where the source-drain current changes rapidly. In this region it is possible to effect a large change in current between source and drain when a small signal is applied to the gate. An important figure of merit for FETs is the subthreshold slope, which is the slope a plot of log(current) vs. V_{gate}. An ideal subthreshold slope is one decade of current per 60 mV of gate bias. Typically, a decade change in source-drain current can be achieved with a change in gate voltage of ~70 mV. The performance of FETs as switches and amplifiers is limited by the subthreshold slope, which in turn is limited by the capacitance of the gate. It is desirable to have a very high gate capacitance, which requires a thin insulating oxide, but also to have a small leakage current, which requires a thick oxide. A current challenge in the semiconductor industry is to continue to scale FETs to even smaller nanoscale dimensions while maintaining acceptable values of these parameters. This is being done by developing new gate insulator materials that have higher dielectric constants than silicon oxide and do not undergo redox reactions with silicon or with metal gate leads. Only a few known materials (such as hafnium oxynitride and hafnium silicates) currently meet these stringent requirements.

Amorphous semiconductors

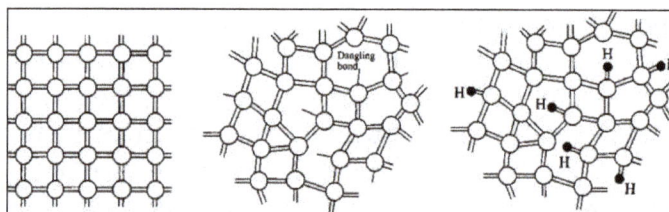

Schematic illustration of the structures of crystalline silicon (left), amorphous silicon (middle), and amorphous hydrogenated silicon (right).

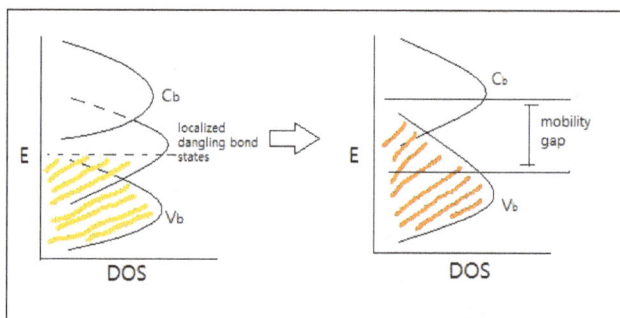

Energy vs. DOS for an amorphous semiconductor. Disorder and dangling bonds result in localized mid-gap states.

Amorphous semiconductors are disordered or glassy forms of crystalline semiconductor materials. Like non-conducting glasses, they are network structures with primarily covalent bonding. Crystalline silicon, which has the diamond structure, is an ordered arrangement of fused six-membered silicon rings, all in the "chair" conformation, The local bonding environment of the silicon atoms is tetrahedral. The silicon atoms in amorphous silicon (a-Si) are also predominantly tetrahedrally coordinated, but there is no long-range order in the structure. In addition to six-membered rings, there are five- and seven-membered rings, as well as some "dangling bond" sites in which Si atoms have only three nearest neighbors.

Two of the most widely studied amorphous semiconductors are a-Si and amorphous selenium, a-Se. Si and Se can both be made in glassy form, usually by sputtering or evaporation at relatively low temperature. In a-Se, as in a-Si, locally, most of the atoms have their "normal" valence, but there are many defects and irregularities in the structure. Dangling bonds in amorphous semiconductors have orbital energies in the middle of gap, and electrons in these states are effectively non-bonding. Because these dangling bond sites are far apart from each other, there is little orbital overlap between them, and they also exist over a range of energies. Electrons in these mid-gap states are therefore localized, a phenomenon known as Anderson localizaton. Amorphous Si is insulating because electrons the Fermi level (in the middle of the gap) are not mobile in the lattice. These localized states create a mobility gap, and only electrons in states that are strongly bonding or antibonding are delocalized. Therefore, unmodified a-Si is not very useful as a semiconductor. However, by hydrogenating the material as it is formed (typically in a plasma of H atoms), the under-coordinated Si atoms are bonded to hydrogen atoms. This generates filled bonding and empty antibonding orbitals, the energies of which are outside the mobility gap. Hydrogenation thus lowers the density of states in the mobility gap. Hydrogenated amorphous silicon (a-Si:H) is insulating in the dark, but is a good photoconductor because light absorption creates electrons and holes in mobile states that are outside the mobility gap.

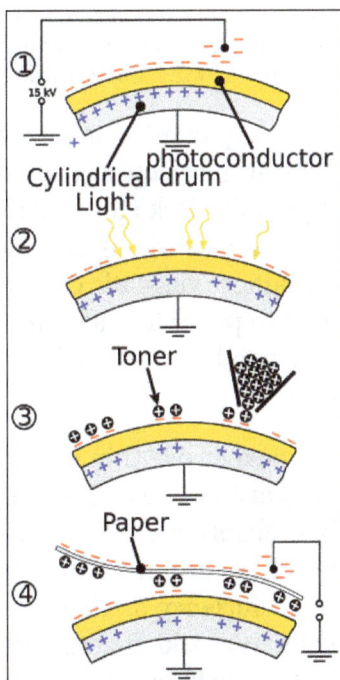

Charging of amorphous Se and pattern transfer in the xerographic cycle.

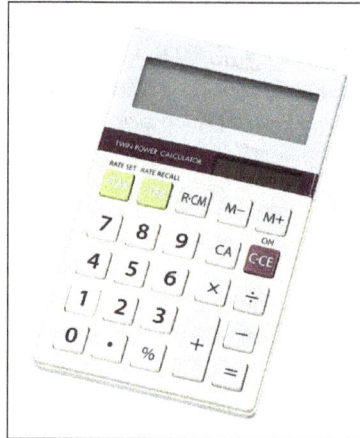

A calculator that runs on solar and battery power.

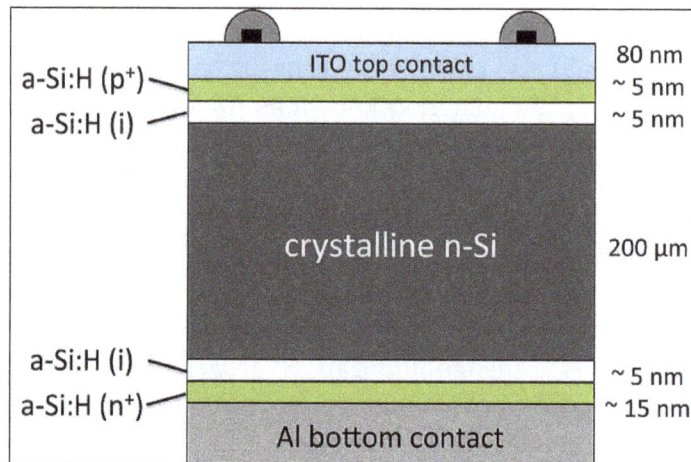

Layered structure of a HIT solar cell. The layers are not drawn to scale. A thick crystalline n-silicon layer is the light absorber, and photogenerated holes, which are the minority carriers, are reflected away from the aluminum back contact by the thin intrinsic a-Si layer there.

The photoconductivity of amorphous Se is exploited in xerography. A conductive drum coated with a-Se, which is insulating, is charged with static electricity by corona discharge from a wire. When the drum is exposed to a pattern of light and dark (the image to be duplicated), the illuminated a-Se areas become conductive and the static charge is dissipated from those parts of the drum. Carbon-containing toner particles then adhere via static charge to the areas that were not exposed to light, and are transferred and bonded to paper to make the copy. The speed of the process and the high resolution of pattern transfer depend on the very low conductivity of a-Se in the dark and its high conductivity under illumination.

Amorphous hydrogenated Si is used in inexpensive thin film solar cells. The mobility gap is about 1.7 eV, which is larger than the bandgap crystalline of Si (1.1 eV). a-Si:H is a direct-gap material, and therefore thin films are good light absorbers. a-Si:H solar cells can be vapor-deposited in large-area sheets. p^+Si-a-Si:H-n^+Si cells have around 10% power conversion efficiency. However amorphous Si solar cells gradually lose efficiency as they are exposed to light. The mechanism of this efficiency loss, called the Staebler-Wronski effect,, involves photogenerated electron-hole pairs which have sufficient energy to cause chemical changes in the material. While the exact mechanism is still unclear, it has been proposed that the energy of electron-hole recombination breaks a weak Si-Si bond, and that one of the resulting dangling bonds abstracts a H atom, leaving a passivated Si-H center and a permanent dangling bond. The effect is minimized by hydrogenating a-Si and can be partially reversed by annealing.

Thin layers of amorphous silicon are used in conjunction with crystalline silicon in heterojunction intrinsic thin-layer (HIT) solar cells. Because the mobility gap of a-Si is wider than the bandgap of c-Si, there is a potential energy barrier at the amorphous-crystalline interface that reflects electrons and holes away from that interface. At the p^+ contact, only holes can tunnel through the barrier, whereas only electrons can tunnel through the barrier to the n^+ contact. The passivation of surface defects that are sites of electron-hole recombination prevents a major loss mechanism in solar cells, increasing both the photovoltage and the photocurrent relative to conventional c-Si p-n junction cells. Panasonic and Sanyo have announced the production of HIT cells with power conversion efficiencies as high as 23%.

Light Emitting Diodes

Light-emitting diode, usually called a LED, is a semiconductor diode that emits incoherent narrow-spectrum of light when electrically biased in the forward direction of the p-n junction, as in the common LED circuit. This effect is a form of electroluminescence. A LED is usually a small area light source, often with optics added to the chip to shape its radiation pattern and assist in reflection. LEDs are often used as small indicator of lights in electronic devices and increasingly in higher power applications such as flashlights and area lighting. The color of the emitted light depends on the composition and condition of the semiconducting material used, and can be infrared, visible, or ultraviolet. LEDs can also be used as a regular household light source. Besides lighting, interesting applications include sterilization of water and disinfection of devices, glass wares and surgical instruments.

Basic Terminology Involved in LEDS Fabrication and Spectral Characterizations

Color Rendering Index (CRI)

The CIE color rendering index (CRI) (or colour rendering index in British-style spelling; sometimes called *color rendition index*), is a quantitative measurement of the ability of a light source to reproduce the colors of various objects faithfully in comparison with an ideal or natural light source. Light sources with a high CRI are desirable in color-critical applications such as photography and cinematography. The color rendering index indicates how well a light sources renders color to the average observer. It is based on color shifts seen in 14 or more colored tiles when the illumination changes from a reference (sunlight or blackbody) to the test sources and ranges from 100 to 0. By definition, day light and black bodies are 100, and every thing else is measured from that points downwards. Poor light sources may have the negative color rendering index. Because the measurement is based on the spectral power distribution, it can be "manipulated" to produce higher color rendering index values. Fluorescent lamp manufacturers often use emission points to their advantage.

Förster Resonance Energy Transfer

Förster resonance energy transfer (abbreviated FRET) describes an energy transfer mechanism between two chromophores. A donor chromophore in its excited state can transfer energy by a non-radiative, long-range dipole-dipole coupling mechanism to an acceptor chromophore in close proximity (typically <10nm). This energy transfer mechanism is termed Förster resonance energy transfer, named after the German scientist de:Theodor Förster. When both molecules are fluorescent, the term "fluorescence resonance energy transfer" is often used, although the energy is not actually transferred by fluorescence. In order to avoid an erroneous interpretation of the phenomenon that, even when occurring between two fluorescent molecules, is always a non-radiative transfer of energy, the name "Förster resonance energy transfer" may be preferred to "Fluorescence resonance energy transfer," although the latter enjoys common usage in scientific literature.

The Color Coordinates

C.I.E. Color Space

The CIE system characterizes colors by a luminance parameter Y and two color coordinates x and y which specify the point on the chromaticity diagram. This system offers more precision in color measurement than do the Munsell and Ostwald systems because the parameters are based on the spectral power distribution (SPD) of the light emitted from a colored object and are factored by sensitivity curves which have been measured for the human eye. Based on the fact that the human eye has three different types of color sensitive cones, the response of the eye is best described in terms of three "tristimulus values". However, once this is accomplished, it is found that any color can be expressed in terms of the two color coordinates x and y. The colors which can be matched by combining a given set of three primary colors (such as the blue, green, and red of a color television screen) are represented on the chromaticity diagram by a triangle joining the coordinates for the three colors.

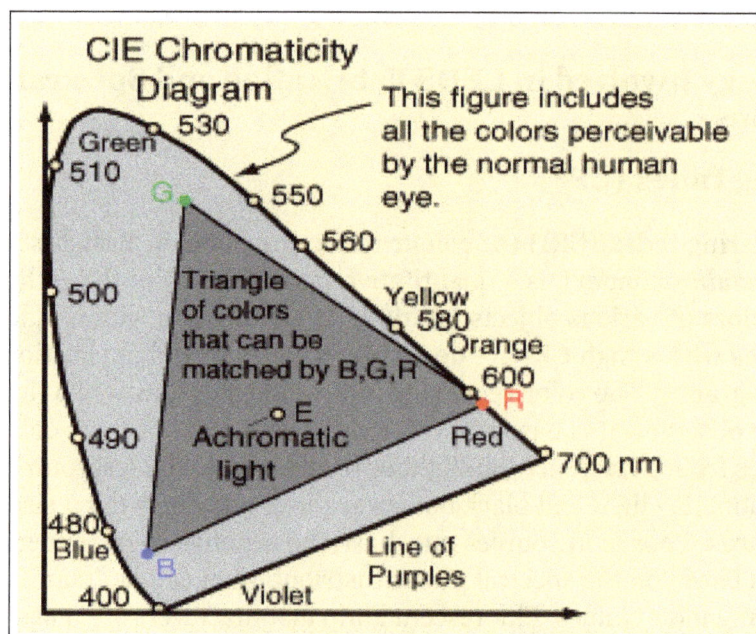

Presents CIE chromaticity diagram.

C.I.E. Chromaticity Diagram

The following display was created by choosing representative RGB values for the color regions from a rendition of the 1976 CIE Chromaticity Diagram provided by Photo Research, Inc. Note that one representative value in about the middle of the hue and saturation ranges was chosen for each section of the diagram. The point chosen was just a visual judgment of a representative color in the range. A different observer would likely have chosen different points to represent the color names, but at least these values might provide a starting point for preferred variations. Approximate color regions on CIE chromaticity diagram is displayed in figure.

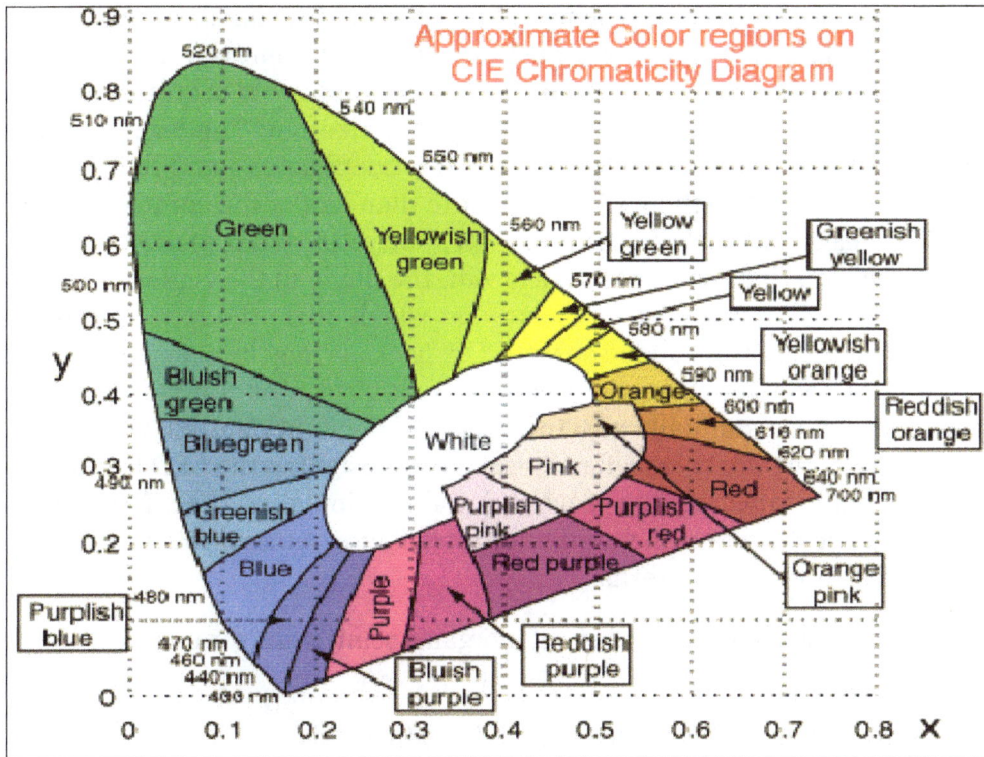

Approximate color regions on CIE chromaticity diagram.

LED Technology

Physical Function

Like a normal diode, the LED consists of a chip of semi conducting material inundated, or doped, with impurities to create a p-n junction. As in other diodes, current flows easily from the p-side, or anode, to the n-side, or cathode, but not in the reverse direction. Chargecarriers— electrons and holes—flow into the junction from electrodes with different voltages. When an electron meets a hole, it falls into a lower energy level, and releases energy in the form of a photon.

The wavelength of the light emitted, and therefore its color, depends on the band gap energy of the materials forming the p-n junction. In silicon or germanium diodes, the electrons and holes recombine by a non-radiative transition which produces no optical emission, because these are indirect band gap materials. The materials used for the LED have a direct band gap with energies corresponding to near-infrared, visible or near-ultraviolet light.

LED development began with infrared and red devices made of gallium arsenide. Advances in materials science have made possible the fabrication of devices with ever-shorter wavelengths, producing light in a variety of colors. LEDs are usually built on an n-type substrate, with an electrode attached to the p-type layer deposited on its surface. P-type substrates, while less common, occur as well. Many commercial LEDs, especially GaN/InGaN, also use sapphire substrate. Substrates that are transparent to the emitted wavelength, and backed by a reflective layer, and light spreading layer, increase the LED efficiency. The refractive index of the package material should match the index of the semiconductor, otherwise the produced light gets partially reflected back into the semiconductor, where it may be absorbed and turned into additional heat, thus lowering the efficiency. This type of reflection also occurs at the surface of the package if the LED is coupled to a medium with a different refractive index such as a glass fiber or air. The refractive index of most LED semiconductors is quite high, so in almost all cases the LED is coupled into a much lower-index medium. The large index difference makes the reflection quite substantial (per the Fresnel coefficients), and this is usually one of the dominant causes of LED inefficiency. Often more than half of the emitted light is reflected back at the LEDpackage and package-air interfaces. The reflection is most commonly reduced by using a dome-shaped (half-sphere) package with the diode in the center so that the outgoing light rays strike the surface perpendicularly, at which angle the reflection is minimized. An antireflection coating may be added as well. The package may be cheap plastic, which may be colored, but this is only for cosmetic reasons or to improve the contrast ratio; the color of the packaging does not substantially affect the color of the light emitted. Other strategies for reducing the impact of the interface reflections include designing the LED to reabsorb and re-emit the reflected light (called photon recycling) and manipulating the microscopic structure of the surface to reduce the reflectance, either by introducing random roughness or by creating programmed moth eye surface patterns.

Conventional LEDs are made from a variety of inorganic semiconductor materials, producing the following colors:

- Aluminium gallium arsenide (AlGaAs): Red and infrared.

- Aluminium gallium phosphide (AlGaP): Green.

- Aluminium gallium indium phosphide (AlGaInP): High-brightness orange-red, orange, yellow, and green.

- Gallium arsenide phosphide (GaAsP): Red, orange-red, orange, and yellow.

- Gallium phosphide (GaP): Red, yellow and green.

- Gallium nitride (GaN): Green, pure green (or emerald green), and blue also white (if it has an AlGaN Quantum Barrier).

- Indium gallium nitride (InGaN): 450–470 nm — near ultraviolet, bluish-green and blue.

- Silicon carbide (SiC) as substrate: Blue.

- Silicon (Si) as substrate: Blue (under development).

- Sapphire (Al_2O_3) as substrate: Blue.

- Zinc selenide (ZnSe): Blue.

- Diamond (C): Ultraviolet.

- Aluminium nitride (AlN), aluminium gallium nitride (AlGaN), aluminium gallium indium nitride (AlGaInN): Near to far ultraviolet (down to 210 nm).

With this wide variety of colors, arrays of multicolor LEDs can be designed to produce unconventional color patterns.

Ultraviolet and Blue LEDs

Blue LEDs are based on the wide band gap semiconductors GaN (gallium nitride) and InGaN (indium gallium nitride). They can be added to existing red and green LEDs to produce the impression of white light, though white LEDs today rarely use this principle. The first blue LEDs were made in 1971 by Jacques Pankove (inventor of the gallium nitride LED) at RCA Laboratories. However, these devices were too feeble to be of much practical use. In the late 1980s, key breakthroughs in GaN epitaxial growth and p-type doping by Isamu Akasaki and Hiroshi Amano (Nagoya, Japan) ushered in the modern era of GaNbased optoelectronic devices. Building upon this foundation, in 1993 high brightness blue LEDs were demonstrated through the work of Shuji Nakamura at Nichia Corporation.

By the late 1990s, blue LEDs had become widely available. They have an active region consisting of one or more InGaN quantum wells sandwiched between thicker layers of GaN, called cladding layers. By varying the relative InN-GaN fraction in the InGaN quantum wells, the light emission can be varied from violet to amber. AlGaN (aluminium gallium nitride) of varying AlN fraction can be used to manufacture the cladding and quantum well layers for ultraviolet LEDs, but these devices have not yet reached the level of efficiency and technological maturity of the InGaN-GaN blue/green devices. If the active quantum well layers are GaN, as opposed to alloyed InGaN or AlGaN, the devices will emit near-ultraviolet light with wavelengths around 350–370 nm. Green LEDs manufactured from the InGaN-GaN system are far more efficient and brighter than green LEDs produced with non-nitride material systems. Figure illustrates some GaN based UV and blue LEDs.

With nitrides containing aluminum, most often AlGaN and AlGaInN, even shorter wavelengths are achievable. Ultraviolet LEDs in a range of wavelengths are becoming available on the market. Near-UV emitters at wavelengths around 375–395 nm are already cheap and often encountered, for example, as black light lamp replacements for inspection of anti-counterfeiting UV watermarks in some documents and paper currencies. Shorter wavelength diodes, while substantially more expensive, are commercially available for wavelengths down to 247 nm. As the photosensitivity of microorganisms approximately matches the absorption spectrum of DNA, with a peak at about 260 nm, UV LEDs emitting at 250–270 nm are to be expected in prospective disinfection and sterilization devices. Recent research has shown that commercially available UVA LEDs (365 nm) are already effective disinfection and sterilization devices.

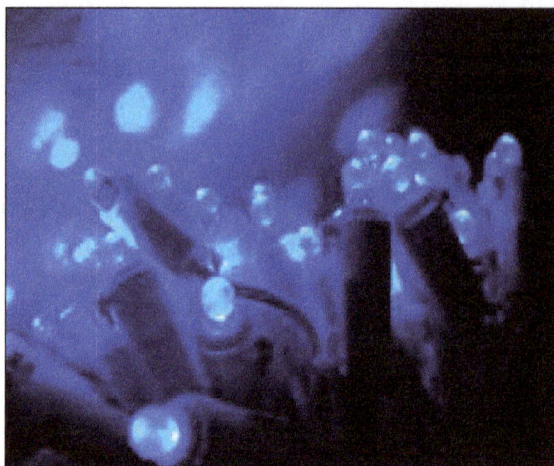

GaN base UV and blue LEDs.

Wavelengths down to 210 nm were obtained in laboratories using aluminum nitride. While not actually LEDs as such, an ordinary n-p-n bipolar transistor will emit violet light if its emitter-base junction is subjected to non-destructive reverse breakdown. This is easy to demonstrate by filing the top off a metal-can transistor (BC107, 2N2222 or similar) and biasing it well above emitter-base breakdown (≥ 20 V) via a current-limiting resistor.

White LEDs

There are two ways of producing high intensity white-light LED. One is to first produce individual LEDs that emit three primary colors – red, green, and blue, and then mix all the colors to produce white light. Hence the product is called multi-colored white LEDs (sometimes referred to as RGB LEDs). Because its mechanism is involved with sophisticated electro-optical design to control the blend and diffusion of different colors, this approach has rarely been used to mass produce white LEDs in the industry. Nevertheless this method is particularly interesting to many researchers and scientists because of the flexibility of mixing different colors. In principle, this mechanism also has higher quantum efficiency in producing white light. On the other hand, the second method of producing white LED is involved with coating a LED of one color (mostly blue LED made of InGaN) with phosphor coating of a different color to produce white light. Depending on the color of the original LED, phosphors of different colors can also be employed. By applying several phosphor layers of distinct colors, we can effectively increase the color rendering index (CRI) value of a given LED. Because this method of producing white LEDs heavily employs the usage of phosphor, the resultant LEDs are called phosphor based white LEDs. Although easier to be manufactured than multi-colored LEDs, phosphor based LEDs have a lower quantum efficiency and other phosphor-related degradation issues. However, it is still the most popular technique of manufacturing high intensity white LEDs as well as high intensity LEDs of other colors because it requires much easier material processing and therefore suits today's applications. Much effort has been spent on optimizing the operating environment, namely temperature and current, for this type of LED.

There are several types of multi-colored white LEDs: di-, tri-, and tetra chromatic white LEDs. Several key factors that play among these different approaches include color stability, color rendering capability, and luminous efficiency. Luminous efficiency is a term expressing the luminous flux per unit electrical input power. It is a key factor in discussing energy efficiency. In principle, if perfect solid-state lighting devices can be fabricated, the same level of luminance can be achieved by using merely 1/20 of the energy that incandescent lighting source requires. Color stability is a self-explanatory term which means the stability of color. Color rendering capability is hard to grasp without being traced back to its origin. In 1977, George Palmer first found that an object's perceived color strongly depends on the illumination source. He discovered that by varying the illumination sources, an object's color appeared differently. Because of their conflicting nature, there is always a trade off between the luminous efficiency and color rendering. For example, the dichromatic white LEDs have the best luminous efficiency (120 lm/W), but the lowest color rendering capability. Oppositely although tetra chromatic white LEDs have excellent color rendering capability, they often have poor luminous efficiency. Tri- chromatic white LEDs are in between, having both good luminous efficiency (>70 lm/W) and fair color rendering capability.

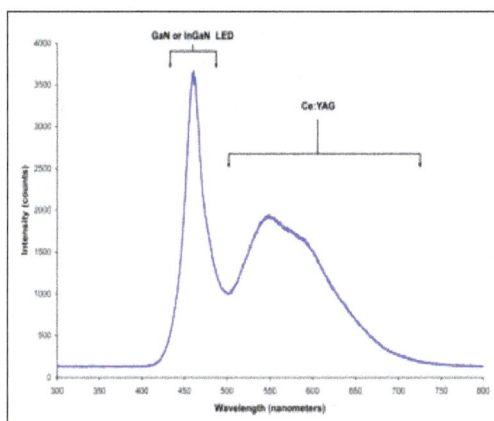

Spectrum of a "white" LED clearly showing blue light which is directly emitted by the GaNbased LED (peak at about 465 nm) and the more broadband Stokes-shifted light emitted by the Ce^{3+}:YAG phosphor which emits at roughly 500–700 nm.

Phosphor based white LEDs encapsulate InGaN blue LEDs inside of a phosphor coated epoxy. A common yellow phosphor material is cerium-doped yttrium aluminum garnet (Ce^{3+}:YAG). Although the phosphor based white LEDs have a relatively easier mechanism, they reach the fundamental limitation due to the unavoidable Stokes energy loss, a loss that occurs when short wavelength photons are converted to long wavelength photons. Regardless this technique of manufacturing is adopted by most of the LED industry because of its low cost and high output. All the high intensity white LEDs now on the market are manufactured by this method. Spectrum of white light generated by combination of GaN blue LED and Ce^{3+}:YAG phosphor is displayed in figure.

What multi-color LEDs offer is not merely another solution of producing white light, but is a whole new technique of producing light of different colors. In principle, all colors in the visible spectrum can be produced by mixing different amount of three primary colors, and this makes it possible to produce precise dynamic color control as well. As more effort is devoted to investigating this technique, multi-color LEDs should have profound influence on the fundamental method which we use to produce and control light color. However, before this type of LEDs can truly play a role on the market, several technical problems need to be solved. These certainly include that for this type of LEDs' emission power decays exponentially with increasing temperature, resulting in a substantial change in color stability. Such problem is not acceptable for industrial usage. Therefore, many new package designs aiming to solve this problem have been proposed, and their results are being reproduced by researchers and scientists.

On the other hand, phosphor based white LEDs are the optimal solution to produce high intensity white light. Since its simplified mechanism, this type of LEDs has attracted much interest from the lighting industry. Because of their more stable performance over a range of temperatures, prototypes as well as products based on this phosphor based mechanism have already appeared on the market, and more high intensity white LEDs are expected to be produced in the near future. However, the biggest challenge these phosphor based white LEDs face is solving the seemingly unavoidable Stokes energy loss. Again this can be done by adapting a better package design or by replacing a more suitable type of phosphor. Philips Lumileds patented conformal coating process addresses the issue of varying phosphor thickness, giving the white LEDs a more consistent spectrum of white light.

White LEDs can also be made by coating near ultraviolet (NUV) emitting LEDs with a mixture of high efficiency europium-based red and blue emitting phosphors plus green emitting copper and aluminum doped zinc sulfide ($ZnS:Cu, Al$). This is a method analogous to the way fluorescent lamps work. However, the ultraviolet light causes photo degradation to the epoxy resin and many other materials used in LED packaging, causing manufacturing challenges and shorter lifetimes. This method is less efficient than the blue LED with YAG:Ce phosphor, as the Stokes shift is larger and more energy is therefore converted to heat, but yields light with better spectral characteristics, which render color better. Due to the higher radiative output of the ultraviolet LEDs than of the blue ones, both approaches offer comparable brightness.

The newest method used to produce white light LEDs uses no phosphors at all and is based on homoepitaxially grown zinc selenide (ZnSe) on a ZnSe substrate which simultaneously emits blue light from its active region and yellow light from the substrate.

A new technique developed by Michael Bowers, a graduate student at Vanderbilt University in Nashville, involves coating a blue LED with quantum dots that glow white in response to the blue light from the LED. This technique produces a warm, yellowish-white light similar to that produced by incandescent bulbs.

Quantum Dot LEDs

Quantum Dots are semiconductor nanocrystals that possess unique optical properties. Their emission color can be tuned from the visible throughout the infrared spectrum.

This allows quantum dot LEDs to create almost any color on the CIE diagram. This provides more color options and better color rendering white LEDs. Quantum dot LEDs are available in the same package types as traditional phosphor based LEDs.

Advantages of Quantum Dots over Organic Fluorophers

Quantum dots have size tunable band gaps, high photoluminescence quantum efficiencies (QEs), good photo stability, narrow emission line width, large spin orbit coupling and good compatibility with solution processing methods. They have shown following advantages over organic materials.

- High quantum yield; often 20 times brighter.

- Narrower and more symmetric emission spectra.

- 100-1000 times more stable to photo-bleaching.

- High resistance to photo-/chemical degradation.

- Tunable wave length range 400-4000 nm.

ZnO-based Leds have Full-colour Potential

The attractiveness of zinc oxide (ZnO) LEDs stems from the potential for phosphor-free spectral coverage from the deep ultraviolet (UV) to the red, coupled with a quantum efficiency that could approach 90% and a compatibility with high-yield low-cost volume production. These LEDs have capability to outperform their GaN-based cousins (which offer a narrower spectral range) due to its three key characteristics such as superior material quality, an effective dopant and the availability of better alloys. The superior material quality is seen in the low defect densities of ZnO layers. The p-type dopant has provided holeconducting layers for ZnO-based devices and growth of BeZnO layers has shown that it is possible to fabricate ZnO-based high-quality heterostructures ("The advantages of ZnO over GaN"). ZnO also promises very high quantum efficiencies, and UV detectors based on this material have produced external quantum efficiencies (EQE) of 90%, three times that of equivalent GaN-based detectors. The physical processes associated with detection suggest that similarly high efficiency values should be possible for the conversion of electrical carriers to photons. So it is plausible that ZnO LEDs will have an EQE upper limit that is three times higher than that of GaN-based devices.

Advantages of ZnO based Leds over Gan based

There are three major advantages of ZnO over GaN:

- Superior material quality, which has been demonstrated by growth of high purity ZnO with defect densities below $105cm^{-2}$, a value typically associated with the best GaN films.

- Improve doping performance, which results from the arsenic p-type dopant that has activation energy of 119 meV in ZnO films, which is less than that of 215 meV for magnesium doped p type GaN. This lower activation energy produces a ten fold increase in the proportion of the activated accepter atoms that are needed for electrical conduction and also reduces the number of defects for a given hole carrier density.

- The availability of better alloys, due to our recent development of high quality BeZnO films. These layers have driven the fabrication of LEDs, lasers and transistors that have less disorder than the structure produced using the AlGaN/GaN material system. The reduced disorder is a consequence of the large difference in band gap between ZnO and BeO and enables only small changes in the alloy's composition to produce relatively large changes in band gap. In comparison, a much larger shift in aluminum composition is required

Organic Light Emitting Diodes

There are two main classes of organic light-emitting diodes: OLEDs (small-molecule based light emitting diodes) and PLEDs (polymer light emitting diods). A typical doubleheterostructure small-molecule OLED consists of three organic layers sandwiched between electrodes. The organic layers adjacent to cathode and anode are the electron transport layer (ETL) and the hole transport layer (HTL), respectively. Emissive layer (EML) usually consists of light-emitting dyes or dopants dispersed in a suitable host material (often same as HTL or ETL material).

PLEDs have relatively simple architectures, with the light-emitting polymer (LEP) layer combining host, emitter and charge transport functions in a single solution-processed layer of the device. Schematic illustration of various layers in OLED is shown in figure.

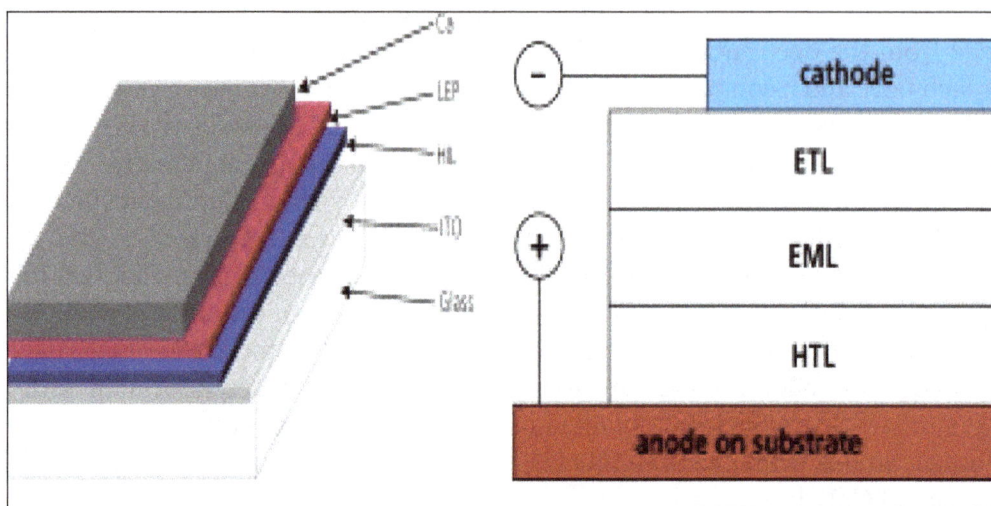

Schematic illustration are as HIL = hole injection layer, LEP= Light Emitting polymer, HTL= Hole transport layer, ETL= Electron transport layer, EML= Emissive material layer.

To function as a semiconductor, the organic emitting material must have conjugated pi bonds. The emitting material can be a small organic molecule in a crystalline phase, or a polymer. Polymer materials can be flexible; such LEDs are known as PLEDs or FLEDs.

Compared with regular LEDs, OLEDs are lighter, and polymer LEDs can have the added benefit of being flexible. Some possible future applications of OLEDs could be in Inexpensive and flexible displays, Light sources, Wall decorations and luminous cloth etc. OLEDs are being widely used to produce visual displays for portable electronic devices such as cellphones, digital cameras, and MP3 players. Larger displays have been demonstrated, but their life expectancy is still far too short (<1,000 hours) to be practical.

Working Principle

As depicted above a typical OLED is composed of an emissive layer, a conductive layer, a substrate, an anode and cathode terminals. The layers are made of special organic molecules that conduct electricity. Their levels of conductivity range from those of insulators to those of conductors, and so they are called organic semiconductors. The first, most basic OLEDs consisted of a single organic layer of poly(p-phenylene vinylene, synthesized by Burroughs et al. Multilayer OLEDs can have more than two layers to improve device efficiency. As well as conductive properties are concrened, some organic layers may be chosen to aid charge injection at electrodes by providing a more gradual electronic profile, or to block a charge from reaching the opposite electrode and being wasted. Combination of electron and hole for the generation of photon in the OLED is illustrated in the figure.

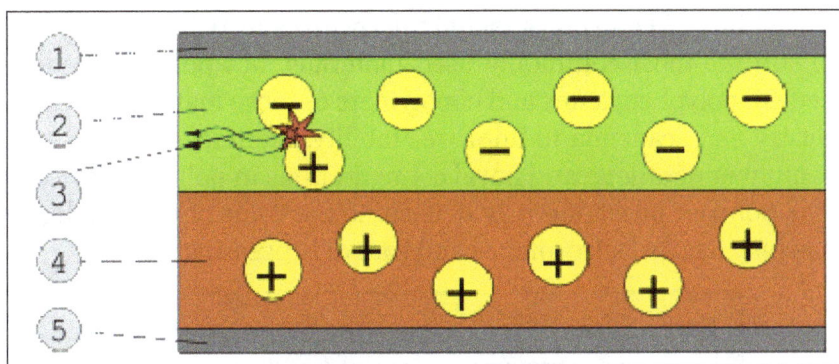

Schematic of a 2-layer OLED: 1. Cathode (−), 2.Emissive Layer,
3.Emission of radiation, 4. Conductive Layer, 5. Anode (+).

A voltage is applied across the OLED such that the anode is positive with respect to the cathode. This causes a current of electrons to flow through the device from cathode to anode. Thus, the cathode gives electrons to the emissive layer and the anode withdraw electrons from the conductive layer; in other words, the anode gives electron holes to the conductive layer.

Soon, the emissive layer becomes negatively charged, while the conductive layer becomes rich in positively charged holes. Electrostatic forces bring the electrons and the holes towards each other and they recombine. This happens closer to the emissive layer, because in organic semiconductors holes are more mobile than electrons (unlike in inorganic semiconductors). The recombination causes a drop in the energy levels of electrons, accompanied by an emission of radiation whose frequency is in the visible region. That is why this layer is called emissive.

The device does not work when the anode is put at a negative potential with respect to the cathode. In this condition, holes move to the anode and electrons to the cathode, so they are moving away from each other and do not recombine.

Indium tin oxide (ITO) is commonly used as the anode material. It is transparent to visible light and has a high work function which promotes injection of holes into the polymer layer. Metals such as aluminum or calcium are often used for the cathode as they have low work functions which promote injection of electrons into the polymer layer. Molecules commonly used in OLEDs include organo-metallic chelates (for example Alq3, used in the first organic light-emitting device) and conjugated dendrimers.

Recently a hybrid light-emitting layer has been developed that uses nonconductive polymers doped with light-emitting, conductive molecules. The polymer is used for its production and mechanical advantages without worrying about optical properties. The small molecules then emit the light and have the same longevity that they have in the small molecule OLEDs.

Polymer light-emitting diodes (PLED), also light-emitting polymers (LEP), involve an electroluminescent conductive polymer that emits light when connected to an external voltage source. They are used as a thin film for full-spectrum color displays and require a relatively small amount of power for the light produced. No vacuum is required, and the emissive materials can be applied on the substrate by a technique derived from commercial inkjet printing. The substrate used can be flexible, such as PET. Thus flexible PLED displays, also called Flexible OLED (FOLED), may be produced inexpensively.

Typical polymers used in PLED displays include derivatives of poly(p-phenylene vinylene) and its derivatives are still the most commonly used materials, but polythiophenes, polypyridines, poly(pyridyl vinylenes), polyphenylenes and copolymers of these materials are now being used. A large effort in industry and academia to improve stability, lifetime, and efficiency has prompted the study of a vast number of unique and novel configurations in polymer devices. Substitution of side chains onto the polymer backbone may determine the color of emitted light or the stability and solubility of the polymer for performance and ease of processing.

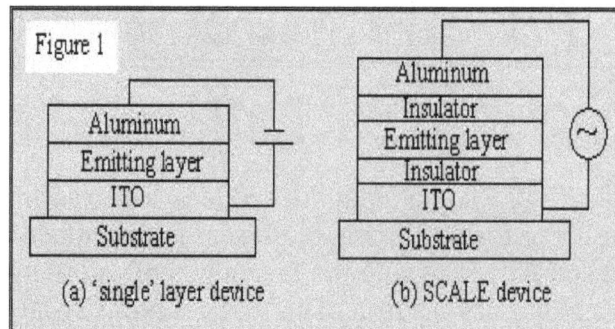

Figure 1

(a) 'single' layer device   (b) SCALE device

Shows the structure of a simple (single layer) device with only the electrodes and a single active layer. Simple 'single' layer devices typically work only under a forward DC bias. Figure shows a novel configuration: a symmetrically configured alternating current light emitting (SCALE) device, which works under AC as well as forward and reverse DC bias.

Potential of Oleds

Today, OLEDs operate at substantially lower efficiency than inorganic (crystalline) LEDs. The best luminous efficiency of an OLED so far is about 10% of the theoretical maximum of 683 for "white" light, or about 68 lm/W. These claim to be much cheaper to fabricate than inorganic LEDs, and large arrays of them can be deposited on a screen using simple printing methods to create a color graphical display.

Bio-LEDs

There are a number of phosphor materials found in the nature and can be used in the fabrication of LEDs and other photonic devices. LEDs having light emissive layer made from biological materials are called Bio LEDs. As these materials are bio degradable and are environmentally green, therefore considerable attention is paid now days for the development of bio LEDs. It is reported that the scales on the wing of the African swalotail butterfly Papilio nireus resemble high-efficiency LEDs. The blue and blue-green scales hey found incorporate natural two dimensional photonic crystals and distributed brag reflectors to boost the output of a fluorescent pigment in the wing. Andrew J. Steckl from Nano-electronics Laboratory, University of Cincinnati have fabricated DNA based LEDs. DNA based solid state lighting devices can be fabricated by making a thin film of DNA. DNA films are fabricated by solution methods, where a reaction between the DNA and a cationic surfactant (such as Cetyltrimethyl Ammonium Bromide CTAB) produces a DNA–lipid complex that is insoluble in water but soluble in alcohols. Using this complex one can make DNA thin film by casting or spin coating, which do not dissolve other organic layers on contact and can thus be integrated into organic devices. In addition, there is also a dry process using ultrahigh vacuum molecular beam deposition to produce nanometer-scale films with very controllable properties. Adjustment in the DNA molecular weight and concentration of reagent will determine the properties of the DNA-CTAB thin film. Electrical resistivity of the film, which plays important role in the LED fabrication, can be reduced by decreasing DNA molecular weight.

DNA can also be used for enhancing the efficiency of OLEDs and PLEDs. Incorporating DNA into OLEDs as an electron-blocking layer (EBL) has been reported to result in BioLEDs that are as much as ten times more efficient and thirty times brighter than their OLED counterparts. Enhanced efficiency using DNA–CTAB nanometer thick film as EBL material is demonstrated in both green and blue-emitting devices, which is presented in figure. The resulting green and blue BioLEDs showed a maximum luminous efficiency of about 8 and 1 cd A^{-1}, respectively. Typical turn-on and operating voltages for these devices were about 4–5 V and about 10–25 V, respectively. The green BioLED achieved the highest luminiscence of about 21,000 cd m^{-2}, whereas the blue BioLED showed a maximum luminance of about 1,500 cd m^{-2}.

In addition, by adding lumophores or chromophores to DNA molecules it is possible to create an exciting new range of optically active materials. Lumophores can be attached to DNA at several locations: intercalated between base pairs, or bound to the minor or major grooves of the double helix. Lumophore intercalation is known to result in a very sensitive signature in the presence of DNA molecules in solution. Lumophore intercalation is particularly important as it has been reported that DNA films can act as a far better host for lumophores than conventional polymer hosts, resulting in much higher luminescence. For example, DNA–CTAB thin-films doped with the luminescent dye sulphorhodamine (SRh) have been reported to exhibit a photoluminescence intensity more than an order of magnitude higher than that of SRh in poly(methyl methacrylate), which is a popular polymer host. Other lumophores have also been reported to luminescence very efficiently in DNA thin-films.

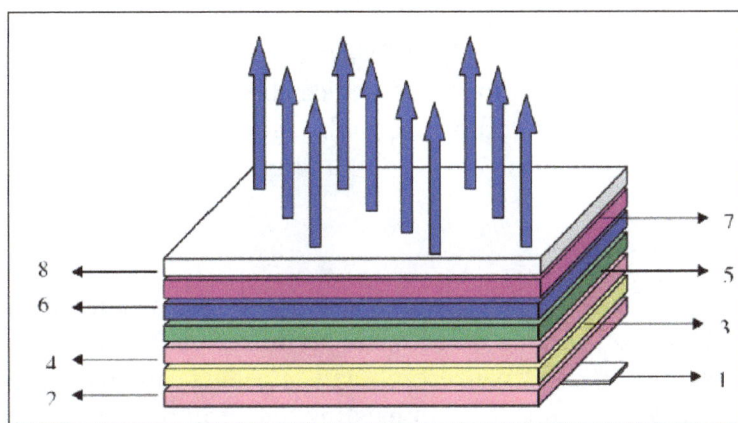

Schematic view of DNA based Bio LED, a layered structure; 1. Al cathode 2. Alq3 (aluminum tris (8-hydroxyquinoline)) 3. BCP (2,9-dimethyl-4,7-diphenyl-1,10- phenanthroline) 4. Alq3 (aluminum tris(8-hydroxyquinoline)) 5. NPB ((N,N'-bis(naphthalene- 1-yl)-N,N'-bis(phenyl)benzidine) 6. DNA-CTAB layer 7. PEDOT is poly(3,4 ethylenedioxythiophene) 8. ITO coated glass plate anode.

Although the mechanisms are not yet completely understood, it is speculated that excited molecules, which are intercalated between base pairs in the DNA structure, are essentially shielded from non-radiative relaxation centers in the host material, thus opening the door to efficient photon emission. Another possible explanation is related to the tight spatial fit between interca-lated molecules and the base-pair structure, which may prevent the conformational relaxation of excited lumophores and thereby enhance the process of radiative relaxation.

Efficiency and Operational Parameters of LEDs

Most typical LEDs are designed to operate with no more than 30–60 milli-watts (mW) of electrical power. Around 1999, Philips Lumileds introduced power LEDs capable of continuous use at one watt (W). These LEDs used much larger semiconductor die sizes to handle the large power inputs. Also, the semiconductor dies were mounted onto metal slugs to allow for heat removal from the LED die.

One of the key advantages of LED-based lighting is its high efficiency, as measured by its light output per unit power input. White LEDs quickly matched and overtook the efficiency of standard incandescent lighting systems. In 2002, Lumileds made five-watt LEDs available with a luminous efficacy of 18–22 lumens per watt (lm/W). For comparison, a conventional 60–100 W incandes-cent light bulb produces around 15 lm/W, and standard fluorescent lights produce up to 100 lm/W.

In September 2003, a new type of blue LED was demonstrated by the company Cree, Inc. to pro-vide 24 mW at 20 milli amperes (mA). This produced a commercially packaged white light giving 65 lm/W at 20 mA, becoming the brightest white LED commercially available at the time, and more than four times as efficient as standard incandescent. In 2006 they demonstrated a proto-type with a record white LED luminous efficacy of 131 lm/W at 20 mA. Also, Seoul Semiconductor has plans for 135 lm/W by 2007 and 145 lm/W by 2008, which would be approaching an order of magnitude improvement over standard incandescent and better even than standard fluorescents. Nichia Corporation has developed a white light LED with luminous efficacy of 150 lm/W at a for-ward current of 20 mA. In May 2008, 130 lm/W is available from Chinese LED manufacturers.

It should be noted that high-power ($\geq$ 1 W) LEDs are necessary for practical general lighting ap-plications. Typical operating currents for these devices begin at 350 mA. The highest efficiency high-power white LED is claimed by Philips Lumileds Lighting Co. with a luminous efficacy of 115 lm/W (350 mA).

Failure Modes

The most common way for LEDs (and diode lasers) to fail is the gradual lowering of light output and loss of efficiency. However, sudden failures can occur as well. The mechanism of degradation of the active region, where the radiative recombination occurs, involves nucleation and growth of dislocations; this requires a presence of an existing defect in the crystal and is accelerated by heat, high current density, and emitted light. Gallium arsenide and aluminum gallium arsenide are more susceptible to this mechanism than gallium arsenide phosphide and indium phosphide. Due to different properties of the active regions, gallium nitride and indium gallium nitride are virtu-ally insensitive to this kind of defect; however, high current density can cause electro-migration of atoms out of the active regions, leading to emergence of dislocations and point defects, acting as non-radiative recombination centers and producing heat instead of light. Ionizing radiation can lead to the creation of such defects as well, which leads to issues with radiation hardening of cir-cuits containing LEDs (e.g., in opto-isolators). Early red LEDs were notable for their short lifetime.

White LEDs often use one or more phosphors. The phosphors tend to degrade with heat and age, losing efficiency and causing changes in the produced light color. Pink LEDs often use an organic phosphor formulation which may degrade after just a few hours of operation causing a major shift in output color. High electrical currents at elevated temperatures can cause diffusion of metal atoms from the electrodes into the active region. Some materials, notably indium tin oxide and silver, are subject to electro-migration with the consequence of leakage current and non radiative recombination along the chip edges. In some cases, especially with GaN/InGaN diodes, a barrier metal layer is used to hinder the electromigration effects. Mechanical stresses, high currents, and corrosive environment can lead to formation of whiskers, causing short circuits.

High-power LEDs are susceptible to current crowding, non-homogenous distribution of the current density over the junction. This may lead to creation of localized hot spots, which poses risk of thermal runaway. Non-homogeneities in the substrate, causing localized loss of thermal conductivity, intensify the situation; most common ones are voids caused by incomplete soldering, or by electro-migration effects and Kirkendall voiding. Thermal runaway is a common cause of LED failures.

Laser diodes may be subject to catastrophic optical damage, when the light output exceeds a critical level and causes melting of the facet. Some materials of the plastic package tend to yellow when subjected to heat, causing partial absorption (and therefore loss of efficiency) of the affected wavelengths. Sudden failures are most often caused by thermal stresses. When the epoxy resin used in packaging reaches its glass transition temperature, it starts rapidly expanding, causing mechanical stresses on the semiconductor and the bonded contact, weakening it or even tearing it off. Conversely, very low temperatures can cause cracking of the packaging.

Electrostatic discharge (ESD) may cause immediate failure of the semiconductor junction, a permanent shift of its parameters, or latent damage causing increased rate of degradation. LEDs and lasers grown on sapphire substrate are more susceptible to ESD damage.

Considerations in Use

Unlike incandescent light bulbs, which light up regardless of the electrical polarity, LEDs will only light with correct electrical polarity. When the voltage across the p-n junction is in the correct direction, a significant current flows and the device is said to be forward-biased. If the voltage is of the wrong polarity, the device is said to be reverse biased, and very little current flows, therefore no light is emitted. LEDs can be operated on an alternating current voltage, but they will only light with positive voltage, causing the LED to turn on and off at the frequency of the AC supply. While the only definitive way to determine the polarity of the LED is to examine its datasheet, these methods are usually reliable:

sign:	+	−
terminal:	anode (A)	cathode (K)
leads:	long	short
exterior:	round	flat
interior:	small	large
wiring:	red	black

Less reliable methods of determining polarity are:

sign:	+	–
marking:	none	stripe
pin:	1	2
PCB:	round	square

While it is not an officially reliable method, it is almost universally true that the cup that holds the LED die corresponds to the cathode. It is strongly recommended to apply a safe voltage and observe the illumination as a test regardless of what method is used to determine the polarity.

Because the voltage versus current characteristics of the LED are much like any diode (that is, current approximately an exponential function of voltage), a small voltage change results in a huge change in current. Added to deviations in the process this means that a voltage source may barely make one LED light while taking another of the same type beyond its maximum ratings and potentially destroying it.

Since the voltage is logarithmically related to the current it can be considered to remain largely constant over the LED's operating range. Thus the power can be considered to be essentially proportional to the current. In order to keep power nearly constant with variations in supply and LED characteristics, the power supply should be a "current source", that is, it should supply an almost constant current. If high efficiency is not required (e.g., in most indicator applications), an approximation to a current source is made by connecting the LED in series with a current limiting resistor to a regulated voltage source.

Most LEDs have low reverse breakdown voltage ratings, so they will also be damaged by an applied reverse voltage of more than a few volts. Since some manufacturers don't follow the indicator standards above, if possible the data sheet should be consulted before hooking up the LED, or the LED may be tested in series with a resistor on a sufficiently low voltage supply to avoid the reverse breakdown. If it is desired to drive the LED directly from an AC supply of more than the reverse breakdown voltage then it may be protected by placing a diode (or another LED) in inverse parallel.

LEDs can be purchased with built in series resistors. These can save PCB space and are especially useful when building prototypes or populating a PCB in a way other than its designers intended. However, the resistor value is set at the time of manufacture, removing one of the key methods of setting the LED's intensity. To increase efficiency (or to allow intensity control without the complexity of a DAC), the power may be applied periodically or intermittently; so long as the flicker rate is greater than the human flicker fusion threshold, the LED will appear to be continuously lit.

Multiple LEDs can be connected in series with a single current limiting resistor provided the source voltage is greater than the sum of the individual LED threshold voltages. Parallel operation is also possible but can be more problematic. Parallel LEDs must have closely matched forward voltages (V_f) in order to have equal branch currents and, therefore, equal light output. Variations in the manufacturing process can make it difficult to obtain satisfactory operation when connecting some types of LEDs in parallel.

Bi-color LED units contain two diodes, one in each direction (that is, two diodes in inverse parallel) and each a different color (typically red and green), allowing two-color operation or a range of apparent colors to be created by altering the percentage of time the voltage is in each polarity. Other LED units contain two or more diodes (of different colors) arranged in either a common anode or common cathode configuration. These can be driven to different colors without reversing the polarity, however, more than two electrodes (leads) are required.

LEDs are usually constantly illuminated when a current passes through them, but flashing LEDs are also available. Flashing LEDs resemble standard LEDs but they contain an integrated multivibrator circuit inside, which causes the LED to flash with a typical period of one second. This type of LED comes most commonly as red, yellow, or green. Most flashing LEDs emit light of a single wavelength, but multicolored flashing LEDs are available too.

Generally, for newer common standard LEDs in 3 mm or 5 mm packages, the following forward DC potential differences are typically measured. The forward potential difference depending on the LED's chemistry, temperature, and on the current (values here are for approx. 20 mA, a commonly-found maximum value).

Color	Potential Difference (Vf)
Infrared	1.6 V
Red	1.8–2.1 V
Orange	2.2 V
Yellow	2.4 V
Green	2.6 V
Blue	3.0–3.5 V
White	3.0–3.5 V
Ultraviolet	3.5 V

Many LEDs are rated at 3 V maximum reverse potential.

LEDs also behave as photocells, and will generate a current depending on the ambient light. They are not efficient as photocells, and will only produce a few microamperes (µA), but will produce a surprising electrical potential—as much as 2 or 3 V. This is enough to operate an amplifier or a CMOS logic gate. This effect can be used to make an inexpensive light sensor, for example to decide when to turn on the LED illuminator.

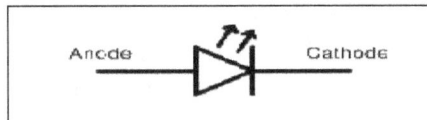
LED schematic symbol.

Advantages of using LEDs

- LEDs produce more light per watt than incandescent bulbs; this is useful in battery powered or energy-saving devices.

- LEDs can emit light of an intended color without the use of color filters that traditional lighting methods require. This is more efficient and can lower initial costs.

- The solid package of the LED can be designed to focus its light. Incandescent and fluorescent sources often require an external reflector to collect light and direct it in a usable manner.

- When used in applications where dimming is required, LEDs do not change their color tint as the current passing through them is lowered, unlike incandescent lamps, which turn yellow.

- LEDs are ideal for use in applications that are subject to frequent on-off cycling, unlike fluorescent lamps that burn out more quickly when cycled frequently, or HID lamps that require a long time before restarting.

- LEDs, being solid state components, are difficult to damage with external shock. Fluorescent and incandescent bulbs are easily broken if dropped on the ground.

- LEDs can have a relatively long useful life. One report estimates 35,000 to 50,000 hours of useful life, though time to complete failure may be longer. Fluorescent tubes typically are rated at about 30,000 hours, and incandescent light bulbs at 1,000–2,000 hours.

- LEDs mostly fail by dimming over time, rather than the abrupt burn-out of incandescent bulbs.

LEDs light up very quickly. A typical red indicator LED will achieve full brightness in microseconds; Philips Lumileds technical datasheet DS23 for the Luxeon Star states "less than 100ns". LEDs used in communications devices can have even faster response times. LEDs can be very small and are easily populated onto printed circuit boards.

LEDs do not contain mercury, unlike compact fluorescent lamps.

Due to the human eye's visual persistence LED's can be pulse width or duty cycle modulated in order to save power or achieve an apparent higher brightness for a given power input. The eye will tend to perceive the peak current light level rather than the average current light level when the modulation rate is higher than approximately 1000 hertz and the duty cycle is greater than 15 to 20%. This is also useful when applied to the multiplexing used in 7-segment displays.

LEDs are produced in an array of shapes and sizes. The 5 mm cylindrical package (red, fifth from the left) is the most common, estimated at 80% of world production. The color of the plastic lens is often the same as the actual color of light emitted, but not always. For instance, purple plastic is often used for infrared LEDs, and most blue devices have clear housings. There are also LEDs in extremely tiny packages, such as those found on blinkies and on cell phone keypads.

Disadvantages of using LEDs

LEDs are currently more expensive, price per lumen, on an initial capital cost basis, than more conventional lighting technologies. The additional expense partially stems from the relatively low lumen output and the drive circuitry and power supplies needed. However, when considering the total cost of ownership (including energy and maintenance costs), LEDs far surpass incandescent or halogen sources and begin to threaten compact fluorescent lamps. In December 2007, scientists at Glasgow University claimed to have found a way to make Light Emitting Diodes brighter and use less power than energy efficient light bulbs currently on the market by imprinting holes into billions of LEDs in a new and cost effective method using a process known as nano-imprint lithography.

LED performance largely depends on the ambient temperature of the operating environment. Over-driving the LED in high ambient temperatures may result in overheating of the LED package, eventually leading to device failure. Adequate heat-sinking is required to maintain long life. This is especially important when considering automotive, medical, and military applications where the device must operate over a large range of temperatures, and is required to have a low failure rate.

LEDs must be supplied with the correct current. This can involve series resistors or current-regulated power supplies.

The spectrum of some white LEDs differs significantly from a black body radiator, such as the sun or an incandescent light. The spike at 460 nm and dip at 500 nm can cause the color of objects to be perceived differently under LED illumination than sunlight or incandescent sources, due to metamerism. Color rendering properties of common fluorescent lamps are often inferior to what is now available in state-of-art white LEDs.

LEDs do not approximate a "point source" of light, so cannot be used in applications needing a highly collimated beam. LEDs are not capable of providing divergence below a few degrees. This is contrasted with lasers, which can produce beams with divergences of 0.2 degrees or less.

There is increasing concern that blue LEDs and white LEDs are now capable of exceeding safe limits of the so-called blue-light hazard as defined in eye safety specifications such as ANSI/IESNA RP-27.1-05: Recommended Practice for Photo biological Safety for Lamp and Lamp Systems.

Types of LEDs

LEDs are classified into three main categories namely; miniature, alphanumeric, and illumination. All these are described as follows:

- Miniature LEDs

These are mostly single-die LEDs used as indicators, and they manufactured in varioussize packages:

> Surface mount
>
> 2 mm
>
> 3 mm (T1)
>
> 5 mm (T1¾)
>
> Other sizes are also available, but less common.

Common package shapes are round, dome top, round, flat top, rectangular, flat top (often seen in LED bar-graph displays), Triangular or square and flat top etc. LEDs in different packaging are illustrated in the figure.

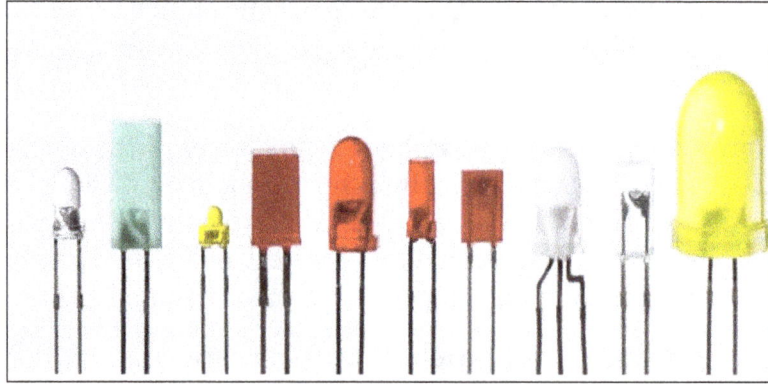

LEDs in different shapes and designs.

The encapsulation may also be clear or semi opaque to improve contrast and viewing angle. There are three main categories of miniature single die LEDs on the basis of operating current:

Low current: typically rated for 2 mA at around 2 V (approximately 4 mW consumption).

Standard: 20 mA LEDs at around 2 V (approximately 40 mW) for red, orange, yellow & green, and 20 mA at 4–5 V (approximately 100 mW) for blue, violet and white.

Ultra-high output: 20 mA at approximately 2 V or 4–5 V, designed for viewing in direct sunlight.

- Multi-color LEDs

  A "bi-color LED" is actually two different LEDs in one case. It consists of two dies connected to the same two leads but in opposite directions. Current flow in one direction produces one color, and current in the other direction produces other color. Alternating the two colors with sufficient frequency causes the appearance of a third color.

  A "tri-color LED" is also two LEDs in one case, but the two LEDs are connected to separate leads so that the two LEDs can be controlled independently and lit simultaneously.

  RGB LEDs contain red, green and blue emitters, generally using a four-wire connection with one common (anode or cathode).

- Five- and twelve-volt LEDs

  These are miniature LEDs incorporating a series resistor, and may be connected directly to a 5 V or 12 V supply.

- Flashing LEDs

  These miniature LEDs flash when connected to 5 V or 12 V. Used as attention seeking indicators where it is desired to avoid the complexity of external electronics.

- Alphanumeric LEDs

  LED displays are available in seven-segment and starburst format. Seven-segment displays handle all numbers and a limited set of letters. Starburst displays can display all letters.

Seven-segment LED displays were in widespread use in the 1970s and 1980s, but increasing use of liquid crystal displays, with their lower power consumption and greater display flexibility, has reduced the popularity of numeric and alphanumeric LED displays.

- Lighting LEDs

LED lamps (also called LED bars or Illuminators) are usually clusters of LEDs in a suitable housing. They come in different shapes, among them the light bulb shape with a large E27 Edison screw and MR16 shape with a bi-pin base. Other models might have a small Edison E14 fitting, GU5.3 (Bipin cap) or GU10 (bayonet socket). This includes low-voltage (typically 12 V halogen-like) varieties and replacements for regular AC mains (120-240 V AC) lighting. Currently the latter are less widely available but this is changing rapidly. Seoul Semiconductor Co., Ltd produces LEDs that can run directly from mains power without the need for a DC converter. For each half cycle part of the LED diode emits light and part is dark, and this is reversed during the next half cycle. Current efficiency is 80 lm/W.

Testing and Characterizations of Leds

Dramatic improvement in the performance of LEDs and other solid state lighting sources over the last few years have produce the unexpected growth for this sector of photonics industry. Advancement in output performance and operating efficiencies as well as innovations in the volume production has resulted in aggressive price competition. Manufacturer offer standard products with the output upward of 100 lm/W for single devices and even higher for multiple LED component engine.

LEDs can be found in applications ranging from architectural and entertainment lighting to the flat panel displays that account for almost one third of their recent growth. With this increased demand, accurate and precise tools and methods for measuring the performance of high brightness LEDs have become all the more critical. There arte a lot of uncertainties and inconsistencies associated with confusion about the methodology, design of measurement fixtures and accessories used in performing optical measurements.

A variety of approaches exist for measuring LEDs that yield different type of information; photometric, radiometric and spectroradiometric. Photometric quantities are measured relative to the response of human eye and typically are identified as "Photopic" or "Luminous" as in "luminous angular intensity". Radiometric quantities are based solely on the energy observed, uncorrected for human response, and they are described as "radiometric" or "radiant" as in radiant flux. Spectroradiometric quantities provide information about how the energy is distributed across the wavelength spectrum and are denoted by the use of "spectral" as in "spectral irradiance".

The type of measurement determines the quantity that is measured. For example as "total flux" is concerned with all the light emergent from the source, whereas "angular intensity" is only the light emitted in specified direction and at particular angles. Light falling onto an area is indicated as "irradiance" or "illuminance". The former being a radiometric quantity and the later being photometric. Similarly the light emitted from an area on the source is termed radiometrically as "radiance" and photometrically as "luminance".

An LED emission depends on temperature, both on the ambient temperature and that of the device, therefore heat sinking is required, which includes how and where electrical connections are made. As the ambient temperature changes in flux, luminance/ radiance and illuminance/ irradiance measurement, so does the LED wavelength, typically from 0.1 to 0.2 nm/°C, depending on the type of LED. Conversely junction temperature of the LED can be extrapolated by recording the shift in the peak wavelength with temperature. As high output LEDs run hotter, therefore active cooling and passive heat sinking is required. Scientists from Rensselaer Polytechnique Institute's, Lighting Research Centre in Troy, N.Y. has reported some compelling results in terms of improving the output efficiency of white LEDs by as much as 60 % using scattered photon extraction techniques. Such arrangement may offset the thermal management problems associated with increasing output. LED output also depends on the supplied current. It is always advisable to use current regulated source rather than that of voltage regulated.

Packages and Standardization of LED Measurements

LED chips are virtually ideal monochromatic light sources. As very small emitters, they approximate point sources and are reasonably lambertian, except at high angles. All types and techniques of optical measurement are easily employed for these devices. When packaged, however LEDs do not behave in such a way. They generally are non uniform and provide highly angular emission, therefore inherent and unique measurement difficulty results. For packaged devices standard conditions are necessary to enable agreement in the measurement results between different laboratories. For luminous, radiant and spectroradiometric intensity, the Commission Internationale de I'Eclairage (CIE) has established guidelines primarily for measuring the luminous intensity of LEDs. As luminous intensity is a quantity that describes a point source, but most of the LEDs are not point sources. Hence the measured luminous intensity varies, depending on the distance and size of the photometer's aperture. To solve this problem the committee introduced standardized geometries. The size of the photometer aperture is seta t 1 cm2 (circular) and the distance between LED and photometer is either 316 or 100 mm. The quantity measured is not a true luminous intensity, so it is called average LED intensity and is designated I_{LEDA} and I_{LEDB}, depending on the geometry used. (In condition A, d = 316mm, solid angle 0.001 sr and in condition B, d = 100 mm, solid angle = 0.01 sr, detector area is set at 1 cm2 for both conditions).

This recommendation has proved successful in solving the problem in luminous intensity measurements and is widely used in the industry. To extend this recommendation further CIE Technical committee 2-46 is preparing a CIE/ISO standard on the measurement of averaged light intensity. However, the sphere geometries described have flaws that could lead to inaccurate measurements. A revision to CIE publication 127 currently in the works will include improve geometries for total luminous flux measurements. In addition partial LED flux (flux emitted within a given cone angle) is being considered. Such a quantity is needed for applications where only forward emission of LED is utilized. The revision also is to include recommendations on using spectroradiometers for photometric and color measurements, including those used with an integrating spheres.

There are defined criterions in its conditions A and B for measurement of averaged intensity in its publication 127. It establishes two conditions for LED position relative to the measurement axis, the measuring instrument's aperture and the distance to the instrument for measuring the average LED intensity.

Flux Measurements of LEDs

Total luminous, radiant and spectral flux of an LED source may be obtained incrementally through series of goniometric measurements or in their entirety by means of an integrating sphere. In the later approach the LED is placed in the centre of the sphere and, depending on the design, a baffle may be placed relative to the emitter to keep direct light from hitting a cosine collection detector at the sphere wall. The walls and the baffle, if used, are coated with a highly diffused broad band reflector such as polytetrafluoroethylene or BaSO4. Because a sphere has area of uniform and non-uniform response, a highly directional source should be pointed at an area of uniform response to obtain the best results.

The LED flux is calculated by comparing the signals detected from the LED and from a standard (known) flux source in the setup. However, anything placed in the sphere affects the throughput, and the lamp or LED used for calibration and the LED to be measured are rarely the same. The difference in the throughput, and the lamp or LED used for calibration and the Led to be measured are rarely the same. The difference in the throughput resulting from changes in the experimental setup will produce incorrect results unless those changes are also incorporated. In well designed system, this is also done using an auxiliary lamp that is housed permanently in the sphere. The lamp is powered while the standard source or test LED is mounted in the sphere but switched off to produce a reference signal level for each. The ratio of the "standard" and "test" reference signals indicates the change in the sphere throughput caused by the swapping of the standard and test devices. This often is incorporated into the calibration procedure of most quality measurement system. Under such circumstances, the optical characteristics of LEDs present no problem into this type of flux measurement.

Standard conditions for flux measurement have not yet been defined, however so there have been some confusion between the measurement of total flux and of forward looking or 2p flux, a quantity measured with the LED placed at the sphere wall. CIE technical committee 2-45 is working on the recommended conditions for this type of measurement and is soliciting industry input. Example as, there is general agreement that the evolution of brighter sources will necessitate larger integrating spheres to insure uniformity of response, but such technical demand must be balanced against the practical realities of the industrial testing laboratory and the production unit.

Colorimetry, Color Temperature Measurement

Colorimetry is science of measuring colors as perceived by three types of colorreceptive cones of human eye, and all measurements are based on the CIE tables of relative responsively values.

From these tables three numbers (the tristimulus values X, Y and Z) can be derived for any perceived color. The familiar CIE x, y color space is graphical model that attempts to map out the chromaticity values, which are relative values derived from tristimulus values. Boundaries represent pure single wavelengths, because all light is combination of one or more wavelengths, all colors lie within the boundary. The CIE uv and u'v' uniform color spaces attempt to make the chromaticity plane more visually uniform so that the distance between two colors is always proportional to the difference in coordinates. These color spaces are strictly two-dimensional, however, and exclude the dimension: brightness.

The CIEUV and CIELAB color spaces were created to include brightness as part of color perception. The CIEUV model is good for emitted color, such as from cathode ray tube monitors and other displays. The CIELAB space is useful for subtractive primary color mixing, such as in inks and in computer printer/plotter outputs.

Glowing or incandescent sources that emit radiation with 100 percent efficiency are called blackbody radiators, or Plankian sources. As the temperature of a black body increases, it changes from red to orange to white and, ultimately, to a pale blue. The correlated color temperature of a source is simply the temperature of blackbody that is closest to the source in CIE UV color space. However, the temperature in degree Kelvin should be attributed only to a source when the coordinates are similar to that of a blackbody. Red and white LEDs can lie close to the blackbody locus, the curve formed by blackbodies' chromaticity values at different temperatures, but green and blue LEDs are too far away for correlated color temperature to be meaningful. Changes in illumination can make objects appear to change color, even if they do not change chromaticity. This is because the sources can have the same chromaticity without sharing spectral characteristics. A fluorescent light may have same chromaticity as sunlight, for example but a fabric that appears green in the shop may appear brown outdoor.

CRI Measurements

Most specifications provided by LED manufacturers do not list chromaticity coordinates, but there is a trend toward supplying colorimetric parameters as product performance improves. Typically manufacturers specify the dominant wavelength, which is defines by the intercept of a line from a white point (Center of the CIE color spaces) through the chromaticity value onto the monochromatic boundary in nanometers. The peak wavelength, the one at the maximum spectral intensity, is easy to obtain. It is the most common value specified by LED manufactures. However, it has little practical significance for applications that are viewed with the human eye; for example, two LED may have the same peak wavelength but may be perceived having different colors.

The most accurate method for measuring color is spectroradiometer. It performs a complete spectral distribution measurement of a source from which all photometric, radiometric and colorimetric parameters can be mathematically calculated and reported in real time. The wavelength accuracy of the equipment should be better than 0.5nm.

Junction Temperature Measurements

Junction temperature affects light-emitting diode (LED) and laser diode performance in many ways. Light output center wavelength, spectrum, power magnitude, and diode reliability are all directly dependent on the junction temperature. The very high power densities found in laser diodes —which reach 1000 W/cm^2 or more in the junction area – provide additional thermal management issues. Thus, thermal design of the diode itself and the packaging in which it is encased becomes crucial to the overall performance of the device. Validation of thermal design and assembly repeatability requires the ability to measure junction temperature. There are basically three different methods for making laser diode junction temperature measurements. Two of the methods use the light output of the device for an indirect measurement of device junction temperature. The third method is more traditional, considering its application to other types of diodes, and is based on one of the device's electrical characteristics for an indirect measurement of junction temperature.

Test of LEDs for Automotive Lighting

LEDs are fast becoming the light source of choice for automotive signal lamps. Before an LED can be used in an automotive signal lamp, however, a study must be made of the LEDs that are available for these applications. This is required to determine the correct LED for the particular lamp being designed. Significant parameters that must be evaluated to determine this LED's suitability for the selected application are the wavelength of the emitted light, the radiation pattern of the light and the usable Lumen output. One additional criterion that must also be considered is the ratio of the cost per Lumen of the targeted LED. These criteria are a few of the parameters that are measured in and out of the laboratory to determine if the LED being considered can be used in an exterior automotive signal lamp application. The selection process of the LED involves the characterization by accurately measuring the light output of the LED at a specified electrical current and thermal resistance. Once this has been completed and it has been established that the selected LED is a possible candidate for an automotive exterior signal lamp, the LED is subjected to durability performance testing. This involves testing the LED for its ability to meet the environmental, electrical and optical requirements. Since there are many manufacturers of LEDs it is necessary to choose an LED that has the highest quality along with the greatest amount of usable light output. It is this testing and characterization process that will enable us to select only the brightest and the highest quality LEDs for the automotive signal lamp. In addition each customer has their own unique set of requirements that the light source must meet. Therefore, the light source must meet or exceed the most stringent requirements that have been set forth.

Measurements have inherent corresponding uncertainties. As they reference a standard that is more and more removed from the National Institute of Standard and Technology (NIST) or another national agency, their associated uncertainty increases. A table of components contributing to uncertainty typically is known as an "uncertainty budget" Total uncertainty is expressed as "expanded uncertainty" or as "percent confidence". As might be expected, uncertainty budget should include the variation in the results, the scan to scan repeatability, the need for realignment, and the drift in samples or in the measurement system. Conformity of the measurement procedure and equipment to the specifications and requirements of the standard, the effects resulting from the differences between, the standard lamp and LED environmental effects, and the accuracy and stability of operating conditions also should be expressed. Many critical issues regarding measurement conditions for LEDs remain unresolved. The CIE and other standards bodies have not yet provided conclusive criteria, but there is vigorous discussion and debate about the appropriate solutions.

Testing and Characterization of LEDs for Large Displays and TVs

LEDs are widely applicable in large video display screens, which show passenger information, video advertising and entertainments. LEDs are useful for such applications, as they have specific and narrow spectral emission characteristics, and can be turned on and off very quickly. LEDs offer much more saturated colors than other display methods, and very high contrast. In these displays each pixel usually consists of three different monochromatic LEDs, emitting red, green and blue light. The light from each must be combined in order to display a full palette of colors along with a series of filters to reflect and transmit the light from different LED sources, combining the different colors into one single light beam. Hence the light from LEDs can be combined to give a display screen with the full spectrum of colors available. However, large screens, which can contain thousands of LEDs, have problems that the intensity of individual LED in each pixel can vary considerably, meaning the display doesn't look uniform when seen from distance. Therefore intensity of each LED in the display should be calibrated and tested individually, in order to achieve uniform brightness at every point of the display. The Lusa system uses a camera to scan upto 720 pixels at once. The data gathered is sent via FireWire to the controlling system, and from there to the computer controlling the display, which can then adjust each pixel until all the units are giving off the same light intensity. As output of an LED is typically highly directional, therefore the intensity can change significantly with viewing angles. It is very important for display designer to know the angular characteristics of the LEDs being used. Because the angular characteristics of LED are set during the design and production of the unit, the testing must be carried out at this stage. The IS-LI system from Radiant imaging uses an imaging sphere, developed with Philips, to measure far field luminous intensity, luminous intensity, CIE chromaticity coordinates and correlated color temperature, all as a function of angle, for LEDs. With this information to hands the designer of an LED display can choose the LEDs most suitable for the display.

Applications of LEDs

LEDs are now widely used from house to laboratory, from industry to military, from lighting to entertainment and also from environmental cleaning to medical surgery. These are brilliant candidates for displays in sexy mobile phones, laptops and high definition TVs. They are rapidly replacing their liquid crystal and cathode ray based display counterparts. High brightness LEDs offer many potential advantages over traditional incandescent bulbs and fluorescent tubes, they are highly reliable with a long service life, they consume relatively littlie energy, lower operating voltage means that they are safer to handle, a particular concern in outside applications where moisture has to be taken into account; and they project less heat into the illuminated field is another important consideration.

A field where LEDs show their great promise is in architectural applications. The highly directional nature of LED illumination, along side the saturated color output they offer, provides architects with lighting options, both internal and external, that simply are not possible using conventional lighting. This also makes LEDs particularly suitable for so called task lighting, highlighting particular architectural or decorative features, or provides local spot lighting. Followings are some recent applications of LEDs.

Applications of LEDs in Display and Televisions

Recent improvements in light output have enabled new applications for LEDs such as automotive lighting, traffic signaling, and more recently television displays. Improvements in manufacturing the epitaxial material and packaging have enhanced thermal conductivity and increased the overall light extraction efficiency of LEDs. Quantum efficiency gains and unique photon extraction techniques also have helped improve brightness. As LED continue to advance in brightness and efficiency, their effect on television applications could be significant. Displays based on cathode ray tubes (CRTs) are heavy, space consumable, power hungry, limited resolution and their replacement with flat panel displays is well publicized. High brightness, full color display made from organic light emitting diodes (OLEDs) can produce brightness and clarity that proponents claim is not available from existing flat panel displays. Adding this with the possibility of low cost fabrication on flexible plastic sheets provide a potential market that no other technology hope to match i.e. one can fold up his computer monitor and sticking it in his pocket.

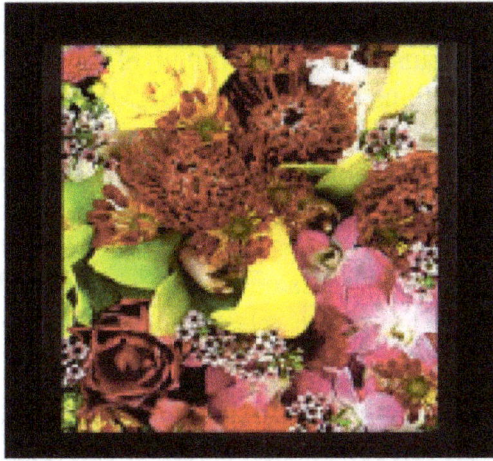

High resolution and high contrast display based on LEDs.

Existing flat panel displays are lit from behind and produce a given color using liquid crystals and filters to subtract the complementary color from white light. All the pixels must continuously illuminate, even when they appear dark. LEDs do not require backlighting and are inherently more efficient, thinner and robust. The appearance of OLED display is essentially independent of viewing angle, and equals a cathode ray tube in brightness, both significant advantages over other flat panel displays. The response time is three orders of magnitude faster than a LCD displays, and OLED display require less than 10 volt to operate. It may even be possible to manufacture large area displays using an inkjet printing in place of vacuum deposition processing. As mentioned above OLED structure consist of a number of thin layers stacked between a transparent anode- must often indium tin oxide (ITO) and a metallic cathode. The organic layers are typically composed of an electron transport layer and a hole transport layer. The active layer may emit light when forward biased by a few volts.

The emission band width of OLEDs is typically on the order of 50-70 nm at full width half maximum. Mixing the output of three such red, green and blue (RGB) diodes will not accurately reproduce an arbitrary color of dynamic full color display. The layers are thin enough (< 1μm) so that thin film interference effect can be used advantageously to narrow the individual emission bandwidth and achieve reliable color reproduction. An RGB display can be made of OLED pixels by overlapping three colored "sub pixels" at each pixel location, enabling a higher than having the color sources side by side. The upper sub pixels are transparent are transparent to the colors produced underneath them. Trace contaminants from the ITO anode have caused problem with this approach.

Organic light emitting diodes are made into both passive matrix and active matrix displays. The passive matrix design is simpler and is in commercial use for low cost applications such as cellular phone displays. It is formed by providing each column of pixels in an array with a data signal as the rows of pixels are scanned with a drive current. Rows must be refreshed by the current in a time less than that detectable as flicker by the human eye. Active matrix displays are used in high resolution and high information applications. They are formed over a pattern of thin film transistors (TFTs) that is itself deposited on a substrate. The TFT backplane can address each pixel independently, and allows for high speeds and high currents (brightness). This active matrix structure is ideal for a device in which each pixel is self illuminating.

A seeming disadvantage to the active matrix approach is the number of drive transistors required, one for each color at each pixel, plus one to turn the pixel off. For VGA resolution (640 rows × 480 columns), this amount to millions of TFTs. However, the same technology is already in use in the high resolution LCD based displays. It has been possible to deposit the TFT back plane over a flexible plastic substrate, in order to fabricate flexible TVs and computer monitor. While crystalline organic materials require vacuum deposition onto a glass substrate. It may be possible to layer polymers over the flexible TFT substrates using spin casting or other inexpensive techniques such as inkjet printing. The spin cast processing is expected to produce OLED arrays no thicker than a piece of paper in about 2 − 3 hours, much less than the the fabrication time of an equivalent inorganic structure.

An LED offer advantage over LCD is to directly generate light. This capability eliminates the needs for a backlight and the associated extra thickness, w.0eight and power drain, which is particularly important in cell phones and other mobile applications. Organic LEDs have wider viewing angles, lower overall power consumption, faster image response, higher contrast, and greater immunity to temperature and lower manufacturing cost that LCDs.

Instead of large displays used in computers and televisions and medium sized displays used in cell phones and mobile applications LEDs are also applicable in micro displays.

Combination of OLEDs with complementary metal oxide semiconductor (CMOS) silicon technology is primary enabler of an important new generation micro displays. The OLEDs are easily formed from evaporated thin films laid on top of the silicon, the required voltages are low enough to allow low cost, high circuit density silicon devices, and the optical characteristics are well for use of low cost, large field of view magnifiers. They are efficient resulting for low power consumption, and the fact that they emit light rather than modulate transmitted or reflected light results in smaller and lighter viewing modules. The integration of OLEDs with integrated circuits (ICs) permits any computer and video electronic system functions to be built directly onto the silicon IC under the OLED layer, resulting in an ultra compact system with lower overall system cost. The technology is often used in near-eye micro displays; a type of flat panel display used in small optically viewed devices such as video head sets, camcorders, viewfinders, and other portable devices. Fabrication of micro displays is similar to that of direct view displays. In the case of direct view displays hole transport layers (HTL) is coated over transparent ITO coated glass plate acting as anode followed by an active emissive layer. An electron injecting layer is deposited over the surface of emissive layer, which is covered by aluminum cathode. As in micro displays hole injecting layer is deposited over the silicon substrate, which is transparent, therefore cathode layer should be semitransparent in order to emit light. Figure depicts layout for direct displays and that of (b) for micro displays.

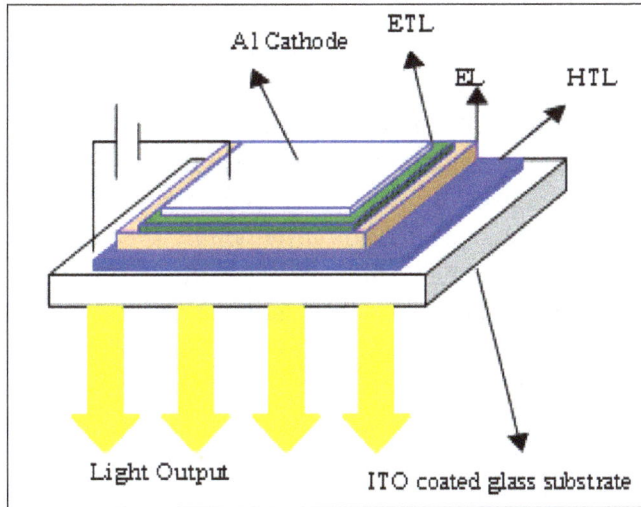

(a) Schematic view of application of LED in Display and Telivision.

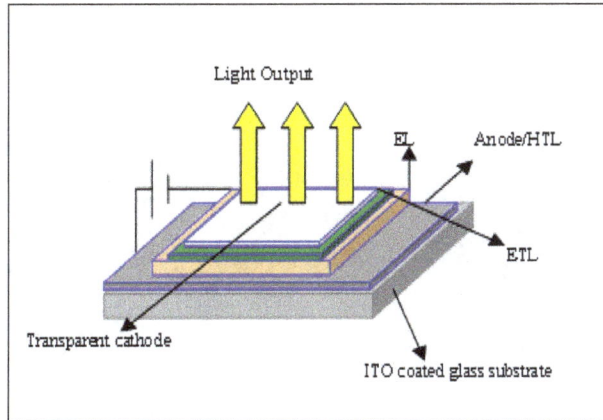

(b) Schematic view of application of LED in microdisplay and mobile display applications.

Applications of LEDs in General Lightening

Light emitting diodes are widely used in general, decorative as well as architectural lightening. This is more expensive per lumen as compared to the incandescent bulb but more durable and long lasting. Grow lights composed of LEDs are more efficient, both because LEDs produce more lumens per watt than other alternatives, and also because they can be tuned to the specific wavelengths plants can make the most use of Light bulbs, Lanterns, Streetlights, Large scale video displays, Architectural lighting, Light source for machine vision systems, requiring bright, focused, homogeneous and possibly strobed illumination. Motorcycle and Bicycle lights, Flashlights, including some mechanically powered models. Emergency vehicle lighting, Backlighting for LCD televisions and displays. The availability of LEDs in specific colors (RGB) enables a full-spectrum light source which expands the color gamut by as much as 45%. Stage lights using banks of LED's as replacement for incandescent bulbs. LED's produce less heat so LED stage lighting is cheaper to operate and reduces the risk of fire considerably. LED-based Christmas lights have been available since 2002, but are only now beginning to gain in popularity and acceptance due to their higher initial purchase cost when compared to similar incandescent-based Christmas lights. For example, as of 2006, a set of 50 incandescent lights might cost US$2, while a similar set of 50 LED lights might cost US$10. The purchase cost can be even higher for single-color sets of LED lights with rare or recently-introduced colors, such as purple, pink or white. Regardless of the higher initial purchase price, the total cost of ownership for LED Christmas lights would eventually be lower than the TCO for similar incandescent Christmas lights since the LED requires much less power to output the same amount of light as a similar incandescent bulb. More to the point, LEDs have practically unlimited life and are hard-wired rather than using unreliable sockets as do replaceable bulbs. So a set of LED lights can be expected to outlive many incandescent sets, and without any maintenance. Following figure depicts lightening applications of light emitting diodes. LED lamps (also called LED bars or Illuminators) are usually clusters of LEDs in a suitable housing. There are different models and shape, among them the light bulb shape with a large E27 Edison screw and MR16 shape with a bi-pin base. Other models might have a small Edison E14 fitting, GU5.3 (Bipin cap) or GU10 (bayonet socket). This includes low-voltage (typically 12 V halogen-like) varieties and replacements for regular AC mains (120-240 V AC) lighting. Currently the latter are less widely available but this is changing rapidly.

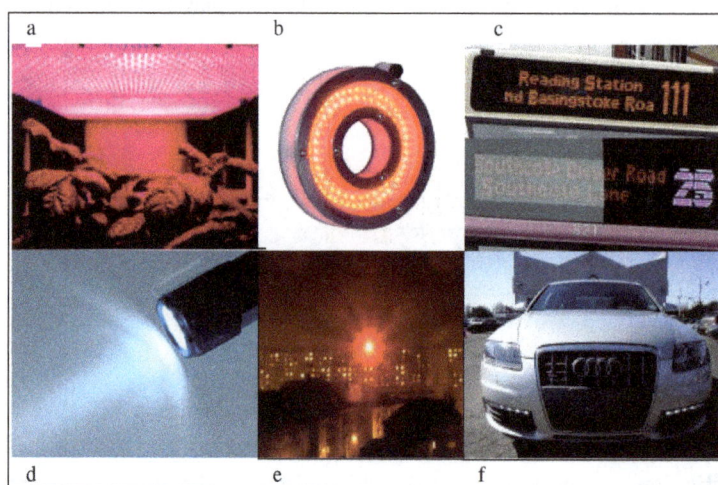

Applications of LEDs in general lightening (a) LED panel light source used in an experiment on plant growth. The findings of such experiments may be used to grow food in space on long duration missions, (b) Light sources for machine vision systems, (c) LED destination displays on buses, one with a colored route number (d) Flashlights and lanterns that utilize white LEDs are becoming increasingly popular due to their durability and longer battery life (e) A single high-intensity LED with a glass lens creates a bright carrier beam that can stream DVD-quality video over considerable distances, (f) LED lights on an Audi S6 car.

Seoul Semiconductor Co., Ltd has developed LEDs that can run directly from mains power without the need for a DC converter. For each half cycle part of the LED diode emits light and part is dark, and this is reversed during the next half cycle. Current efficiency is 80 lm/W.

Laboratory and Medical Applications of LEDs

Light emitting diodes have become key components in the research and developments now days. They are widely used in displays in multimeters, indicators in power supplies, water purification, sterilization, diode lasers and several medical applications. There are several applications of LEDs of different wavelength ranges, which are illustrated in the following table.

S. No.	Applications	Wavelength required (nm)
1	Detection of micro contamination	225nm/ 365nm
2	Water purification	225 – 280 nm
3	UV curing	365-375 nm
4	Scientific analysis	265-415nm
5	Bank note verification	365-385 nm
6	Phototherapy	310 – 420 nm
7	Crime Scene inspection	340 – 420 nm
8	UV dental whitening	345 – 420 nm

LED phototherapy for acne using blue or red LEDs has been proven to significantly reduce acne over a three-month period. Sterilization of water and other substances using UV light is frequently used for biological experiments.

Applications in Information and Telecommunication

LEDs have shown their potential applications in the field of information broadcasting and tele-communications. They are widely used in computer mices, scanners, sensors etc. IR LEDs are most important components in the remote controls system, which are main components of current era of modern technology. IR LEDs are also used in optical fiber and free space optical communication. Some flatbed scanners use an array of red, green, and blue LEDs rather than the typical cold-cathode fluorescent lamp as the light source. Having independent control of three illuminated colors allows the scanner to calibrate itself for more accurate color balance, and there is no need for warm-up. LEDs are also used in computers, for hard drive activity and power on. Some custom computers feature LED accent lighting to draw attention to a given component. Many computer manufacturers use LEDs to tell the user its current state.

Exogenous Physical Irradiation on Titania Semiconductors

As one of the mostly explored multidisciplinary research frontiers, nanomedicine has attracted the broad attention of scientific community ranging from material science, chemistry, pharmacy, biology, and biomedicine. It has shown the intriguing performance and application prospect in molecular imaging for disease diagnosis, targeted drug delivery for enhanced chemotherapy, some physically triggered novel therapeutic modalities, diagnostic biosensing, and even tissue engineering. Various nanoparticles with their intrinsic desirable composition, nanostructure, physiochemical property, and biological effects have been explored to achieve the efficient therapeutic performance and outcome in the past decades, among which inorganic nanoplatforms have been attracting the high research interest very recently because of their unique physiochemical property, multifunctionality (e.g., optical property, magnetism, electronic behavior and acoustic property) and relatively high biocompatibility.

Organic nanosystems have been broadly investigated and some of them have entered the clinical stage for benefiting the patients. It is noted that the organic nanosystems typically lack the functionality, which means that they cannot be easily designed for some unique and specific theranostic purposes. Comparatively, inorganic nanosystems can be facilely endowed with specific properties of magnetism, fluorescence, ultrasound responsiveness, electronic conductivity, etc. They can also be designed with some intriguing nanostructures and topologies. For instance, the mostly explored mesoporous silica nanoparticles (MSNs) are fabricated with well-defined mesoporous nanostructure, which provide the large reservoirs for the efficient loading and delivery and therapeutic guest molecules. Another paradigm of inorganic nanoparticles is the mostly studied superparamagnetic iron oxide nanoparticles (SPIONs) for contrast-enhanced magnetic resonance imaging (MRI), magnetically targeted drug delivery and magnetic hyperthermia, which are all based on their intriguing magnetic properties. Especially, the plasmonic resonance property of gold (Au) nanoparticles has been adopted for photo-triggered hyperthermia, computed tomography (CT) imaging, and biosensing applications.

Compared to mostly explored metal oxides such as silica, manganese oxide, and iron oxide nanoparticles, titania nanosystems have emerged as a novel inorganic nanoplatform with their intrinsic physiochemical properties suitable for biomedical applications. Titanium (Ti) has been demonstrated as one of the biocompatible elements. For instance, titanium oxide (TiO_2) has been extensively used in colorant in food, cosmetics, and sunscreen. Especially, Ti-containing metal alloys have been employed as the medical implantation devices. As one of the mostly explored semiconductors, traditional TiO_2 nanoparticles have a bandgap of 3.2 eV, which can be excited by ultraviolet (UV) light. The UV light with radiative energy higher than its bandgap of 3.2 eV (387 nm) excites an electron to the conduction band (CB) and creates an electron hole pair. The produced electrons can reduce the absorbed molecular oxygen to superoxide radicals, and the holes are capable of oxidizing the water molecules into hydrogel radicals (•OH). Therefore, TiO_2 nanoparticles can function as the inorganic photosensitizers to produce large amounts of reactive oxygen species (ROS) for photodynamic therapy (PDT). Especially, the semiconductor nature and unique photoresponsiveness of TiO_2 have been employed for the degradation of organic substrates and deactivation of microorganisms/viruses.

However, traditional titania nanoparticles only respond to UV light, which unfortunately has potential phototoxicity and low tissue-penetrating depth, severely hindering their further clinical translation. Very recently, the fast development of theranostic nanomedicine has explored more effective exogenous physical triggers for activating titania nanoparticles for achieving some specific but intriguing therapeutic modalities, such as near infrared (NIR)-triggered photothermal therapy (PTT), NIR-activated PDT, X-ray/Cerenkov radiation (CR)-activated deep-seated PDT, and US-triggered SDT. All these exogenous physical triggers are featured with their intrinsic characteristics but also with some drawbacks, which are mainly based on the semiconductor nature of titania nanosystems accompanied with the unique oxygen-defect modulation for photothermal hyperthermia. In addition, the critical issue of biocompatibility and biosafety of these titania nanoplatforms has also been discussed to potentially guarantee the further clinical translation. Finally, the critical challenges and further future developments of the novel physical-triggered titania-based nanoplatforms and the corresponding intriguing therapeutic modalities are also deeply discussed to promote the progress of this novel inorganic nanoplatform in theranostic nanomedicine.

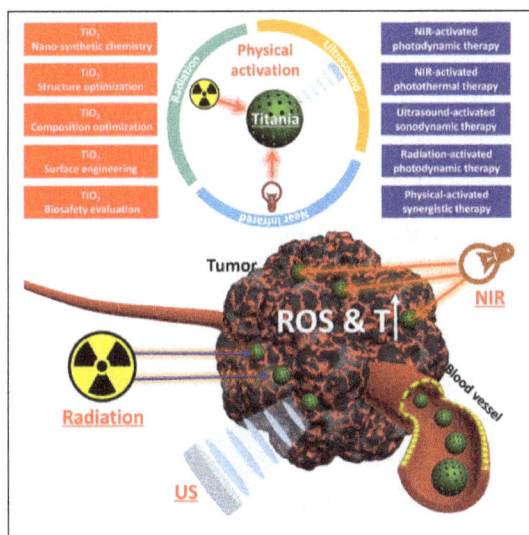

Schematic illustration of exogenous physical activation of titania nanoparticles for tumor-specific therapy. It includes NIR-activated PDT/PTT, radiation-activated PDT, US-activated SDT and physical activation-based synergistic therapy. The related research frontiers are also summarized in the figure, such as nano-synthetic chemistry for titania fabrication, structure/composition optimization, surface engineering, and biosafety evaluation.

Synthesis, Multifunctionalization and Surface Engineering of Titania Nanoparticles in Biomedicine

The rational design and successful construction of titania-based nanoplatforms are the bases for achieving the high theranostic performance in biomedicine, which is mostly based on the advances of nanosynthetic chemistry and material chemistry. We recently synthesized highly dispersed mesoporous titania nanoparticles (MTNs) with monodispersity and uniformity by a method of prehydrolysis of titanium precursors combined with solvothermal treatment, which provided both the high crystallized framework and well-defined mesoporous nanostructure. In addition, we recently synthesized oxygen-deficient core/shell-structured black TiO_{2-x}. nanoparticles by a facile aluminum reduction methodology for enhanced sonodynamic therapy (SDT) and simultaneous NIR-triggered photothermal hyperthermia at NIR-II biowindow. Especially, mesoporous black TiO_{2-x}. nanoparticles were fabricated by a two-step procedure where the mesoporous white TiO_2 nanoparticles were initially synthesized, followed by reduction under hydrogen atmosphere at 500 °C for 1 h to turn white TiO_2 into black TiO_{2-x}.

Especially, the advances of nanosynthetic chemistry make the precise controlling of titania's composition, nanostructure and functionality possible where these titania nanocomponents can also be integrated with other functional moieties or nanoparticles to achieve some specific purposes. For instance, the $NaYF_4$:Yb,Tm nanoparticles were initially coated by a silica layer with the grafting of (3-aminopropyl)-trimethoxysilane, which provided the positively charged amino groups for efficient binding titanium precursors, guaranteeing the gradual epitaxial growth of uniform TiO_2 layer onto the surface of initially synthesized nanoparticles. $NaYF_4$:Yb^{3+},Tm^{3+}@NaGdF$_4$:Yb^{3+}@TiO_2 (UCNPs@TiO_2) core/shell nanoparticles were synthesized by direct in situ growth protocol. $NaYF_4$:Yb^{3+},Tm^{3+}@NaGdF$_4$:Yb^{3+} were initially synthesized, followed by surface modification with polyvinylpyrilidone (PVP). Then, TiF_4 acted as the Ti precursors for the direct formation of TiO_2 nanoshells on the surface of UCNPs by the hydrolysis and condensation process. This one-step PVP-mediated methodology is facile and generic for the construction of TiO_2-based nanocomposites, especially for the construction of UCNPs@TiO_2 composite nanoplatforms.

For TiO_2-based functionalization, we also directly grew TiO_2 nanoparticles onto the surface of graphene oxide (GO) by a hydrothermal treatment of the cosolvent solution of TiO_2 nanoparticles and GO suspension, based on which TiO_2 nanoparticles were uniformed dispersed onto the surface of 2D planar GO. $Au - TiO_2$ nanocomposites were synthesized by a facile photoreduction of Au^{3+} ions, which could be deposited on the surface of TiO_2 nanoparticles to grow Au nanoparticles under UV irradiation. The particle size of deposited Au nanoparticles could be controlled by adopting the UV-irradiation durations. Especially, the dumbbell-like $Au - TiO_2$ nanoparticles were synthesized by a seed-mediated growth approach. Au nanoparticles were initially synthesized, acting as the growing sites for the TiO_2 generation in an anisotropic manner by controlling the hydrolysis degree of introduced Ti precursor.

The surface chemistry of titania-based nanoplatforms is also of high significance in biomedicine. For instance, the adequate surface modification can either improve the stability of these nanoparticles in physiological solution or achieve the positive-targeting accumulation into tumor cells/tissues. The surface of $Au - TiO_2$ nanocomposites was modified with biocompatible carboxymethyl dextran (CMD) for achieving prolonged systemic circulation and subsequent enhanced tumor-homing capability. We modified the surface of black TiO_{2-x} nanoparticles with $NH_2 - PEG_{2000}$ molecules by a simple sonication procedure, which was based on the coordination interaction between N component of PEG molecules and Ti atoms of black TiO_2 nanoparticles. We also modified the surface of TiO_2-loaded GO by PVP molecules for enhanced stability in physiological condition, which could guarantee the further in vivo biomedical applications on combating cancer. In addition, the precoating of SiO_2 layer onto the surface of TiO_2-based nanocomposite could provide the anchoring sites of silane group of maleimide-PEG-silane, achieving the efficient PEGylation of TiO_2-based nanocomposites. Especially, anti-cAngptl4 Ab was conjugated onto the surface of N- TiO_2/$NaYF_4$:Yb,Tm nanocomposites for targeted cancer-cell PDT on killing cancer cells as induced by NIR irradiation.

Light Irradiation on Titania for PDT

PDT on combating cancer is featured with noninvasiveness and tumor specificity, which typically employs the external physical light source for activating photosensitizers to produce toxic ROS and consequently kill the cancer cells. Light-excited PDT has been clinically used for the treatment of cancers on skin and other epidermal tissues. As compared to traditional organic photosensitizers, TiO_2-based inorganic nano-photosensitizers are featured with high stability and nontoxicity, which has shown broad application potentials in PDT-based cancer treatment.

Based on titania nanoparticles, a polychromatic visible light-activated nano-biohydrid system was constructed by covalently binding an antibody via a dihydroxybenzene bivalent linker that could selectively recognize glioblastoma multiforme (GBM) cells. This targeting strategy enhanced the intracellular uptake of TiO_2 nanoparticles and produced large amounts for ROS for damaging the cell membrane, inducing the cancer-cell death under the visible-light irradiation. The PDT efficiency of titania could also be achieved by heterogeneous atom doping. For instance, the Fe-doping of

TiO$_2$ nanotubes was demonstrated to realize near-visible light-driven (2.30 mW cm^{-2}, $\approx$ 405 nm) PDT on killing cervical cancer cells, and the phototoxicity of Fe-doped TiO$_2$ nanotubes was much higher than that of undoped TiO$_2$ nanotubes. Furthermore, nitrogen-doped $TiO_2(N?TiO_2)$ also showed visible light-triggered PDT against HeLa cancer cells where the N-doped TiO$_2$ nanoparticles were featured with higher PDT efficiency as compared to that of pure TiO$_2$ nanoparticles. The related mechanism investigation revealed that $N?TiO_2$ induced more loss of mitochondrial membrane potential and higher increase of intracellular Ca^{2+} and nitrogen monoxide in HeLa cancer cells than pure TiO$_2$ nanoparticles.

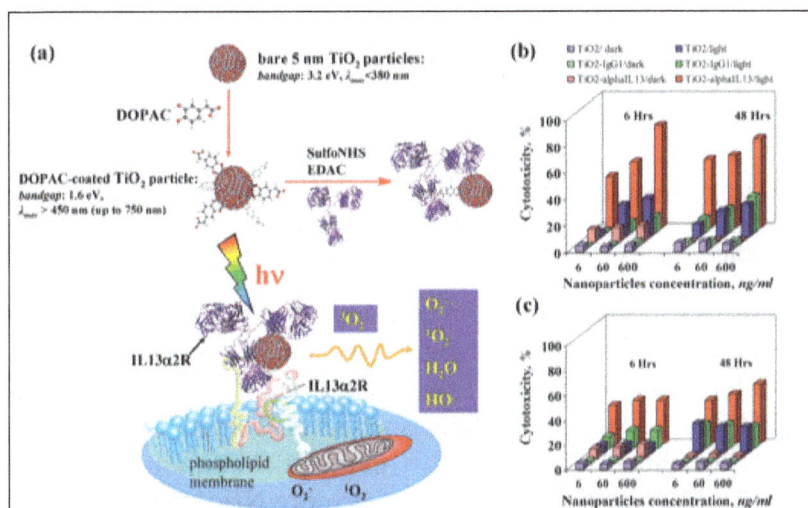

a) Schematic illustration of the fabrication of titania nanoparticles with surface-linked IL13R-recognizing antibody, and their further recognition and binding to surface IL13R of cancer cell for visible light-activated ROS generation and subsequently inducing cancer-cell death. In vitro therapeutic efficiency by evaluating the phototoxicity of TiO$_2$-mAb against b) A172 GBM cells (high ILα2R expression) and c) U87 GBM cells (low ILα2R expression).

To further enhance the PDT efficiency of TiO$_2$-based photosensitizers, TiO$_2$ nanoparticles were conjugated with ruthenium complex (N3) for improved and synergistic production of ROS in both hypoxic and normoxic conditions. By light irradiation (365 nm), the N3 injected electrons into TiO$_2$ nanoparticles, resulting in the production of three- and fourfold more hydroxyl radicals ($\cdot$OH) and hydrogen peroxide (H_2O_2) as compared to bare TiO$_2$ nanoparticles, respectively. Bare TiO$_2$ nanoparticles could oxidize water molecules to produce hydroxyl radical ($\cdot$OH, figure). The presence of light-induced electron–hole pair in TiO$_2$ facilitated the reduction of molecular oxygen to superoxide and then transformation to single oxygen ($^1O_2$, figure). Especially, under the hypoxic condition, the N3 facilitated the electron–hole reduction of absorbed water molecules to enhance the hydroxyl radical production with nearly threefold increase. This strategy could transform TiO$_2$ photosensitizer from a dual type I and II PDT nanoagents into a mainly type I photosensitizer independent of the oxygen level. This work provides an efficient strategy to enhance the TiO$_2$-based PDT efficiency in hypoxic condition by N3 hybridization. Coating a homogenous TiO$_2$ layer onto the surface of ZnTPyP self-assembly nanocrystal achieved the photoelectron transfer at ZnTPyP self-assembly/TiO$_2$ interfaces, which further enhanced the two-photon PDT against HeLa cancer cells via type-1-like PDT process. This titania-based composite nanoplatform is very intriguing because the achieved two-photon PDT is highly desirable for deep-tissue disease treatment.

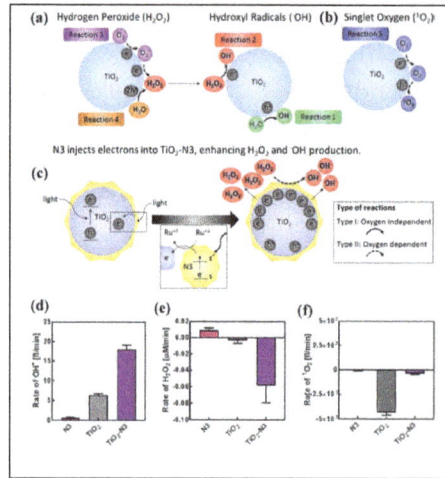

Schematic illustration of ROS generation, including a) H_2O_2 production and transformation into hydroxyl radical, b) singlet oxygen ($^1O_2$) production, and c) electron injection by N3 into TiO_2–N3 to enhance H_2O_2 and hydroxyl radical production. The generation rate of d) hydroxyl radical, e) H_2O_2, and f) singlet oxygen under hypoxic conditions.

The major challenge of TiO_2-based PDT is the light responsiveness only in the wavelength range of UV or visible light, which has the low tissue-penetrating distance and causes the failure in the treatment of deep-seated tumor. Upconversion nanoparticles (UCNPs) are capable of generating high energy light from the low energy light such as NIR light. Therefore, lanthanide-doped UCNPs can convert NIR light into UV or visible photons via an anti-Stokes emission process, which potentially acts as the "nano-transducers" to achieve NIR-triggered PDT. On this ground, core/shell-structured UNCPs ($NaYF_4$:Yb^{3+},Tm^{3+}@$NaGdF_4$:Yb^{3+}) with enhanced upconverting UV emission were initially synthesized, followed by coating with TiO_2 shells using TiF_4 as the Ti precursor to in situ grow TiO_2 shells onto the surface of UNCPs under mild hydrolysis condition. The UCNPs core emitted upconverting light in UV/visible range by 980 nm NIR irradiation, which was substantially diminished by the absorbance of TiO_2 shells in such a wavelength range. Such an energy-transferring process induced the extracellular and intracellular generation of ROS for causing the cancer-cell death by inducing the cell apoptosis. HeLa tumor-bearing model results showed that the intratumoral injection of $NaYF_4$:Yb^{3+},Tm^{3+}@$NaGdF_4$:Yb^{3+}@TiO_2 (UCNPs@TiO_2) core/shell nanoparticles followed by 980 nm laser irradiation achieved the substantial tumor-growth suppression with high therapeutic efficiency/outcome, which was further demonstrated by immunohistochemical staining for caspase 3.

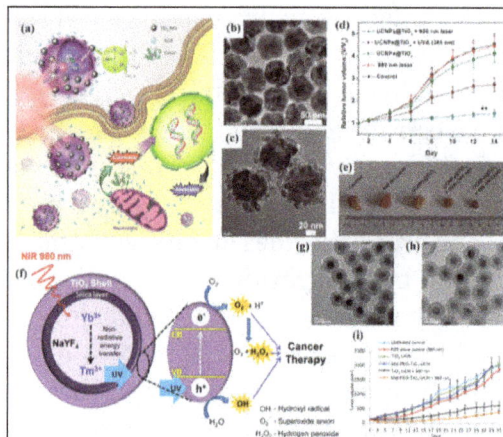

a) Schematic illustration of NIR-triggered PDT based on UCNPs@TiO_2-based nano-photosensitizers and the underlying therapeutic mechanism including the induced apoptosis of cancer cells. TEM images of b) NaYF$_4$:Yb^{3+},Tm^{3+}@NaGdF$_4$:Yb^{3+} core/shell nanoparticles and c) NaYF$_4$:Yb^{3+},Tm^{3+}@NaGdF$_4$:Yb^{3+}@TiO_2 (UCNPs@TiO_2) core/shell nanoparticles. d) The tumor-volume changes of tumor-bearing mice in varied treatment groups as shown in the figure, and e) the corresponding digital photographic images of excised tumors at the end of treatments. f) The scheme of UCNP-mediated activation of surface-coated TiO_2 layer for ROS production to kill the cancer cells. TEM images of g) TiO_2-UCNPs and h) Mal-PEG- TiO_2-UCNPs. i) The in vivo OSCC tumor-growth volumes within the 35 d duration after varied treatments as indicated in the figure.

Similarly, Zhang and co-workers coated a TiO_2 layer onto the surface of SiO_2-coated UCNPs (NaYF$_4$:20%Yb,0.5%Tm) for 980 nm NIR-triggered PDT. The UCNPs core converted NIR irradiation into UV light, which photoexcited electrons in the valence band (VB) of TiO_2 shell to the CB, forming the photo-induced hole–electron pairs. The postgenerated hole–electron pairs reacted with surrounding molecular oxygen and water molecules to generate ROS and then induce the cancer-cell death. The TiO_2-coated UCNPs were clearly characterized by TEM image. After the intratumoral injection of these composite nanoparticles followed by 980 nm NIR irradiation, the significant tumor-growth suppression was achieved. In fact, the TiO_2-coated UCNPs themselves are highly biocompatible, and only the NIR-irradiated tumor region can produce toxic ROS, therefore their impact to normal cells and tissues are low, leading to high therapeutic biosafety. Furthermore, anti-EGRF-affibody was conjugated to PEGylated TiO_2–UCNPs nanocomposites for targeting epithelial growth factor receptor (EGFR) overexpressing oral cancer cells and the subsequent NIR-excited PDT with the therapeutic outcome of significantly suppressed tumor growth and improved survival rate of tumor-bearing mice.

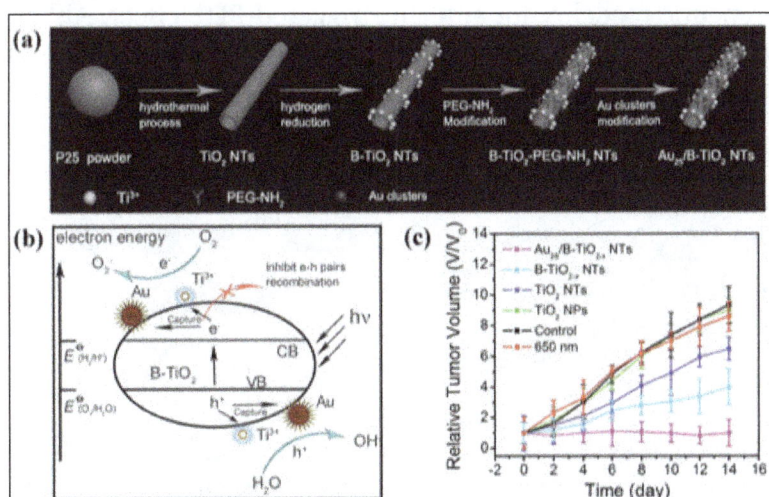

a) Schematic illustration of the synthetic process for Au$_{25}$/B–TiO_{2-x} nanotubes. b) The scheme of photocatalytic mechanism for ROS production as assisted by the introduced Au$_{25}$/B– TiO_{2-x} nanotubes. c) In vivo therapeutic outcome as indicated by the relative tumor-volume changes after varied treatments.

The photoresponsive wavelength range could also be controlled by rational design of the composition and nanostructure of titania-based nanoplatforms. For instance, 808 nm NIR-activation of black TiO_2 nanoparticles with a narrow bandgap of around 2.32 eV was demonstrated to absorb NIR light and subsequently produce abundant ROS for photodynamic killing of bladder cancer cells. In addition, Au cluster-anchored black anatase TiO_{2-x} nanotubes (designated as

$Au_{25}/B?TiO_{2-x}$) were stepwise synthesized by gaseous hydrogen reduction of TiO_2 nanotubes followed by the deposition of Au clusters. These $Au_{25}/B?TiO_{2-x}$ exhibited the photo-responsiveness in NIR range (650 nm) for PDT against cancer. The surface modification of Au clusters changed the electrical distribution in the composite nanosystem, which could reduce the recombination of electrons and holes as triggered by NIR irradiation. Importantly, the hydrogen reduction generated large amount of Ti^{3+} ions in the matrix of black TiO_{2-x}, which extended the light response of anatase TiO_2 nanoparticles from UV light to NIR light. In vivo therapeutic evaluation on tumor-bearing xenograft revealed that the significantly enhanced therapeutic efficacy was achieved based on the photocatalytic synergistic effect, which substantially suppressed the tumor growth after the injection of $Au_{25}/B?TiO_{2-x}$ followed by NIR irradiation.

To achieve simulated sunlight-irradiated PDT, TiO_2–Au–graphene (designated as TAG) heterogeneous nanocomposites were designed and fabricated for employing simulated sunlight as physical triggering source to kill melanoma skin cancer cells by photodynamic effect. The narrow bandgap of Au nanoclusters and staggered energy bands of Au–TiO_2–graphene resulted in the efficient use of simulated sunlight, which also enhanced the separation efficiency of electron–hole pairs for producing large amounts of hydroxyl and superoxide radicals. Typically, the sunlight-excited electrons from HOMO to LUMO of Au nanoclusters were transferred to the conductive band of titania nanoparticles and then to the graphene matrix, which further acted as the free electrons and further generated superoxide radicals by reacting with oxygen molecules. The holes from both HOMO of Au nanoclusters and valance band of titania nanoparticles accumulated on HOMO of Au nanoclusters, which further reacted with water molecules to produce hydroxyl radicals. These TGA nanocomposites have been demonstrated to trigger a series of toxicological effects on killing B16F1 melanoma cells against B16F1 tumor xenograft, indicating high photodynamic efficiency of this prominent therapeutic modality for sunlight-triggered PDT effect. Although paradigms are effective on phototriggered PDT for cancer therapy based on TiO_2-based photosensitizers, this therapeutic modality is still suffering from the low tissue-penetrating capability of light as the irradiation source.

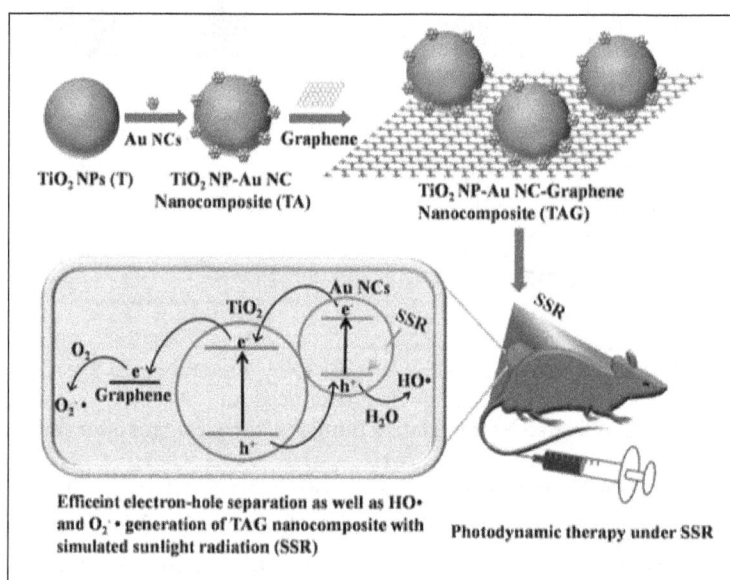

Schematic illustration of the fabrication of TAG composite nanoplatform and the related PDT mechanism for cancer therapy.

Laser Irradiation on Titania for PTT

In addition to NIR-triggered PDT for combating cancer, NIR-induced PTT has emerged as an efficient therapeutic modality for tumor treatment. Typically, the exogenous NIR laser can penetrate through the skin and activate the photothermal agents for converting NIR energy into heat and then ablating the tumor tissue by simply elevating the tumor temperature subsequently. Therefore, the development of desirable photothermal-conversion agents plays the determining role for achieving the efficient and desirable PTT outcome.

Based on an ambient heterogeneous spark discharge, $Au - TiO_2$ heterodimers were fabricated by incorporating Au component into TiO_2 nanoparticles, which exhibited the visible light-induced photothermal effect on killing HeLa cancer cells based on the localized surface plasmon resonance of integrated ultrafine Au nanoparticles. Al reduction could transform white P25-type TiO_2 nanoparticles into oxygen-deficient black TiO_{2-x} (B-TiO_{2-x}) nanoparticles, which endowed these black TiO_{2-x} nanoparticles with unique photothermal-conversion capability for efficient photothermal hyperthermia of cancer. Mo et al. modified the surface of B-TiO_{2-x} nanoparticles with PEG molecules for guarantee their high stability in physiological condition. After intravenous administration into HeLa tumor-bearing mice, these PEGylated $B?TiO_{2-x}$ nanoparticles efficiently accumulated into tumor tissue and rapidly elevated the tumor temperature by 808 nm NIR irradiation, causing the complete photothermal eradication of tumor tissue. Besides the Al reduction to fabricate black TiO_{2-x} nanoparticles, the hydrogenated black TiO_2 $(H?TiO_2)$ nanoparticles also exhibited high NIR absorption, which were further developed as the photothermal-conversion nanoagents for efficient tumor photohyperthermia based on their high photothermal-conversion efficiency of as high as 40.8% at the wavelength of 808 nm.

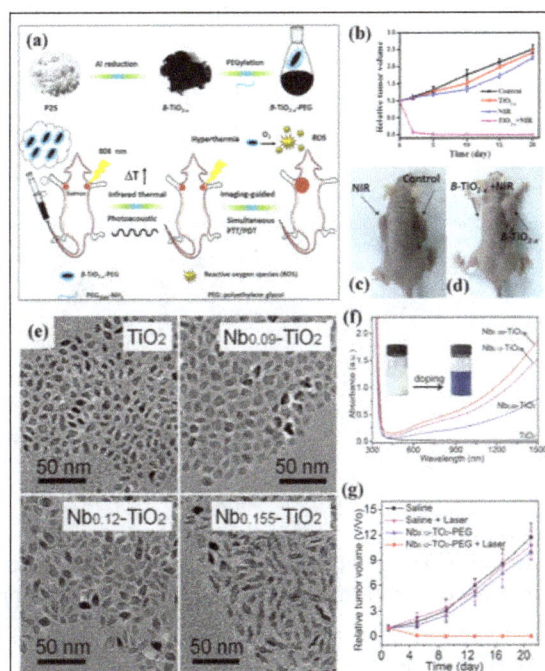

a) Schematic illustration of synthesizing PEGylated black TiO_{2-x} nanoparticles and their unique functionality for PA imaging-guided photothermal hyperthermia of tumor under NIR laser irradiation. b) The relative tumor-volume changes after varied treatments including control group, NIR group, TiO$_{2-x}$ group and TiO$_{2-x}$ combined with NIR irradiation group. c,d) Photographic image of tumor at the end of each treatment. e) TEM images of TiO$_2$ nanoparticles with varied Nb-doping amount. f) UV–vis–NIR absorbance spectra of Nb-doped TiO$_2$ nanoparticles in chloroform. g) The relative HeLa tumor-volume changes as a function of feeding time after different treatments as shown in the figure.

In addition to the mostly explored high-temperature treatment strategy to obtain black TiO$_2$ nanoparticle for PTT against cancer, Chen and co-workers successfully converted UV-responsive TiO$_2$ nanoparticles to blue TiO$_2$ nanocrystals by a simple Nb-doping approach. The different Nb-doping amount induced varied morphology of TiO$_2$ nanocrystals with high dispersity. Especially, the efficient Nb-doping endowed these blue TiO$_2$ with the strong NIR absorbance, which was originated from the localized surface plasmon resonances because of Nb doping-induced considerable free electrons. These blue Nb-doped TiO$_2$ nanocrystals efficiently converted laser at NIR-II biowindow (1064 nm) into heat and induced the photothermal effect on ablating the tumor tissue with high PTT efficiency.

The endowed targeting property of titania nanoparticles potentially enhances the tumor-accumulation efficiency for improved cancer therapy. On this ground, the surface of NIR-responsive TiO$_2$ nanoparticles as the photothermal-conversion nanoagents was conjugated with cyclo(Arg-Gly-Asp-d-Tyr-Lys) peptide c(RGDyK) for targeted photothermal hyperthermia of cancer. Based on the absorption of electron localized on Ti(III) sites and free electrons existing in the conduction bond, these TiO$_2$ nanoparticles showed high photothermal-conversion efficiency of nearly 38.5%. The surface-modified c(RGDyK) peptide selectively targeted the $\alpha_v\beta_3$ integrin on the cancer-cell membrane (U87-MG human glioblastoma cells) for efficiently killing the cancer cells, demonstrating the effectiveness of targeting strategy for improving the therapeutic efficiency of PTT.

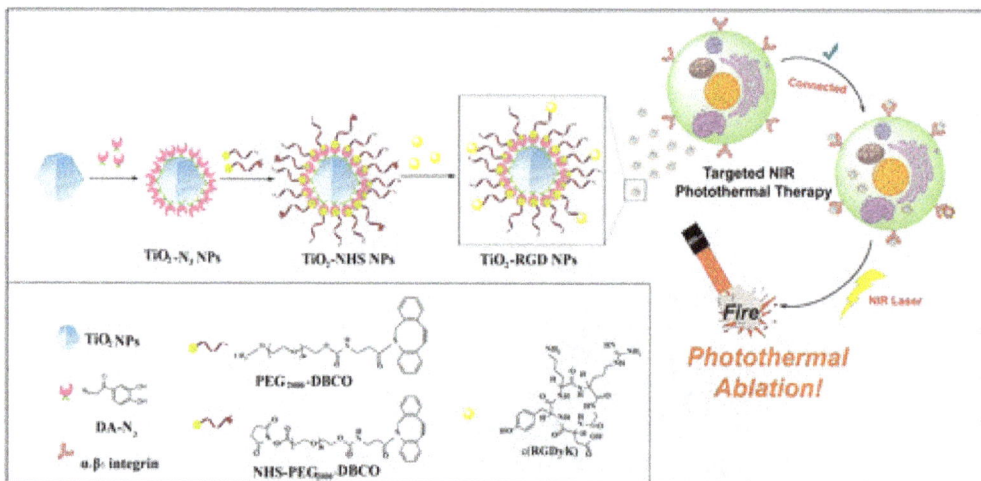

The scheme of the synthetic procedure of targeted TiO$_2$–RGD nanoparticles and their unique functionality for NIR-triggered photothermal hyperthermia against $\alpha_v\beta_3$ integrin-overexpressed cancer cells.

Radiation-activated Titania for PDT

The traditional external laser-activated PDT or PTT still suffers from the low tissue-penetrating depth of laser because of the rapid light attenuation passing through tissue and difficulty for reaching the deep-seated malignant lesions, which only confines the photointerventions for the treatment of superficial diseases. Radiation therapy by using radiation sources can solve critical issue because of the high tissue-penetrating capability of these radiation sources. Especially, the advances of theranostic nanomedicine has demonstrated the augmenting effect of some nanoparticulate radiosensitizers for substantially enhanced radiation-therapy outcome. Therefore, recent advances have also revealed the possibility of exogenous physical radiation sources for activating titania-based nanoplatforms to achieve efficient cancer therapy.

Au and titania anisotropic nanostructure was rationally designed as radio-sensitizers for X-ray-activated radiation therapy. Typically, bare TiO_2 nanoparticles could generate cytotoxic hydroxyl and superoxide radicals by the activation of UV light. Dumbbell-like Au–TiO_2 nanoparticles (DATs) can be activated by ionizing radiation and then produce secondary photons or electrons, which could induce the ROS production and migrate over the interface of DAT to TiO_2 component for further ROS production on the surface of TiO_2. The anisotropic nanostructure of DATs was constructed by stepwise seed-mediated growth. Based on the strong asymmetric electric coupling between Au component and dielectric TiO_2 at the interface, these DATs exhibited a synergistic therapeutic efficiency on X-ray-triggered radiation therapy where the production of secondary electrons and ROS from DATs substantially enhanced the radiation effect, causing the high tumor-suppressing effect and survival rate of tumor-bearing mice. This paradigm demonstrates that the rational integration of TiO_2 nanoparticles with functional nanoparticles can significantly enhance the efficiency of radiation therapy by taking the unique characteristics of each integrated component.

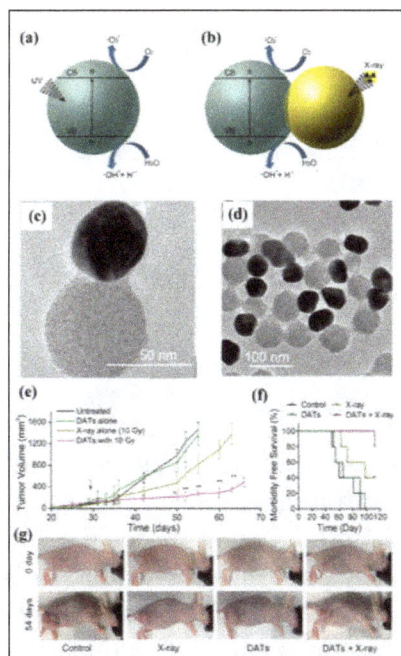

The scheme of ROS production on a) photoactivated TiO_2 nanoparticles and b) X-ray-induced hybrid DATs. TEM images of hybrid DATs at different magnifications. c) In vivo tumor-volume changes of tumor-bearing xenograft after different treatments as indicated in the figure, and d) corresponding survival rate of SUM159-tumor-bearing mice after varied treatments. e) Photographic images of SUM159 tumor-bearing mice before and at the end of treatments. f) The survival rate of tumor-bearing mice after varied treatments, and g) corresponding representative photographic images of mice before and after different treatments.

As an internal light source, CR is featured with high tissue-penetrating depth, which is typically triggered when the charged particles (e.g., β^+ and β^-) pass through a dielectric medium beyond the light speed. The UV can be emitted by CR for triggering UV-responsive photosensitizers for PDT. On this ground, CR-induced therapy was achieved by the radiation of PET radionuclides for activation of TiO_2 nanoparticles to produce hydroxyl and superoxide radicals. Especially, titanocene (Tc) was further anchored onto the surface of TiO_2 nanoparticles for enhancing and complementing CR-irradiated TiO_2 cytotoxicity because it could generate cyclopentadienyl and titanium-centered radicals once exposure to UV light. Furthermore, apo-transferrin (Tf) was modified onto the surface of TiO_2 nanoparticles for enhancing the positive accumulation into the tumor tissue. The results demonstrated that the intravenous administration of Tf-anchored TiO_2 nanoparticles and clinically employed radionuclides efficiently suppressed the tumor growth accompanied with prolonged survival rate of tumor-bearing mice. This paradigm provides a new strategy to develop low-radiance-sensitive nanophotosensitizers for efficient Cerenkov-radiation-activated cancer therapy with the tissue-depth impendence.

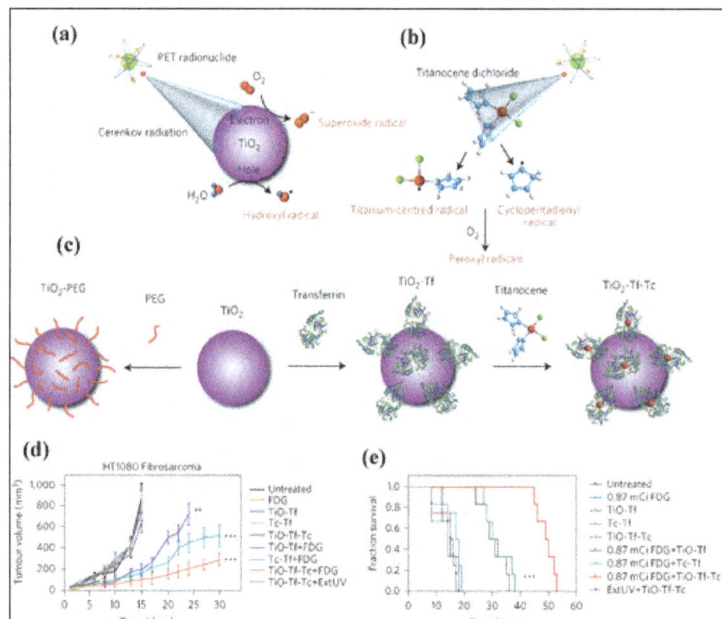

a) Schematic illustration of CR activation of TiO_2 nanoparticles for producing cytotoxic hydroxyl and superoxide radicals by the electron–hole pair generation where CR was generated by PET radionuclides. b) The scheme of CR activation of Tc for the generation of cyclopentadienyl radical and titanium-centered radical by photofragmentation. c) Schematic illustration of the fabrication of TiO_2–PEG, TiO_2–Tf, and TiO_2–Tf–Tc, nanoparticles. d) The tumor-volume changes of HT1080-tumor-bearing nude mice after varied treatments as indicated in the figure. e) The survival rate after the treatments with 0.87 mCi/0.1 mL FDG.

The efficient PDT strongly depends on the ROS production efficiency, which is significantly influenced by the local photon intensity. Gallium-68 (Ga-68) is a promising CR source because of the 30-time higher Cerenkov productivity as compared to fluorine-18 (F-18) such as ^{18}F-fluorodeoxyglucose $\left(^{18}F?FDG\right)$. Therefore, ^{68}Ga-labelled bovine serum albumin $\left(^{68}Ga?BSA\right)$ was employed as the CR source to activate dextran-modified TiO_2 nanoparticles for inhibiting the tumor growth, which could emit UV light to produce electron (e^-) and hole (h^+) from energy band of TiO_2

and generate ROS subsequently. By PET imaging, it has been found that intratumoral injection of $^{68}Ga?BSA$ and $^{18}F?FDG$ showed the similar tumor uptake of $^{68}Ga?BSA$ and $^{18}F?FDG$. Importantly, the tumor-bearing mice after the treatment with ^{68}Ga-BSA and TiO_2 photosensitizer exhibited significantly inhibited tumor volume and prolonged survival time while the mice in the group of $^{18}F?FDG$ and TiO_2 showed much lower tumor-suppressing rate, indicating that Ga-68 could act as the more efficient radionuclide as compared to F-18 for CR-induced in vivo PDT on combating cancer. The effective cancer treatment of deep-seated tumor by radiation-activated TiO_2 nanoparticles is highly promising for clinical use but the potential biosafety risk of radiation source should be seriously considered.

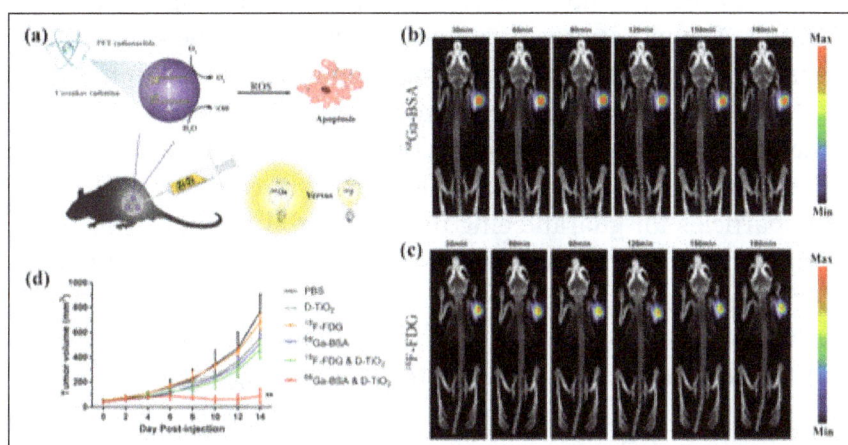

a) Schematic illustration of CR-activated TiO_2 photosensitizers for producing ROS to kill the cancer cells. The comparison of CR-induced PDT by dynamic PET imaging of 4T1 tumor-bearing mice after intratumor injection of equivalent amount of b) ^{68}Ga-BSA and c) ^{18}F-FDG from 30 min to 3 h. d) The tumor-volume changes of tumor-bearing mice after varied treatments as indicated in the figure.

Ultrasound Irradiation on Titania for SDT

Ultrasound (US) has been broadly explored in biomedicine for decades, not only for diagnostic imaging but also for therapeutic applications. For instance, the thermal, mechanical, and cavitation effects of high-intensity focused ultrasound (HIFU) have been used for noninvasive cancer surgery. In addition, the sonosensitizer-involved SDT produces ROS for inducing the cancer-cell death for cancer-dynamic therapy. Especially, US is featured with high tissue-penetrating depth in human bodies, which can reach internal organs such as liver, spleen, and kidney. Therefore, both the tissue-penetrating capability and theranostic biosafety of US make it a promising exogenous physical triggering source for versatile biomedical applications.

As the mostly explored inorganic nanosonosensitizers with high biocompatibility and stability, titania nanoparticles have been extensively employed for US-activated SDT against cancer. PEGylated TiO_2 nanoparticles have been demonstrated to be effective on inducing the cell death of U251 monolayer cells (1.0 MHz, 1.0 W cm^{-2}), and the related therapeutic mechanism was found to be different from that of UV light-induced PDT. Avidin protein-conjugated TiO_2 nanoparticles were designed to preferentially discriminate cancerous cells from healthy cells for targeted SDT. For in vivo assessment, the combination of TiO_2 nanoparticles and US irradiation (1 MHz, 1.0 W cm^{-2}, 2 min) substantially inhibited the tumor growth on subcutaneously implanted C32 xenograft, demonstrating the high in vivo therapeutic efficiency of TiO$_2$-sonosensitized SDT. Especially, the introduction of dual-frequency US for activation of TiO$_2$ nanoparticles as the nano-sonosensitizers was demonstrated to be more efficient for enhancing the hydroxyl radical production, which was verified in vitro on killing HepG2 cells.

We recently synthesized MTNs for US-triggered SDT. These MTNs were featured with ellipsoidal topology and high dispersity. Especially, they showed the highly single-crystalline structure with well-defined mesoporosity, which could enhance the SDT efficiency based on the fact that the high crystallity without defects could avoid the recombination of electrons (e$^-$) and holes (h$^+$) as triggered by US irradiation. The mesoporosity potentially facilitated the encapsulation and delivery of therapeutic agents such as anticancer drugs. After accumulation into tumor tissue of PEGylated MTNs (PEG-MTNs) via the typical enhanced permeability and retention (EPR) effect, the US-triggered SDT effect achieved 40% tumor-suppression rate under the intravenous administration mode. Hydrophilized TiO$_2$(HTiO$_2$) nanoparticles were fabricated by anchoring CMD onto the surface of TiO$_2$ nanoparticles for guaranteeing the high stability in physiological condition, prolonging the blood-circulation duration and enhancing the tumor accumulation. The accumulation of HTiO$_2$ into tumor tissue and further US activation not only enhanced the immune response but also destroyed the tumor microvasculature, which was demonstrated by the gradually decreased tumor volume and the decreased tumor vasculature by US-triggered SDT effect in bright-field images.

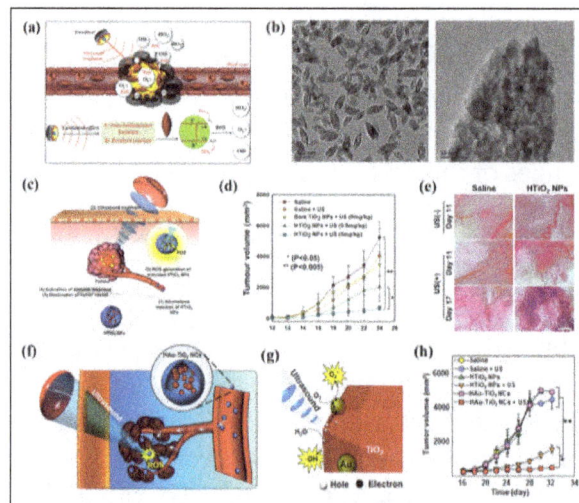

a) Schematic illustration of the accumulation of PEG–MTNs into the tumor tissue, and further US-triggered production of ROS for killing cancer cells. b) TEM images of MTNs at low (left image) and high (right image) magnifications. c) The scheme of HTiO$_2$ nanoparticle-enhanced SDT, including EPR effect-enabled accumulation into tumor and US-triggered ROS production to enhance the immune response and destroy tumor microvasculature. d) The tumor-volume changes of SCC7 tumor-bearing mice in each treatment group. e) The bright-field images of tumor vasculature by US-triggered SDT effect. f) Schematic illustration of in vivo US-triggered activation of HAu–TiO$_2$ nanoparticles for SDT and g) the underlying mechanism regarding the ROS production by US activation of HAu–TiO$_2$ nanoparticles. h) The comparison of tumor-volume changes with respect of feeding time by varied treatments as indicated in the figure.

It is noted that the low quantum yield of nanosonosensitizers resulting from the fast electron–hole recombination hinders the further clinical translation of TiO_2-based sonosensitizers. To address this critical issue, noble metal Au was combined with TiO_2 nanoparticles to prevent the undesirable electron–hole recombination by trapping the sono-excited electrons. This principle has been extensively explored in the typical TiO_2-based photocatalysis. In addition, CMD was also anchored onto the surface of Au– TiO_2 nanoparticles for further in vitro and in vivo evaluations. It is important to find that more ROS could be produced under US activation of Au–TiO_2 composite nanosonosensitizers as compared to pure TiO_2 without Au deposition, demonstrating the effectiveness of Au and TiO_2 combination. This enhanced SDT effect was also revealed in tumor-therapeutic outcome where the Au– TiO_2 composite nanosonosensitizers induced the more significant tumor suppression as compared to TiO_2 nanoparticles upon US activation.

By learning the lessons from typical photocatalysis, we recently fabricated an oxygen-deficient $TiO_{2-?x}$ nanosonosensitizers for enhancing the SDT efficiency against tumor, which was achieved by Al reduction at high temperature to create an oxygen-deficient TiO_{2-x} layer onto the surface of TiO_2 nanoparticles. Such an oxygen-deficient TiO_{2-x} layer facilitated and enhanced the separation of electrons (e^-) and holes (h^+) from the energy-band of TiO_2 semiconductor, which was activated by external physical US irradiation. This effect has been demonstrated to substantially enhance the SDT efficiency at solvent level, in vitro cellular level and in vivo tumor xenograft level. Especially, such a process to create oxygen-deficient TiO_{2-x} (black TiO_{2-x}) endowed this unique TiO_2-based nano-sonosensitizers with unique photothermal-conversion capability at NIR-II biowindow (1064 nm), which synergistically enhanced the SDT efficiency with the therapeutic outcome of complete tumor eradication.

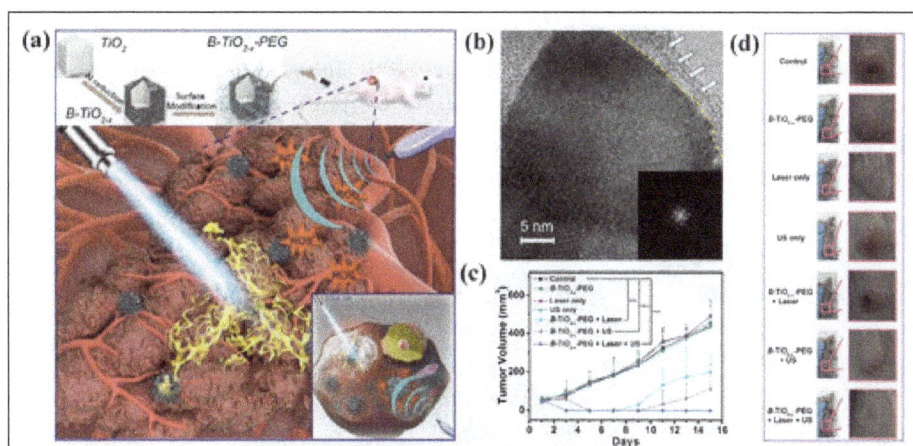

a) Schematic illustration of the fabrication of PEGylated B-TiO_{2-x} nanosonosensitizers and enhanced SDT by ROS production and synergistic NIR-II-triggered photothermal hyperthermia. b) High-resolution TEM image of B-TiO_{2-x} nanosonosensitizers and corresponding SAED patter (inset image). c) The tumor-volume changes of 4T1 tumor-bearing mice after varied treatments as indicated in the figure, and d) corresponding photographic images of tumor at the end of treatments.

Compared to light-triggered TiO_2-based photosensitizer for PDT, US-activated SDT based on TiO_2 nano-sonosensitizers is more applicable for clinical use based on the high tissue-penetrating depth of US as compared to the conventional light as the irradiation source. However, US-activated SDT is still at the preliminary stage, which still requires the further deep understanding of the underlying mechanism on the anticancer effect, which is highly beneficial for further improving the SDT efficiency on combating cancer.

Exogenous Physical Irradiation on Titania for Synergistic Cancer Therapy

Although therapeutic modalities enabled by TiO_2-based nanoplatforms have shown promising clinical-translation potential, each of these therapeutic modalities suffers from its intrinsic drawbacks hindering further broad applications. For instance, the therapeutic efficiency of RT, PDT, and SDT is limited by the hypoxia microenvironment of tumor. The heat shock response of local phototriggered hyperthermia causes the low PTT efficiency. The continuous chemotherapy usually induces the multidrug resistance (MDR) of cancer cells. To solve this critical issue, the combination therapy with involved two or more therapeutic modalities is expected to integrate the features and advantages of each therapeutic modality to achieve synergistic therapeutic outcome, which has been broadly explored in abundant therapeutic-modality combinations including physical-triggering of TiO_2 nanoplatforms.

Based on the high electroconductivity of graphene, we recently loaded TiO_2 nanoparticles onto the surface of graphene (designated as MnO_x/TiO_2–GR–PVP, MnO_x for MR imaging) for enhanced and synergistic SDT and photothermal hyperthermia, which was uniformly distributed onto graphene's surface. On one hand, the high electroconductivity of graphene facilitates the separation of electron (e^-) and hole (h^+) pairs from the energy-band structure of TiO_2 nanosonosensitizers upon external US irradiation, which could significantly enhance the SDT efficiency on killing the cancer cells. On the other hand, the graphene matrix showed the high photothermal-conversion performance for hyperthermia, which further synergistically enhanced the SDT efficiency of loaded TiO_2 nanosonosensitizers with high therapeutic biosafety as indicated by neglectable body-weight changes. This paradigm demonstrates that the rational combination of nano-TiO_2 semiconductors with high electroconductivity nanosystems could enhance the SDT efficiency by facilitating the US-triggered separation of electron and hole pairs.

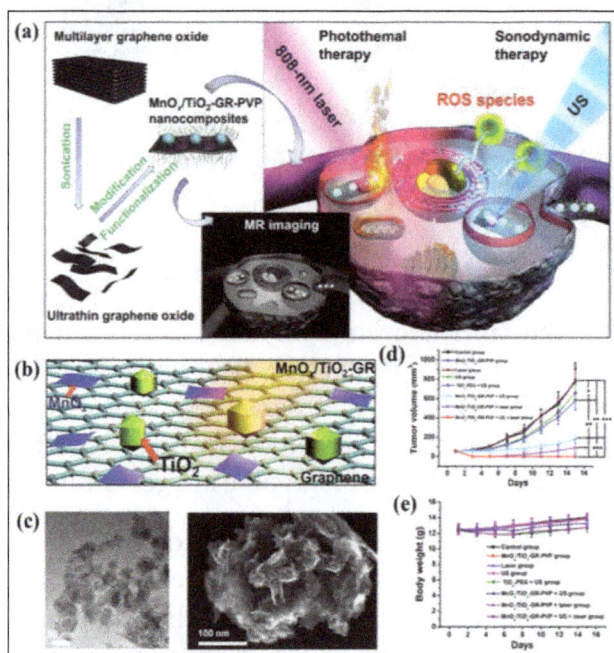

a) Schematic illustration of the fabrication of MnO_x/TiO_2–GR–PVP composite nanosheets, and their synergistic therapy based on MR/CT/PA multiple imaging-guided photothermal hyperthermia (808 nm) and enhanced SDT. b) The scheme of loading MnO_x and TiO_2 onto the surface of graphene nanosheets, and corresponding c) TEM (left image) and SEM (right image) images. d) The tumor-volume changes with the prolonged feeding time after varied treatments as shown in the figure, and e) corresponding body-weight changes in each therapeutic group.

To reverse the MDR of cancer cells, a "nano-bomb" was designed for US-triggered multiple and synergistic cancer therapy based on hollow MTNs. Chemotherapeutic drug doxorubicin acting as the ammunition was loaded into MTNs as the ammunition depot, and the surface of MTNs was coated by dsDNA as the safe device to avoid the prerelease of loaded doxorubicin. Especially, the US irradiation on drug-loaded MTNs achieved multiple effects, including US-triggered SDT for MTN-sonosensitized ROS generation, US-activated drug release, reversal of MDR, and final synergistic cancer treatment. The reversal of MDR of MCF-7/ADR cancer cells was based on the inhibition of mitochondrial energy supply by the US-triggered "explosion" of MTNs, causing the substantially suppressed tumor growth.

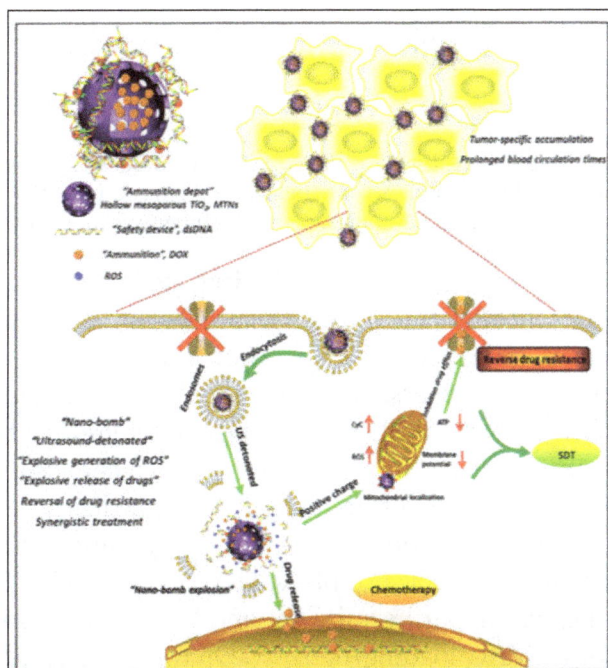

Schematic illustration of US-triggered combinatorial therapy using a "nano-bomb," including US-triggered SDT for ROS generation, US-activated drug release, reversal of MDR, and final synergistic cancer treatment.

In addition, envelope-type mesoporous titanium dioxide nanoparticles (MTN) were fabricated with the subsequent loading of docetaxel (DTX) accompanied with a high drug-loading capacity of ≈26%. Furthermore, β-cyclodextrin (β-CD) was anchored onto the surface of MTN as a bulky gatekeeper, which was based on a ROS-sensitive linker to seal DTX within the mesopores. Upon US irradiation, large amounts of ROS were produced by SDT effect to break the ROS-sensitive linker and then trigger the DTX release from the mesopores. Therefore, the US irradiation not only induced ROS generation for SDT cancer therapy, but also triggered DTX releasing from mesopores for synergistic chemotherapy, which was demonstrated by the synergistic therapeutic outcome where the tumor growth in the synergistic group got the maximum suppression. Magnetic core shell structured Fe_3O_4-NaYF@TiO_2 nanocomposites were constructed for synergistic chemotherapy by loaded doxorubicin and SDT by the TiO_2 component. The further surface engineering with hyaluronic acid (HA) enabled targeted intracellular transportation, which induced high tumor-inhibition rate of 88.36% in synergistic group, much higher than that of the single therapeutic modality such as chemotherapy (28.36%) and SDT (38.91%).

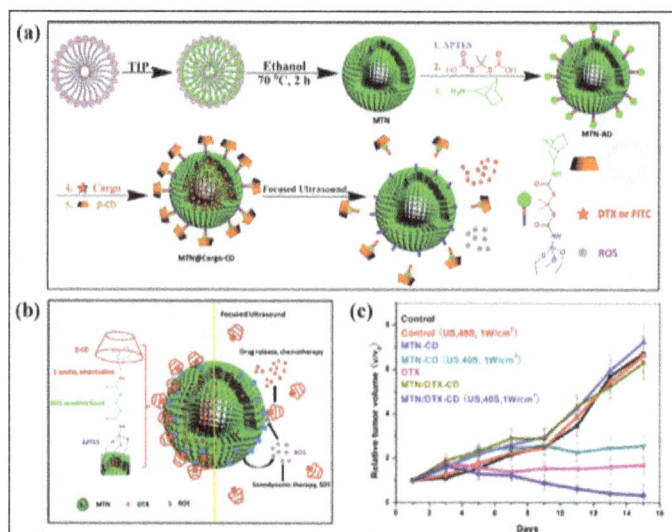

a) Schematic illustration of stepwise synthesis of MTN@DTX-CD composite nanosystem. b) The scheme of microstructure of MTN@DTX-CD and the functionality of each component. c) The tumor-volume changes of S180 tumor-bearing mice after varied treatments as indicated in the figure.

TiO_2-based nanoparticles have been explored as the drug-delivery nanosystems for chemotherapy. For instance, ZnPc@TiO_2 (ZnPc: zinc phthalocyanine) hybrid nanoparticles were employed for the intracellular delivery of anticancer drug doxorubicin based on the electrostatic interaction between drug molecules and nanocarriers. Doxorubicin was also loaded into Fe_3O_4@TiO_2 core/shell nanoparticles for synergistic chemotherapy by loaded chemotherapeutic drug and US-activated SDT by the TiO_2 component as sonosensitizer. The typical strategy only loaded the chemotherapeutic drug molecules onto the surface of TiO_2 nanoparticles, therefore the drug-loading amount was low and the release in uncontrollable. On this ground, Wu and co-workers coated a mesoporous silica layer onto the surface of black TiO_2 nanoparticles for achieving synergistic chemotherapy and photothermal hyperthermia. On one hand, the mesoporous silica layer provided the large reservoir for the drug loading/delivery. On the other hand, the black TiO_2 acted as the photothermal-conversion nanoagents for PTT. By the integration with folic-acid targeting, the mesoporous silica-coated black TiO_2-enabled produced the synergistic therapeutic outcome on suppressing the tumor growth against MCF-7 breast cancer xenograft.

TiO_2 nanoparticles could integrate with other functional nanosystems for achieving some specific synergistic therapeutic purposes. For instance, praseodymium (Pr)-doped TiO_2 on GO nanoplatforms were fabricated by a facile hydrothermal synthesis. First, the sp^2 carbonaceous framework of GO converted NIR light into heat for photothermal hyperthermia. Second, the Pr-doped TiO_2 nanoparticles could absorb more hydroxide ions onto the surface to promote the generation of hydroxyl radicals and suppress the electron–hole recombination. Third, the 4f electron transition of doped Pr achieved the incorporation of additional energy levels in the bandgap of TiO_2, which induced the enhanced photocatalytic activity on killing cancer cells under visible light (450 nm).

Fourth, this composite nanosystem could store therapeutic anticancer agents (doxorubicin) for enhanced chemotherapy. Especially, the synergistic chemotherapy, PTT and phototriggered PDT (triple-therapeutic modality) significantly induced the cancer-cell death as compared to either monomodal or dual-modal therapy.

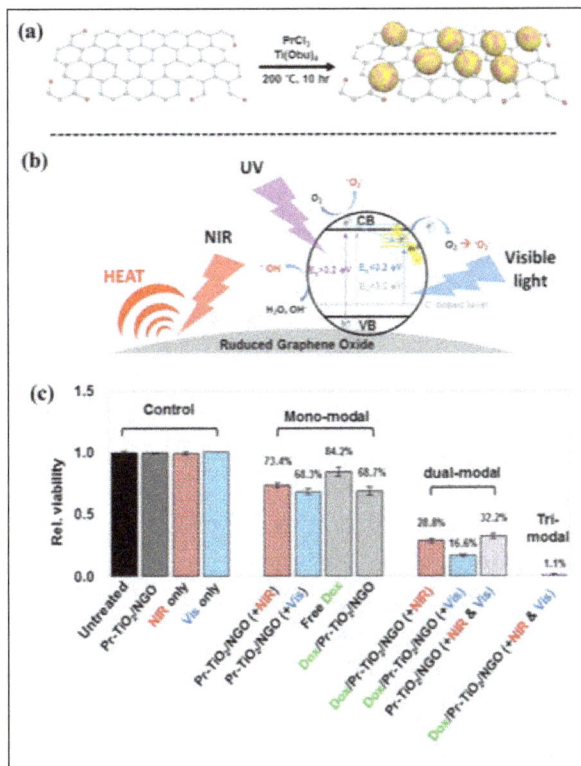

The scheme of a) synthesizing Pr-TiO$_2$/NGO nanocomposites and b) light-activated photocatalytic process for ROS production. c) In vitro therapeutic efficacy of different treatments, including mono, dual, and triple-modality cancer-cell treatment such as chemotherapy (loaded Dox), PTT, and phototriggered dynamic therapy.

That the UV wavelength range of 320 to 400 nm might cause the phototoxicity and have the low-penetrating capability. To solve the critical issue of UV-responsiveness of traditional TiO$_2$ nanoparticles, zinc phthalocyanine as the intriguing photochemical molecule with high stability, efficiently extended the light window of TiO$_2$ nanoparticles from UV region to NIR region for phototreatment, which was based on an intercomponent electron transfer between zinc phthalocyanine and titania nanoparticles. Especially, the ROS-sensitive compound BCBL was conjugated to zinc phthalocyanine-modified TiO$_2$ nanoparticles for ROS-triggered chemotherapy. Upon NIR irradiation, the generated large amounts of ROS triggered the release of loaded BCBL for chemotherapy, which also acted as the toxic species for PDT, inducing the synergistic chemotherapeutic and PDT efficiency. The UV-activated TiO$_2$ nanoparticles have been previously demonstrated to reverse the MDR of cancer cells. To further overcome critical issue of UV light for reversal of MDR, doxorubicin was loaded into NaYF$_4$:Yb/Tm-TiO$_2$ inorganic photosensitizers for simultaneous 980 nm NIR-activated PDT and intracellular drug delivery. The surface folic acid modification enhanced intracellular uptake of the nano-photosensitizer and accelerated the doxorubicin release in both drug-sensitive MCF-7 and drug-resistant MCF-7/ADR cancer cells, inducing the synergistic MCF-7/ADR tumor-inhibition rate of up to 90.33%, significantly higher than that of free doxorubicin.

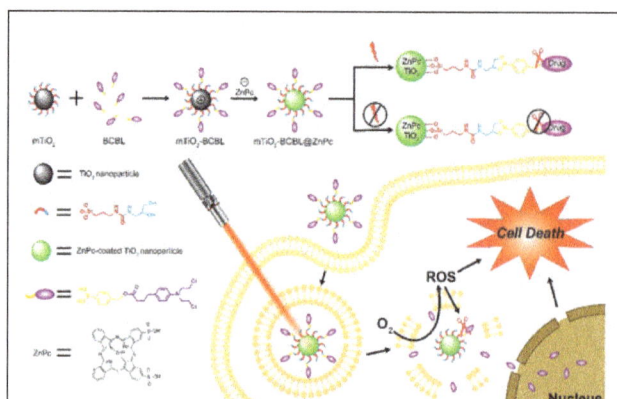

Schematic illustration of synthesizing mTiO$_2$–BCBL@ZnPc nanoparticles and their unique therapeutic functionality for cancer therapy, which was based on ROS-responsive drug-releasing performance under NIR irradiation by breaking up the phenylboronic acid ether.

The construction of mesoporous titania-coated UCNPs is expected to achieve NIR-triggered PDT and simultaneous chemotherapy for synergistic therapy, originating from UNCPs core and mesoporous titania shell. Mesoporous TiO$_2$ upconverting nanoparticles (abbreviated as MTUN) were synthesized by direct coating of a mesoporous TiO$_2$ layer onto the surface of NaGdF$_4$:Yb25%,Tm0.3% as mediated by a middle silica layer. The UCNPs converted NIR irradiation to UV light, which further activated mesoporous TiO$_2$ layer to generate ROS for inducing cancer-cell apoptosis. Especially, the well-defined mesopores of surface TiO$_2$ layer acted as the drug-storage reservoirs for drug delivery and chemotherapy, inducing the synergistic NIR-activated PDT and chemotherapy. Importantly, the HA was anchored onto the surface of this composite nanosystem for targeting cluster determinant 44 (CD44) that was overexpressed on cancer-cell membrane and achieving controlled drug releasing as triggered by the specific enzyme in tumor region. Based on mesoporous TiO$_2$-coated UCNPs, a photolabile o-nitrobenzyl derivative was incorporated to act as the gate by forming a sensitive linker for avoiding the drug releasing. The NIR-triggered ROS production not only induced the PDT effect, but also cause the breaking of the sensitive linker for on-demand drug releasing, leading to synergistic chemotherapy and PDT against cancer cells. To further enhance the drug-loading capability, rattle-type UCNPs@Void@ TiO$_2$ nanocomposites were fabricated with large voids between UCNPs core and mesoporous TiO$_2$ shell, producing TiO$_2$-based PDT by NIR irradiation and doxorubicin-induced synergistic chemotherapy.

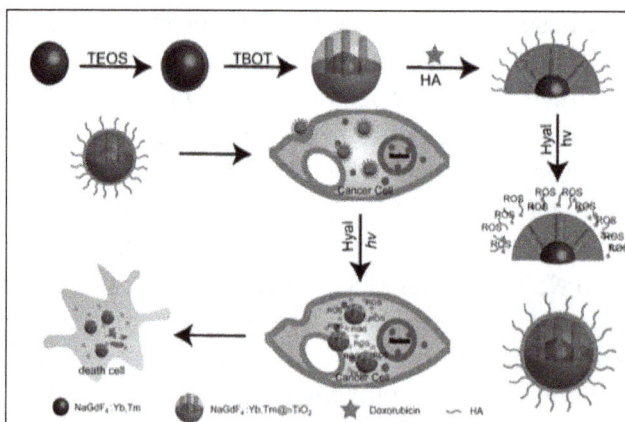

The scheme of the fabrication of NaGdF$_4$:Yb,Tm@mTiO$_2$ core/shell nanoparticles the novel drug-delivery system for simultaneous and synergistic chemotherapy and NIR-triggered PDT.

The rational structure design of TiO_2-based nanoplatforms could endow them with more therapeutic functionalities. It has been demonstrated that anticancer drug doxorubicin-loaded TiO_2 nanoparticles overcame the MDR of breast cancer cells (MCF-7/ADR) by bypassing the P-glycoprotein-mediated doxorubicin-pumping system. Furthermore, TiO_2-based composite nanosystems (DOX@TiO_{2-x}@PAD-Cy5.5, PDA: poly dopamine) were stepwise synthesized for simultaneous fluorescent/PAT bimodal tumor imaging and NIR-activated chemo/photodynamic/photothermal combinatorial therapy. Because of the high photothermal-conversion capability of TiO_{2-x} matrix, the tumor temperature was rapidly elevated upon NIR irradiation (808 nm, figure). Especially, the NIR irradiation of DOX@TiO_{2-x}@PAD-Cy5.5 generated ROS for efficient PDT, and the presence of mesopores in TiO_{2-x} matrix provided the reservoirs for the encapsulation and controllable delivery of therapeutic anticancer drugs (doxorubicin) with unique responsiveness to endogenous mild acidity of TME and exogenous NIR irradiation. The simultaneous and synergistic triple therapy induced the high tumor-suppressing outcome with almost complete tumor eradication, which was caused by DOX-induced DNA damage and PDT/PTT-induced mitochondrial dysfunction/change of membrane.

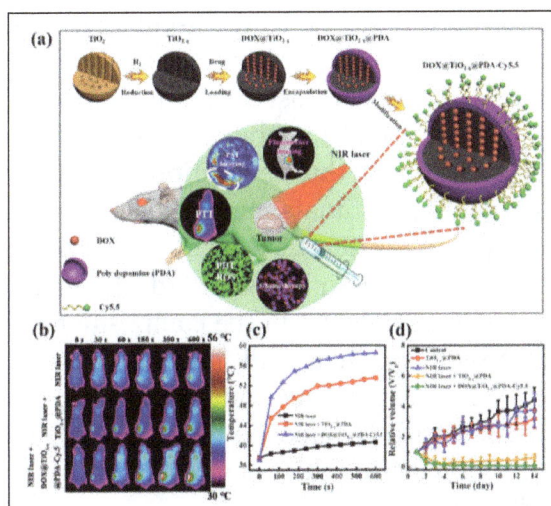

a) Schematic illustration of synthesizing DOX@TiO_{2-x}@PAD-Cy5.5 nanocomposites, and their unique functionality for dual-mode imaging-guided synergistic chemotherapy, PTT, and PDT. b) Infrared thermal images of MDA-MB-231 tumor-bearing mice after the administration of DOX@TiO_{2-x}@PAD-Cy5.5 nanocomposites followed by NIR irradiation, and c) corresponding temperature-elevating profiles. d) The tumor-volume change as a function of feeding time after the different treatments as indicated in the figure.

Both the microstructure and functionality of TiO_2-based nanoplatforms could be rationally designed for achieving synergistic cancer-therapeutic outcome, which is highly unique in TiO_2 nanosystems because of their easy synthesis and specific semiconductor nature. It should be noted that the surface inertness of TiO_2 nanoparticles makes their surface engineering difficult, which typically seeks the help from other organic or inorganic functionality by some unique synthetic strategies, such as inorganic silica coating or hydrophobic–hydrophobic interaction. It should be noted that the multifunctionalization design of TiO_2 nanoparticles should be based on the practical requirements of clinical use because the complex design of these composite nanosystems usually causes the difficulty for potential clinical translation.

Diagnostic-imaging of Titania for Therapeutic Guidance and Monitoring

It has been well demonstrated that some metal oxides nanoparticles can act as the contrast agents for enhancing the diagnostic-imaging resolution and sensitivity of diverse imaging modalities, such as manganese oxide (T_1-weighted MR imaging), gadolinium oxide (T_1-weighted MR imaging), iron oxide (T_2-weighted MR imaging), and tantalum oxide (CT imaging). Titania nanoparticles have been seldom explored for enhancing the contrast of various diagnostic-imaging modalities because of lacking the characteristic physiochemical properties. Fortunately, the fast advances of material-synthetic chemistry and nanomedicine make it possible based on two typical strategies. On one hand, the structure of titania nanoparticles can be tuned with contrast-enhanced imaging functionality. On the other hand, these titania nanoparticles can be integrated with some imaging contrast agents for achieving some specific imaging purposes. It is intriguing that the diagnostic-imaging capability of titania nanoparticles can play the specific role for precise therapeutic guidance and monitoring, which is promising for enhancing the therapeutic efficiency and mitigating the damage to the surrounding normal tissue/cell.

This property has been generally developed for contrast-enhanced PA imaging, such as 2D MXene, black phosphorous, MoS_2, and Au nanoparticles. On this ground, oxygen-deficient black titania nanoparticles were explored as the contrast agents for PA imaging after the injection into tumor-bearing mice. It has been found that the obvious contrast enhancement was observed in tumor region after intratumoral injection of these black titania nanoparticles, demonstrating their imaging capability. As another paradigm, titania nanoparticles were integrated with manganese oxide nanoparticles to construct a composite nanosystem, where the integrated manganese oxide nanoparticles acted as the contrast agents for T_1-weighted MR imaging and guided the SDT of cancer as contributed by the titania component in the composite nanosystem. Especially, titania nanoparticles were simultaneously conjugated with fluorescent moieties and Gd-based chelates for labeling HeLa cancer cells by both fluorescence microscopy and MR imaging, showing the dual-imaging capability of the titania-based composite nanosystem. Additionally, the construction of magnetic $Fe_3O_4 - TiO_2$ nanocomposites achieved simultaneous T_2-weighted MR imaging and PDT against MCF-7 cancer cells.

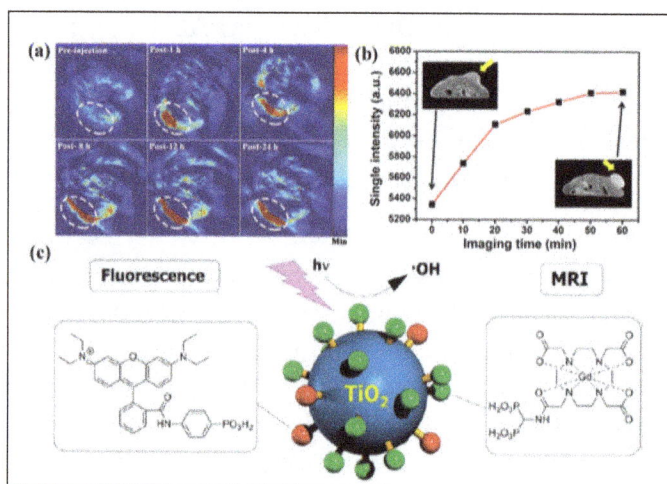

a) In vivo PA imaging of tumor-bearing mice before and after intratumoral administration of DOX@TiO$_{2-x}$@PDA-Cy5.5 nanocomposites for prolonged durations. b) In vivo T_1-weighted MRI signal-intensity variations after intravenous administration of MnO_x / TiO_2–GR–PVP for different time intervals. c) Schematic illustration of surface conjugation of fluorescent moieties and Gd chelates for concurrent fluorescence and MR imaging, and the further UV-activated production of hydroxyl radicals for killing the cancer cells.

Biocompatibility and Biosafety of Theranostic Titania

Titanium (Ti) element is one of the most biocompatible elements present in nature, as demonstrated by the fact that TiO_2-based micro/nanoparticles have been broadly used in food, cosmetics, and sunscreen and Ti-containing metal alloys has been used as the medical implantation devices. It is highly expected that these Ti-based nanoparticles. However, it is generally accepted that the particles would induce some abnormal biological behaviors and effects or even toxicity when their particle size are reduced into nanoscale. Therefore, systematic investigation of these titania nanoparticles should be further conducted to guarantee their high biocompatibility and biosafety for further clinical translation.

Actually, the biological effects and biocompatibility of titania-based compound or micro/nanoparticles have been broadly investigated in the past decade, which has also been summarized and discussed in some excellent reviews. Therefore, this review herein focuses more on the biocompatibility and biosafety of some rationally designed novel titania-based nanoplatforms with unique responsiveness to exogenous physical irradiations. Furthermore, we systematically assessed the in vivo biocompatibility of these MTNs on healthy mice. It has been found that either single high dose at 150 mg kg^{-1} or repeated dose at as high as total 400 mg kg^{-1} exhibited no obvious in vivo toxicity, as demonstrated by the hematology markers and blood biomedical parameters where no significant changes were monitored as compared to control groups without any treatments, indicating the high biocompatibility of these MTNs. For titania-based nanocomposites, it has been demonstrated that Au_{25}/B-B- TiO_{2-x} nanotubes not only showed low hemolytic effect on red blood cells, but also revealed their low cytotoxicity to L929 cells (mouse fibroblast cell line) and HeLa cells (human cervical cancer cell line). The targeting titania-based nanocomposites anti-EGFR–PEG– TiO_2–UCNPs were demonstrated to have no major sub-acute or long-term toxicity as revealed in no significant blood biomedical, hematological or histopathological changes at the dose of 50 mg kg^{-1}.

One of the unique advantages of titania-based nanoplatforms with responsiveness to exogenous physical irradiation is the high therapeutic biosafety. These nanoplatforms can only induce the toxic effect under the tumor sites as irradiated by external diverse physical triggers while other organs or tissues without physical irradiation will not be damaged even if these titania nanoparticles are accumulated into them. This high therapeutic biosafety is expected to significantly mitigate the side effects of traditional therapeutic modalities such as chemotherapy where the toxic drugs or substances are usually introduced, causing the severe side effects. Our results have demonstrated that SDT against cancer with the assistance of black TiO_{2-x} nanosonosensitizers induced no obvious pathological changes of the major organs after the therapeutic process, demonstrating the high therapeutic biosafety of this SDT modality. In addition, the combinatorial and synergistic SDT and chemotherapy of DTX-loaded MTN were demonstrated to be featured with sustainably decreased side effects of loaded chemotherapeutic drug DTX by avoiding the spleen and hematologic toxicity to tumor-bearing mice.

Conclusions and Outlook

As one of the mostly explored biocompatible metal oxides in biomedicine, TiO_2 nanosystems are featured with their intrinsic physiochemical properties for some specific theranostic applications, which is mainly originated from their semiconductor nature. Traditional strategies mainly focus on the UV light irradiation of TiO_2 nanoparticles for PDT by forming the electrons (e^-) and holes (h^+) pairs from the energy-band structure and then inducing the ROS generation for killing the cancer cells. The potential phototoxicity and low tissue-penetrating depth of UV light severely limit the further in vivo biomedical applications of these TiO_2 nanosystems. The fast development of nanosynthetic material chemistry enables the fine tuning of the composition, nanostructure, and property of TiO_2 nanosystems possible. Importantly, the intriguing development of theranostic nanomedicine promotes the generation of diverse novel therapeutic modalities, which can be easily extended to TiO_2 nanosystems for achieving physiochemical property-oriented bioapplications, especially for cancer treatment. On this ground, this review mainly focuses on the very-recent development of TiO_2-based nanoplatforms for cancer treatment with specific focuses on the NIR-triggered photothermal hyperthermia, NIR-activated PDT, X-ray/CR-activated deep-seated PDT, US-triggered sonodynamic therapy, and some synergistic therapeutic paradigms. Most of these novel therapeutic modalities are based on the semiconductor nature of TiO_2 nanoplatforms, together with the defect modulation for PTT.

Table: Paradigms of nanotitania semiconductors for exogenous physical irradiation-activated tumor-specific therapy.

Nanotitania	Irradiation source	Therapeutic modality	Performance
Targeted TiO_2	Visible light	Photodynamic therapy	Enhanced intracellular uptake and visible light-activated PDT for damaging the cell membrane.
N3-TiO_2	Light irradiation (365 nm)	Photodynamic therapy	Enhanced hydroxyl radical production under the hypoxic condition.
UCNPs@TiO_2	NIR irradiation (980 nm)	Photodynamic therapy	Inducing the substantial tumor suppression with high therapeutic efficiency by NIR irradiation.
UCNPs@TiO_2	NIR irradiation (980 nm)	Photodynamic therapy	Efficiently killing the cancer cells both in vitro and in vivo by NIR activation.
$Au_{25}/B-TiO_{2-x}$	NIR irradiation (650 nm)	Photodynamic therapy	Improved tumor-suppressing effect based on photocatalytic synergistic effect by NIR irradiation.
TiO_2–Au–graphene	Simulated sunlight	Photodynamic therapy	Triggering a series of toxicological effects on killing B16F1 melanoma cells against B16F1 tumor xenograft.
Black TiO_{2-x}	NIR irradiation (808 nm)	Photothermal therapy	Elevating the tumor temperature and inducing tumor-tissue hyperthermia.
Nb-doped TiO_2	NIR irradiation (1064 nm)	Photothermal therapy	Ablating the tumor tissue and suppressing tumor growth at NIR-II biowindow.
Au–TiO_2	X-ray	Photodynamic therapy	Inducing a synergistic therapeutic outcome with high tumor-suppressing effect and improved survival rate of mice.
TiO_2–Tc-Tf	Cerenkov radiation	Photodynamic therapy	Suppressing tumor growth and improved survival rate with deep tissue-penetrating depth.
Dextran–TiO_2	Cerenkov radiation	Photodynamic therapy	Efficiently killing the cancer cells and improving the survival rate of tumor-bearing mice.

Mesoporous TiO_2	Ultrasound	Sonodynamic therapy	Inducing tumor-suppressing effect against 4T1 tumor xenograft.
Hydrophilized TiO_2	Ultrasound	Sonodynamic therapy	Enhancing immune response, suppressing tumor growth and destroying tumor microvasculature.
Au–TiO_2	Ultrasound	Sonodynamic therapy	Improved SDT effect against cancer by trapping the sono-excited electrons.
Black TiO_{2-x}	Ultrasound and NIR (1064 nm)	Sonodynamic therapy and photothermal therapy	Synergistic SDT and PTT on killing the cancer cells accompanied by enhanced SDT effect by oxygen-deficient titania layer.
MnO_x/TiO_2-GR	Ultrasound and NIR (808 nm)	Sonodynamic therapy and photothermal therapy	Decreasing the re-combination of electrons and holes for enhanced SDT effect on suppressing the tumor growth.
Targeted TiO_2	Visible light	Photodynamic therapy	Enhanced intracellular uptake and visible light-activated PDT for damaging the cell membrane.

The unique physiochemical property of TiO_2 nanosystems has achieved high therapeutic efficacy of aforementioned cancer-therapeutic modalities, which is difficult to be achieved in other metal oxides such as SiO_2 nanoparticles and superparamagnetic Fe_3O_4 nanosystems. Although these TiO_2-based novel therapeutic modalities are highly promising, it should be noted that they are still in infancy and at the preliminary stage.

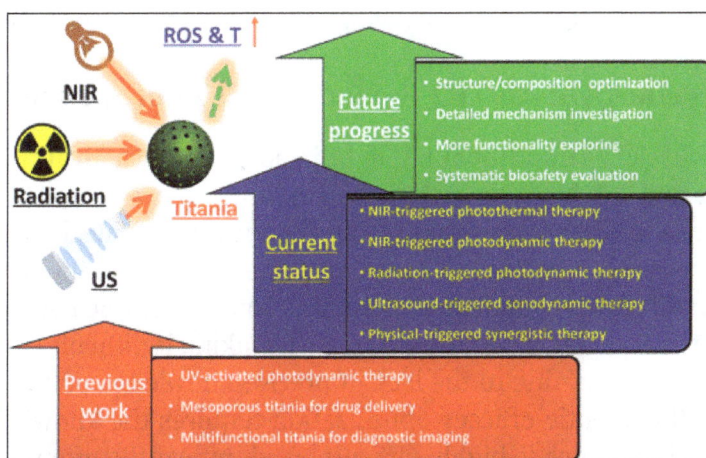

The summative scheme of previous work, current status, and future progress on physical irradiation-activated titania nanoparticles for versatile biomedical applications, especially on combating cancer.

Fabrication of TiO_2 Nanosystems

The synthetic process for desirable TiO_2 nanosystems is a bit more difficult as compared to other metal oxides such as SiO_2 and Fe_3O_4 because the hydrolysis of titanium precursors is very fast in most cases. Therefore, their morphology and nanostructure are difficult to be precisely controlled. Especially, some fabrication process requires high-temperature treatment such as the metal reduction to synthesize oxygen-deficient black TiO_{2-x} nanoparticles, which avoidably causes the aggregation of TiO_2 nanoparticles with low dispersity. In addition, there lacks the specific surface chemistry for the surface modification of these TiO_2 nanoparticles, which, however, is necessary for guaranteeing their high stability in physiological solution or achieving the high tumor accumulation by targeting modification. These fabrication difficulties might lower the therapeutic efficacy of TiO_2-based therapeutic modalities or cause the biosafety issue. Therefore, two strategies are suggested to fabricate desirable TiO_2 nanosystems. On one hand, the precise controlling of the sol–gel process during the synthesis of TiO_2 nanosystems should be taken into consideration, which can fabricate TiO_2 nanoparticles with desirable composition, nanostructure, and physiochemical property. On the other hand, the surface of as-synthesized TiO_2 nanoparticles could be endowed with some functional groups for further surface modification, which can be achieved by some specific treatments such as PEGylation or silica/mesoporous silica coating. It means that the surface chemistry can be controlled by using some "mediators" for satisfying the surface-modification requirements.

Biocompatibility Issue of TiO_2 Nanosystems

As compared to SiO_2 and Fe_3O_4 metal oxides, the biological effects and biocompatibility of TiO_2 nanoparticles are significantly less explored, which severely hinders their further clinical translation because of the lack of solid biocompatibility data. It is considered that TiO_2 has been used in colorant in food, cosmetics and sunscreen and Ti-containing metal alloys has been broadly used as the medical implantation devices, therefore these TiO_2 nanosystems might possess the relatively high biocompatibility. Our previous results have demonstrated the low in vivo toxicity of either mesoporous TiO_2 nanoparticles or black TiO_2 nanosystems. These preliminary results are encouraging, but the further systematic investigations on the biocompatibility and biosafety issue are still highly urgent and necessary, which is the following research target in the near future.

Not Very Clear Mechanism on TiO_2-based Novel Therapeutic Modalities

To overcome the drawbacks of traditional UV light irradiation for activating TiO_2 nanoparticles, NIR light, X-ray, CR, and US have recently been explored to activate TiO_2 nanoparticles. The therapeutic performance is highly encouraging, but the underlying mechanism has not been fully revealed. Most of the results are mainly based on some phenomena in vitro. The exact in vivo therapeutic process is still highly challenging to monitor and determine because of the lack of adequate techniques and the complex in vivo environment, making it difficult to further optimize and enhance the therapeutic efficacy due to the lack of precise knowledge on the related mechanism. Therefore, the further therapeutic-efficacy optimization requires the knowledge accumulation of the therapeutic mechanism, which is highly difficult but significantly urgent.

Influence of Crystalline Types of TiO_2 Nanoparticles on Therapeutic Performance

It has been well documented that TiO_2 nanoparticles have varied crystalline types with different physiochemical properties. However, at current stage, most therapeutic applications of titania nanoparticles did not consider the influences on the crystalline types of titania nanoparticles because the therapeutic use of titania nanoparticles as the emerging inorganic nanoplatform is still in the infancy, which still requires the following systematic investigations on the detailed underlying mechanism of the influence of crystalline types, precise structure/composition control and the following performance optimization.

References

- Semiconductors: chemistryexplained.com, Retrieved 14 April, 2019

- Semiconductor, museum-exhibition-principle: tel.com, Retrieved 23 January, 2019

- What-are-semiconductors-properties-of-semiconductors, nuclear-engineering-radiation-detection-semiconductor-detectors: nuclear-power.net, Retrieved 18 June, 2019

- Extrinsic-semiconductor, semiconductor-electronics-materials-device-and-simple-circuits: toppr.com, Retrieved 13 May, 2019

- Basics-of-Light-Emitting-diodes-Characterizations-and-Applications-200071029: researchgate.net, Retrieved 19 August, 2019

- Electronic-Properties-of-Materials-Superconductors-and-Semiconductors, Introduction-to-Inorganic-Chemistry: en.wikibooks.org, Retrieved 17 February, 2019

Polymer Chemistry: An Integrated Study

Polymer chemistry studies the synthesis, chemical and physical properties, and structure of polymers and macromolecules. Polymerization mechanisms include addition polymerization, step growth polymerization and catalysis. The topics elaborated in this chapter will help in gaining a better perspective about the concepts and mechanisms associated with polymer chemistry.

Polymers are high molecular weight materials/macromolecules, composed of large number of low molecular weight species. The term polymer can be defined as a macromolecule with high molecular mass arising due to the joining of a large number of simpler/smaller molecules (called monomers). The polymers are characterized by variable molecular weight (depending on the source or mode of synthesis or extraction), low specific gravity, better resistance to erosion, corrosion, insects or fungi etc. The advantageous aspect of polymeric materials over other materials is that they can be tailor-made i.e. polymers can be synthesized as per our property requirement or soft to rigid/tough, transparent to opaque, light to heavy, crystalline to amorphous materials can be synthesized. To cite an example, there are different types of polyethylene $\{(CH_2\text{-}CH_2)n\}$ such as low density, medium density and high density polyethylene (abbreviated as LDPE, MDPE and HDPE). Polymers find applications as major class of materials such as plastics, fibres, rubbers, rexin (artificial leather) and explosives. Polymerization is the union of two or more simpler molecules (effected under the influence of heat, pressure, catalyst etc.), resulting in the formation of new macromolecule with the characteristic C-C linkages. The resultant macromolecule is called polymer and the original/parent simpler molecule is called monomer. The condition for a substance to be polymerized is the functionality. The functionality of a monomer is defined as the number of reactive/bonding sites available in the molecule such as carbon-carbon multiple bond, condensible functional groups such as hydroxyl, carboxyl, amine, halo group etc. E.g. carbon – carbon double bond is termed as bi-functional, because when the double bond is broken, two single bonds become available for combination i.e. ethylene can be polymerized but not ethane.

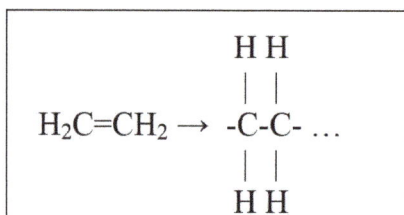

$$
H_2C{=}CH_2 \rightarrow \quad
\begin{array}{c}
H\ H \\
|\ \ | \\
\text{-C-C-} \dots \\
|\ \ | \\
H\ H
\end{array}
$$

Polymers are classified based on their origin as natural and synthetic polymers. Natural polymers are the ones which are isolated from the natural sources. Cotton, silk, wool, rubber etc. belong to this type. There are also polymers modified from these natural polymers viz., cellophane, cellulose rayon, leather etc. Synthetic polymers are the polymers obtained from the low molecular weight compounds such as ethylene, phenol, formaldehyde etc. polyethylene, polyvinyl chloride (PVC), nylon, bakelite etc. are some of the synthetic polymers. Polymers are further classified as homo-chain and heterochain polymers based on whether the polymeric chain is made of only carbon at-oms or also of heteroatoms such as oxygen, nitrogen, sulphur, silicon etc. e.g. polyethylene, Teflon, polystyrene, PVC (polyvinyl chloride) are homochain polymers since their polymeric chain is made of only carbon atoms.

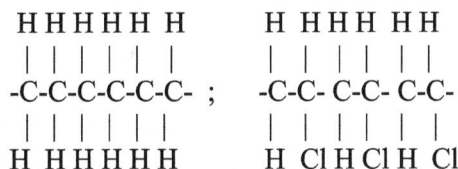

```
    H H H H H  H        H  H H H  H H
    | | | | | |         |  | | |  | |
   -C-C-C-C-C-C- ;     -C-C- C-C- C-C-
    | | | | | |         |  | | |  | |
    H  H H H H H        H  Cl H Cl H  Cl
```

Polyesters, polyamides, polysulfones, silicones are examples of heterochain polymers since their polymeric chain is also made of hetero-atoms such as oxygen, nitrogen, sulphur, silicon etc.

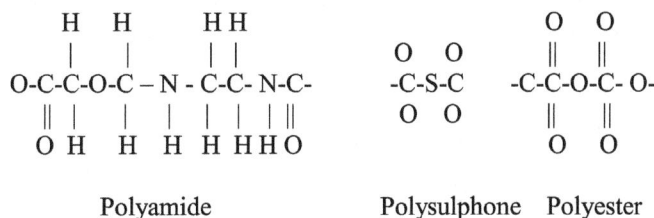

```
      H    H      H H                           O   O
      |    |      | |           O   O           ||  ||
   O-C-C-O-C – N - C-C- N-C-   -C-S-C      -C-C-O-C- O-
     || |    |    | | | | ||    O   O          ||  ||
     O  H    H    H  H H H O                    O   O

          Polyamide            Polysulphone   Polyester
```

Polymers are characterized by their variable molecular weight depending upon the source of ex-traction and synthetic route adopted. Synthetic polymers may be either straight chain polymers or branched chain polymers. The chain length of a polymer and the degree of polymerization are the parameters to be carefully understood in the case of a branched chain polymer. The degree of polymerization (DP) may be defined as the total number of repeating units/monomers present in the given polymer macromolecules whereas the chain length of a polymer is the number of mono-mer units present in the main polymeric chain. Let us consider the following copolymer structure, where 'M' represents the monomer or the repeating unit.

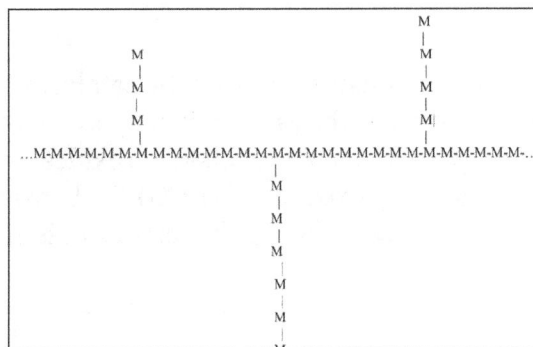

The above structure shows a branched chain polymer (monomer is represented as M) with three braches consisting of 3, 4 and 6 monomers in them. The chain length of this polymer is 29 (mono-mers) whereas the DP value for this structure is 29 + 3 + 4 + 6 = 42 (monomers).

Polymerization is classified based on (i) the mechanism of the process and (ii) practical aspects of the process. Based on the mechanism, polymerization is broadly classified into two types namely

(1) Addition or Chain polymerization and (2) Condensation or Step polymerization. Based on practical considerations, polymerization is of types such as solution polymerization, emulsion polymerization, bulk polymerization etc.

Addition polymerization is the mechanism, which results in products, which are exact multiples of the original monomer molecule. Here, the monomer units, under the influence of heat, pressure, catalyst etc., add together with the conversion of C=C into C-C systems. Addition polymerization is also termed as chain polymerization as the reaction proceeds by chain mechanism. Some examples of addition polymers are polyethylene, polypropylyene, polystyrene etc. It is obvious that any chain reaction involves three major steps namely chain initiation, chain propagation and chain termination. The chain initiation step is the decomposition of the initiator molecule to give rise to reactive species such as free radicals or cations or anion. Decomposition of initiator proceeds by homolytic or heterolytic fission. Consider the breaking of a covalent linkage between two atoms A and B. Homolytic fission of the covalent bond results in the formation of free radicals A. and B. i.e. the atoms A and B are separated with their unpaired electrons. In heterolytic fission, the covalent electron pair is transferred towards the more electronegative atom resulting in the formation of ions.

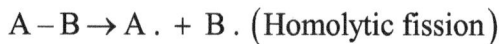

$$A - B \rightarrow A \, . \, + \, B \, . \, \left(\text{Homolytic fission} \right)$$

$$A - B \rightarrow A^+ + B^- \, \left(\text{or} \right) \, A^- + B^+ \, \left(\text{Homolytic fission} \right)$$

Addition or Chain polymerization is of types namely free radical polymerization, cationic polymerization and anionic polymerization based on whether the initiation of polymerization is caused by free radicals or cations or anions. Free radical polymerization is an important type, whose mechanism is discussed below.

Chain Initiation

The initiator molecule decomposes to give free radicals (highly reactive species).

$$I \rightarrow 2 \, R \, .$$

Here, the initiator species has to be distinguished from the catalyst in that the initiator starts the reaction between various species whereas the catalyst just speeds up the reaction that is already taking place but slowly. Organic peroxides such as alkyl peroxides or acyl peroxides are used as initiators for polymerization e.g. acetyl peroxide - $(CH_3CO)_2O_2$, benzoyl peroxide - $(C_6H_5)_2O_2$ etc. Another important initiator used in free radical polymerization is AIBN – Azo bis (Iso Butyro Nitrile).

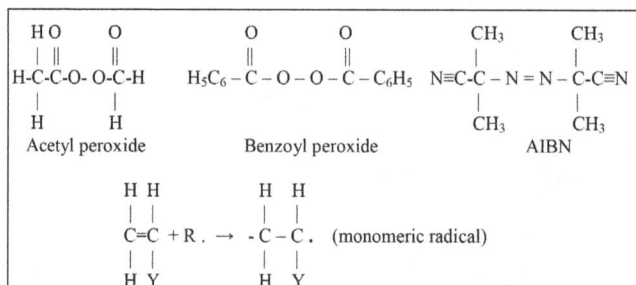

Chain Propagation

The monomer molecule is attacked by the initiator radical to form a new reactive species monomeric radical; the monomeric radical, in turn, attacks another monomer molecule to form dimeric radical etc. this step proceeds continuously as long as polymeric radicals (reactive species) are available.

$$M + R\cdot \rightarrow M\cdot \; ; \; M\cdot + M \rightarrow M_2\cdot \; ; \; M_2\cdot + M \rightarrow M_3 \cdot \; .... \; M_n\cdot + M \rightarrow M_{n+1}$$

Here M_n represents the polymer with degree of polymerization 'n'. Thus, the degree of polymerization is the number of monomer units present in the polymer molecule. The polymeric radicals are also termed as living polymers as they can be further polymerized. Thus living polymers of various degrees of polymerization are formed in the chain propagation step.

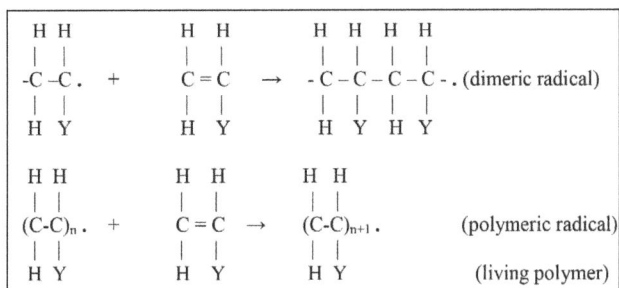

Chain Termination

It involves the recombination of polymer radicals to form dead polymers, which cannot be further polymerized. Chain termination may take place either by disproportionation or by coupling process, both of which are explained below with the example of a vinyl monomer $H_2C=CHY$ (Y is any group such as OH, X, COOH etc).

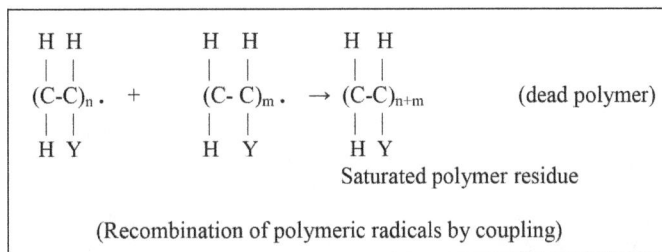

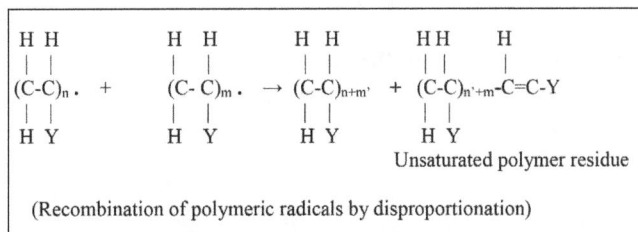

The chain length is the number of monomers present in the main chain (carbon skeleton) of the polymer. (The term main chain signifies the chain with maximum carbon atoms, discarding the other chain branches). Another important salient feature of addition/chain polymerization is that the high molecular weight is formed at once/immediately.

Ionic Polymerization

It is a type of chain reaction or addition polymerization involving cations or the anion intermediates as the chain initiators. The main steps are of course, chain initiation, chain propagation and chain termination. The type of polymerization reaction initiated by the proton and the chain propagated by the carbonium ion is called cationic polymerization.

The chain initiation of cationic polymerization involves the attack of the π electron pair of the monomer. A proton (H^+) pulls the π electron pair towards it and the positive charge of the proton is transferred to the farther end of the monomer molecule, forming the carboinum ion or carbo-cation (C^+), which acts as the cationic initiator. Here, a σ bond is formed between the proton, monomer unit and hence the polymeric chain further grows as a dimeric, trimeric, ..., polymeric cation, which constitutes the chain propagation process. The chain termination may be effected by the collision between the growing polymeric cation and a counter-ion i.e. an anion of other growing chain or a deliberately added anion. Lewis acids such as boron trifluoride (BF_3), aluminium chloride ($AlCl_3$), tin(IV)chloride ($SnCl_4$), titanium(IV)chloride ($TiCl_4$) etc. are used as the initiators / catalysts. Water is used as the cocatalyst. The catalyst and co-catalyst initially combine together forming the proton. The proton, in turn attacks the monomer, forming its cation which further acts as the chain carrier. The sequence of steps in this type of polymerization is represented below:

Chain initiation:

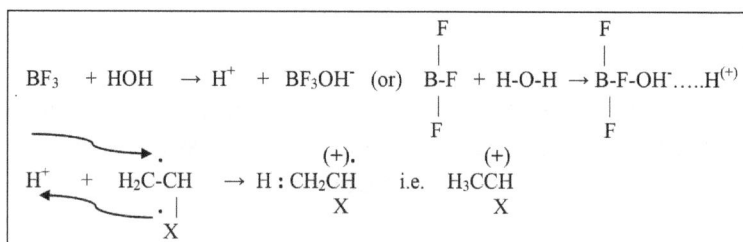

Chain propagation: (attack of second monomer by the carbonium ion of the first monomer, etc).

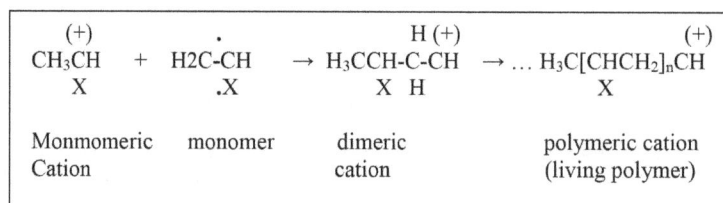

Here the electron pair transfer, as a whole is a unique process rather than one electron transfer. Lewis acid catalysts, in the presence of small amount of water as co-catalyst, form hydrates, which exist as ion pairs, as shown above. The net initiation and the propagation processes can be represented as:

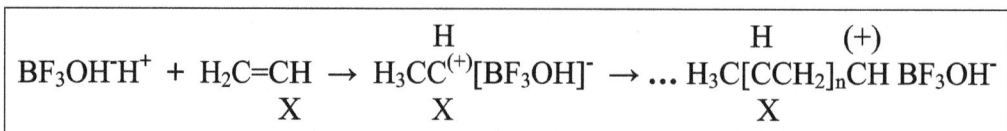

It is to be noted that the π electron pairs of the monomer units are pulled towards a direction opposite to that of the chain growth. The C$^+$ ion keeps on migrating in the direction of chain growth and the counter-ion also moves along with the carbonium ion.

Chain termination: The possible reactions of this step are governed by the two facts mentioned below:

(i) The proton donation to the counter-ion (reversal of the initiation step), resulting in the formation of a double bond at the end of the growing polymer molecule and hence the arrest of the chain growth.

$$\begin{array}{ccccc} & H & & H & & & H & OH \\ & | & & | & & & | & | \\ H_3CC & \wedge\wedge\wedge\wedge & CH_2C^{(+)}[BF_3OH]^- & \rightarrow \dots & H_3C[CCH_2]_nCH_2CH & + & BF_3 \\ & X & & X & & & X & X \end{array}$$

This process of proton donation and the reformation of BF3 hydrate is called ion-pair precipitation.

(ii) Formation of covalent bond between the carbonium ion and the counter-ion and the termination being effected by coupling.

$$\begin{array}{ccccc} & H & & H & & & H & OH \\ & | & & | & & & | & | \\ H_3CC & \wedge\wedge\wedge\wedge & CH_2C^{(+)}[BF_3OH]^- & \rightarrow \dots & H_3C[CCH_2]_nCH_2CH & + & BF_3 \\ & X & & X & & & X & X \end{array}$$

The termination by any of these two processes regenerates the initiator species (catalyst).

The monomers isobutylene, styrene, methyl styrene and vinyl ethers etc. undergo cationic polymerization.

Anionic Polymerization

The type of polymerization reaction initiated and the chain propagated by the carbanion is called cationic polymerization.

Here the attack on the π electron pair of the monomer by the anion takes place at the farthest end of the monomer substrate. At the same time, it forms a σ bond with the monomer molecule. The electron pair is pushed to the farthest end of the molecule, resulting in the formation of the carbanion – the chain carrier.

[Initiation]			[Propagation]	
(+) . HH	H(-)			(-)
R + H2C-CH $\rightarrow$ R:CC:	(or) RCC + n H$_2$C=CH $\rightarrow$ RCH$_2$[CH$_2$CH]$_n$CH$_2$CH			
.X HX	HX	X	X	X
	Monmomeric monomer		polymeric anion	
	anion		(living polymer)	

The carbanion formed in the initiation step attacks the π electron pair of the next monomer molecule and so on, pushing the π electron of the polymeric species to the farthest end and forming a σ bond with the new monomer unit. It is to be noted here that the electron pair movement is in the same direction as that of the chain growth, unlike the cationic mechanism.

The chain termination step of the anionic polymerization is not a spontaneous process. Hence the chain termination has to be made by the deliberate addition of strongly ionic impurities. In the absence of such materials, the anionic polymerization process continuously, once initiated and propagated. Thus the living polymers (species in the chain propagation process) grow as long as the fresh monomers are added. Hence this technique is called living polymerization technique and is very useful in applications such as the preparation of block copolymers.

Condensation polymerization may be defined as the reaction occurring between simple polar-group containing monomers. It proceeds by condensation mechanism i.e. takes place with the removal of minor molecules such as water, ammonia, hydrogen chloride etc e.g. condensation polymerization between adipic acid and hexamethylene diamine.

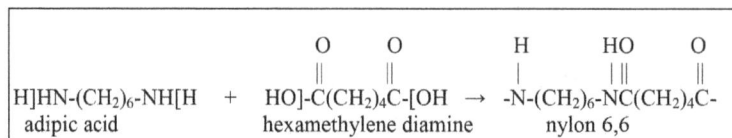

$$
\begin{array}{ccc}
& O \quad\quad O & H \quad\quad HO \quad\quad O \\
& \parallel \quad\quad \parallel & \mid \quad\quad \mid\parallel \quad\quad \parallel \\
\text{H]HN-(CH}_2)_6\text{-NH[H} \;+\; & \text{HO]-C(CH}_2)_4\text{C-[OH} \;\rightarrow\; & \text{-N-(CH}_2)_6\text{-NC(CH}_2)_4\text{C-} \\
\text{adipic acid} & \text{hexamethylene diamine} & \text{nylon 6,6}
\end{array}
$$

Co-polymerization is a type of addition polymerization involving two different monomers. Generally addition polymerization takes place with a similar monomer but addition polymerization can also take place between two different monomers (M_1 and M_2) with the resultant polymeric chain being composed alternatively of the two monomers.

The major differences between addition polymerization and condensation polymerization are tabulated below:

	Addition polymerization	Condensation polymerization
1	Growth reaction adds repeating units one at a time to the polymeric chain.	Any two molecular species can react.
2	Number of monomer units increases steadily throughout the reaction.	Monomer disappears early in the reaction.
3	High polymer is formed at once.	Polymer molecular weight (degree of polymerization) rises steadily throughout the reaction.
4	Longer reaction times have very little effect on molecular weight but gives higher yields.	Longer reaction times are essential to obtain higher molecular weights i.e. reaction time influences molecular weight of the polymer.
5	The reaction mixture contains only monomers, high polymers and very small amount (10^{-8}) of growing chains.	All types of molecular species are present at any stage.
6	E.g. polymerization of ethylene, styrene, vinyl chloride, propylene etc.	Polymerization to get nylon, PET, polycarbonate, polyurethane etc.

The polymeric chain of a copolymer of two monomers (M_1 and M_2) will be as follows:

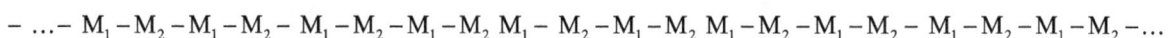

$$- \ldots - M_1 - M_2 - M_1 - M_2 - M_1 - M_2 - M_1 - M_2 \; M_1 - M_2 - M_1 - M_2 \; M_1 - M_2 - M_1 - M_2 - M_1 - M_2 - M_1 - M_2 - \ldots$$

(the above structure is that of a linear co-polymer)

E.g. SBR – styrene butadiene rubber – a copolymer is composed of alternative units of styrene and butadiene.

Based on the mechanism of polymerization, polymers are classified as addition polymers, condensation polymers and copolymers (addition polymers). Based on branching of the polymeric chain, polymers are classified as linear polymers, branched chain polymers and cross-linked polymers. Cross-linked polymers are the polymers, which have the adjacent polymeric chains, connected together by means of covalent linkages such as ether linkage, methylene linkage etc. e.g. Bakelite is a cross-linked polymer. Copolymers may be either linear with alternately placed two different monomers or branched chain polymers where one monomer is attached as a side chain/branch.

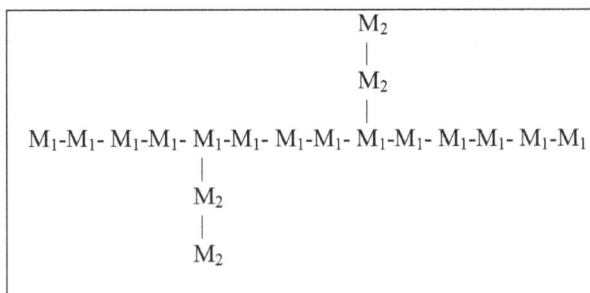

$$
\begin{array}{c}
M_2 \\
| \\
M_2 \\
| \\
M_1\text{-}M_1\text{-}\ M_1\text{-}M_1\text{-}\ M_1\text{-}M_1\text{-}\ M_1\text{-}M_1\text{-}\ M_1\text{-}M_1\text{-}\ M_1\text{-}M_1\text{-}\ M_1\text{-}M_1 \\
| \\
M_2 \\
| \\
M_2
\end{array}
$$

The schematic of a block copolymer is given below:

Properties of polymers

Glass Transition Temperature (T_g)

It is obvious that the polymers can be synthesized for wide range of properties such as those from flexibility to rigidity, crystalline to amorphous nature etc. A polymeric sample on heating may transform itself from a state of softness to a state of hardness or brittleness. Thus the temperature below which a polymer is hard and above this temperature, it becomes so soft that it transforms from glassy (brittle) state to a rubbery (elastic) or a visco-elastic state is termed as glass transition temperature (T_g). On further heating the polymer sample beyond the T_g value, it starts melting, acquiring the flow properties. The temperature at which the polymer becomes a fluid is called flow temperature (T_f). The phase transition of the polymer in these temperature ranges can be represented as:

Glassy State (brittle)	Rubbery state (visco-elastic)	Viscous fluid state (polymer melt)

Depending on whether the polymer is more crystalline or more amorphous, the T_g, T_f or melting temperatures are either sharp or a temperature range respectively. These temperature values of a polymer are indicative of other mechanical properties of the polymer and hence become more significant. Thus if rubber (originally known for its elastic behavior), if cooled below its T_g value of -70 °C, it becomes brittle and resembles glass in its brittleness.

Tacticity

Polymers may be synthesized as stereo regular compounds. Stereo regularity is a phenomenon of regular or orderly, spatial arrangement of groups or radicals of the monomer on either side of the main polymeric chain. Co-ordination polymerization is a type of chain polymerization where Ziegler_Natta catalysts (organo-metallic compounds containing transition metals such as Ti, Mo, Cr, V, Ni, Rh etc.) are used for achieving the stereo regularity in the polymer structure. Based on tacticity/stereo regularity, the polymers are classified as atactic, isotactic and syndiotactic polymer. Atactic polymer is one in which the stereo regularity is lacking or the groups/radicals are randomly/disorderly arranged on either side of the polymeric chain. Isotactic polymer is one in which the stereo regularity exists such that the groups /radicals are orderly arranged on same (one or other) side of the polymeric chain. Syndiotactic polymer is one in which the stereo regularity exists such that the groups/radicals are orderly arranged on either side of the polymeric chain. This can be illustrated with the vinyl (H_2C = CHY) monomer (Y is a substituent radical such as alkyl/R, halogen/X etc). Thus propylene (H_2C = $CHCH_3$) is a vinyl monomer where the alkyl group is the simplest methyl group (R = CH_3). Propylene can be polymerized with Ziegler_Natta catalysts to get stereo regular polymers such as isotatic polypropylene or syndiotactic polypropylene or atactic polypropylene if polymerized with other initiators/catalysts. The relative arrangement of the methyl groups attached to the alternate carbon atoms of the monomers in the polymeric chain decides the type of polypropylene. These three types are shown in fig. below:

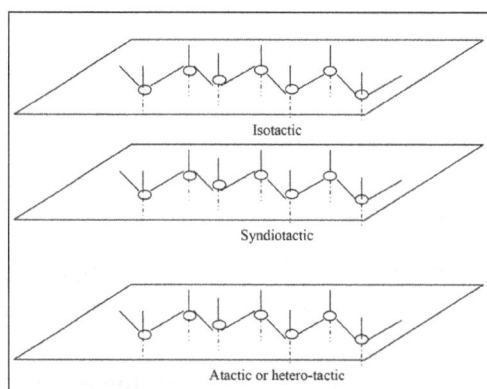

Isotactic and syndiotactic polymers are collectively termed as homo-tactic polymers whereas the atactic polymers are termed as hetero-tactic polymers. Poly buta-dienes and their derivatives, used as repeating units/monomers in synthetic rubber (co-polymers) can be synthesized as stereo regular polymers in this way.

It is obvious that the molecular weight of a polymer is not a fixed parameter but variable dependent on the synthetic method. Hence there are three important versions of representation of polymer molecular weight as number average molecular weight (M_n), weight average molecular weight (M_w), and viscosity average molecular weight (M_v). First two types of parameters are discussed in detail below:

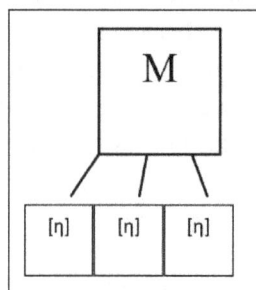

Let us consider a polymer system containing 'n$_1$' numbers of the monomer species with molecular weight 'M$_1$', 'n$_2$' numbers of the monomer species with molecular weight 'M$_2$', 'n$_3$' numbers of the monomer species with molecular weight 'M$_3$', and so on. The total number of monomers in this system is '$\left(\sum n_i\right)$', with 'i' ranging from '1' to 'n'. the total weight (W) of the polymer sample is W $= \left(\sum n_i M_i\right)$.

The number fraction of fraction 1 (i.e. fraction of monomers with molecular weight 'M$_1$') is given by $\left[n_1 / \left(\sum n_i\right)\right]$.

Thus the molecular weight contribution by the fraction 1 is given by $\left(n_1 M_1\right) / \left(\sum n_i\right)$.

Similarly, the molecular weight contribution by the fraction 2 is given by $\left(n_2 M_2\right) / \left(\sum n_i\right)$.

Hence, The number average molecular weight $\left(\overline{M_n}\right)$ is defined as:

$$\left(\overline{M_n}\right) = \left(\Sigma n_i M_i\right) / \left(\Sigma n_i\right),$$

where 'n$_i$' is the number of monomers each with molecular weight 'M$_i$' and 'i' varies from '1' to 'n' i.e. 'n$_1$' is the number of monomers each with molecular weight 'M$_1$', 'n$_2$' is the number of monomers each with molecular weight 'M$_2$', so on.

The weight fraction of fraction 1 (i.e. weight fraction of monomers with molecular weight 'M$_1$') is given by [n$_1$ M$_1$ / (W)] = [n$_1$ M$_1$ / (W)] = $\left[n_1 M_1 / \left(\Sigma n_i M_i\right)\right]$.

Thus the molecular weight contribution by the fraction 1 is given by $\left[n_1 M_1 . M_1 / \left(\Sigma n_i M_i\right)\right]$ i.e = $\left[n_1 M_1^2 / \left(\Sigma n_i M_i\right)\right]$.

Similarly, the molecular weight contribution by the fraction 2 is given by: $\left[n_2 M_2 . M_2 / \left(\Sigma n_i M_i\right)\right]$ i.e = $\left[n_1 M_2^2 / \left(\Sigma n_i M_i\right)\right]$.

Hence, the weight average molecular weight (M$_w$) is defined as:

$\left(\overline{M_w}\right) = \left(\Sigma n_i M_i^2\right) / \left(\Sigma n_i M_i\right)$, where 'n$_i$' is the number of monomers each with molecular weight 'M$_i$' and 'i' varies from '1' to 'n' i.e. 'n$_1$' is the number of monomers each with molecular weight 'M$_1$', 'n$_2$' is the number of monomers each with molecular weight 'M$_2$', so on.

For synthetic polymers, $\overline{M_w}$ is greater than $\overline{M_n}$, i.e. $\overline{M_w} > \overline{M_n}$. For a given polymer, these two molecular weight values may be same or different i.e. more than unity. Accordingly, the polymer is said to be mono-disperse or poly-disperse. Mono-disperse polymers are considered as homogenous and the poly-disperse ones as heterogenous polymers. But in practice, homogenous polymers are not realizable. Thus the Poly Dispersity Index (PDI) of a polymer is defined as the ratio of the weight average molecular weight to the number average molecular weight:

$$PDI = \overline{\left(M_w\right)} / \overline{\left(M_n\right)}$$

Typical PDI values for the synthetic polymers are given below for reference:

Free radical polymers: obtained by solution/suspension/emulsion methods: 1.5 – 2.

: obtained by bulk polymerization method: 2 – 5.

: obtained by auto-acceleration methods: 8 – 10.

Polymers by cationic/anionic mechanism: Using homogenous catalysts: < 1.5 (less than).

Using heterogenous catalysts: > 10 (higher than).

Polymers obtained by poly-condensation / poly-addition / ring opening mechanisms: 2 – 3.

Branched chain polymers: > 20 (greater than 20).

The polymerization techniques used in practice are of types such as bulk polymerization, solution polymerization, emulsion polymerization, suspension polymerization etc.

Bulk Polymerization

In this method, the monomer is taken in the liquid state and the initiator is dissolved in the monomer. The chain transfer reagent, if required, is also dissolved in the monomer. The function of chain train transfer reagent is to control the molecular weight of the final polymer and result in a homogenous phase. The resultant mass is heated or exposed to radiations of particular wavelength for the initiation of the polymerization. As the polymerization reaction proceeds, the viscosity of the medium increases and hence the mixing becomes progressively difficult, which is the disadvantage. Another disadvantage of this process is the restricted diffusibility of the growing polymer chain, because of the higher viscosity of the medium. Moreover, the accumulation of active radical sites causes the enormous increase in the polymerization rate, leading to the auto-acceleration mechanism and ultimately to explosion, if not controlled properly. The advantages of bulk polymerization technique are (i) simplicity, (ii) higher purity of the polymer obtained, (iii) no requirement of additives other than the initiator and the chain transfer reagent and (ii) direct utility of the polymer with no isolation requirement (due to higher purity). Bulk polymerization is used in the free radical polymerization of methyl methacrylate ($H_2C = C(CH_3)COOCH_3$) or styrene ($H_2C = CHC_6H_5$) to get transparent powders of PMMA or polystyrene (thermocole) and also the cast sheets of poly vinyl chloride (PVC).

Emulsion Polymerization

Here, the monomer is dispersed in aqueous phase as emulsion. The emulsion is stabilized by the addition of surfactants (surface active agents), protective colloids and some buffers. Anionic surfactants such as sodium or potassium aryl sulphonates or cationic surfactants such as alkyl amino hydro chlorides or alkyl ammonium halides or non-ionic surfactants such as alkyl glycosides or saccharic esters of higher fatty acids are used. They function by lowering the surface tension at the water-monomer interface and hence facilitate the emulsion of monomer in water. Due to their low solubility, surfactants are molecularly dispersed even at low concentrations. At a particular concentration, the excess un-dissolved ones form as molecular aggregates, called micelles and an equilibrium is established between the dissolved surfactant molecules and the aggregated micelles. The highest concentration of surfactants wherein all the molecules are in the dispersed state and the concentration beyond which the micelle formation is possible is termed as critical micelle concentration (CMC).

The emulsifier molecules are made of two parts – a long, non-polar hydrocarbon (H/C) moiety, to which is attached a polar entity such as –COONa, -SO$_3$Na, -NH$_2$.HCl, -NBr etc. in micelle formation, the emulsifier molecules aggregate in such a way that the polar ends align outward and the H/C ends come close to each other at the interface.

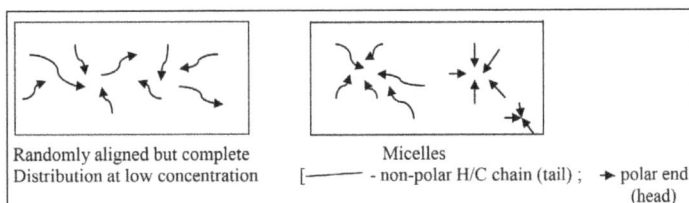

| Randomly aligned but complete Distribution at low concentration | Micelles [——— - non-polar H/C chain (tail) ; → polar end (head)] |

Due to the proximity of the H/C entities of all emulsifier molecules, the interiors of the micelles act as the H/C phases, where the monomers can be made soluble. On further addition of the monomer followed by agitation, emulsification takes place. The resultant emulsion is a complex system. These micelles possess favourable condition for polymerization to occur. The initiator molecules are available at the surface layers whereas the interior of the monomer is filled with the solubilized monomer. Hence the polymerization starts at the surface and proceeds inwards. On consumption of the monomer in the micelle, more amount of monomer diffuses from the aqueous phase, into the micelle. The polymer chain growth continues until another radical species enters and arrests the chain growth. With more and more amount of polymer formed, the polymer chains aggregate into fine particles and get surrounded and stabilized by the emulsifier layer. At the end of polymerization, the fine particles of the polymer are stabilized by the emulsifier layer and dispersed uniformly in the aqueous phase. The resultant milky white dispersion is termed as latex, which can be used as such for applications like adhesives, emulsion paints or the polymer can be isolated from the latex by the de-stabilization of the emulsified polymer mass using electrolytes or by spray drying or by freezing. Depending upon the relative solubility of the monomer and initiator in water and the ratios of amounts of monomer/water and emulsifier water, the polymerization occurs either at the dissolved phase or at the interface or at the surface or inside the monomer droplets.

Emulsion polymerization is the most widely used industrial technique for the polymerization of monomers such as vinyl chloride (H$_2$C = CH.Cl), buta-diene (H$_2$C = CH.CH = CH$_2$), chloroprene (H$_2$C = CCl.CH = CH$_2$), vinyl alcohol (H$_2$C = CH.OH), acrylates (H$_2$C = CH.COO-) and methacrylates (H$_2$C = C(CH$_3$).COO-) etc. It is to be noted that the heat transfer is very difficult with this technique and hence the viscosity buildup of the polymer mass is quite low compared to the bulk and solution polymerization techniques.

Solution Polymerization

In this technique, the monomer and the free radical initiator are dissolved in a suitable solvent, along with the ionic co-ordination catalyst and the chain transfer reagent, if any. The selection of some inert solvent enables viscosity control and prevents heat transfer. The advantage of this method lies in its use when the polymer is used in solution form or when the polymer is insoluble in monomer or any other solvent or precipitates out as slurry and is amenable for easy isolation. The industrial production of poly acrylo nitrile (PAN) by free radical mechanism, that of poly isobutylene by cationic initiation and manufacture of block co-polymers are affected by solution polymerization technique.

Suspension Polymerization

Water-insoluble monomers are polymerized by this technique. The monomer is suspended in water in the form of fine droplets, which are then stabilized against coalescence, using stabilizers, surfactants, protective colloids and by stirring. But the initiators are monomer-soluble. Each monomer droplet is isolated and is independent of other droplets and hence acts as an independent bulk polymerization nucleus (where the polymer chain growth starts and proceeds). The continuous aqueous phase separating the monomer droplets, acts as the efficient heat transfer medium and hence the exothermicity is controlled. The size of the monomer droplets depends upon the monomer/water ratio, type and concentration of stabilizers and the mode and speed of agitation. This technique is more economical compared to solution polymerization technique, since water is used as the heat transfer medium. As the entire bulk of the monomer is divided into numerous tiny droplets, the control of the kinetic chain length (chain length of living polymer; that in chain propagation stage) of the formed polymer is quite good and results in a fairly narrow molecular weight distribution of the product. Thus the polymerization precedes to 100% conversion levels and the product is obtained as 'spherical beads or pearls'. The technique is so termed as bead or pearl polymerization. The isolation of the product is easy as it involves only the filtration of pearls or beads and the removal of surfactants and other additives by mere washing with water. The water-washed and dried polymer sample is as such used for molding purposes or can be dissolved in a suitable medium for use as adhesives/coatings. Expanded polystyrene beads (for making polystyrene foams), styrene-divinyl benzene copolymers (for the preparation of ion exchange resins) and poly vinyl acetates (for further conversion to poly vinyl alcohol) are produced by suspension polymerization technique.

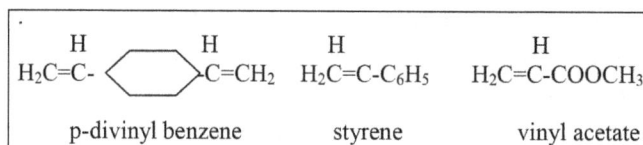

p-divinyl benzene   styrene   vinyl acetate

Preparation, Properties and uses of Nylon 6,6

Nylon 6, 6 is a polyamide prepared by the condensation polymerization between/polycondensation of the monomers hexamethylene diamine and adipic acid. Here two water molecules are removed from each set of monomers.

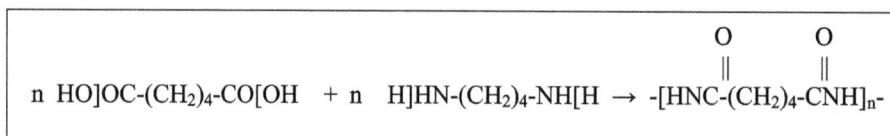

$$\text{n HO]OC-(CH}_2)_4\text{-CO[OH} + \text{n H]HN-(CH}_2)_4\text{-NH[H} \rightarrow \text{-[HNC-(CH}_2)_4\text{-CNH]}_n\text{-}$$

Nylon 6, 6 is a tough, fairly crystalline material with the melting point of 270 °C. It has good tensile strength and abrasion resistance up to 150 °C. It is resistant to the action of solvents but soluble in formic acid, cresols and phenols. Its high tensile strength and rigidity/stiffness are due to the high degree of polymerization.

Nylon 6, 6 is largely used to produce tyre chord and to make filaments and ropes. Due to its toughness, it is used as a good substitute for metals in gears and bearings. Its fibre forming tendency is used to yarn textile fibres and dress materials.

Epoxy Resins

These are basically poly-ethers. The common types of epoxy resins are prepared from the condensation polymerization between epichlorohydrin and bis-phenol A. The reaction is carried out with excess of epichlorohydrin.

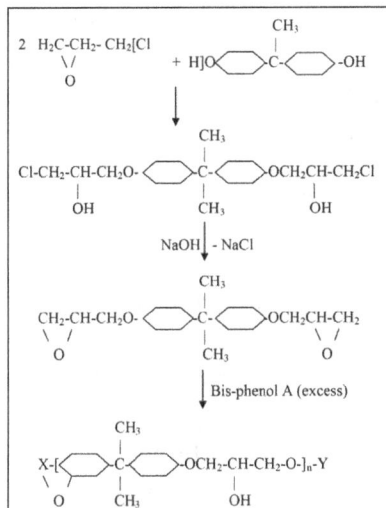

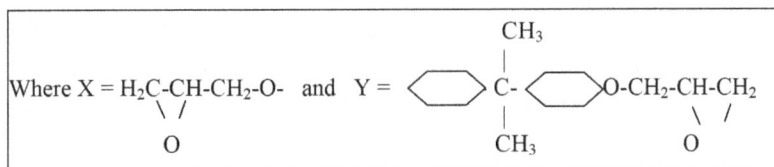

Instead of bis-phenol A, other compounds with OH groups such as glycols, glycerols and resorcinols can also be used. The epoxy resins obtained through these reactions will be either highly viscous liquids or solids with high melting points. The epoxy resins can be further cured with substances such as amines, poly sulfides and poly amides. The thiol (-SH) covalent linkages are established between the polymer chains, during curing, to get the resultant strong, highly cross-linked polymers.

Epoxy resins find enormous applications due to their remarkable chemical resistance and good adhesion properties. They are used as excellent structural adhesives. On proper curing, they yield tough materials as used in industrial floorings, as foaming materials, as pottering materials in electrical insulation etc. A principal constituent of most of the fibre reinforced plastic (FRP) materials is the epoxy resins. Also, the EPI coating in lorries carrying corrosive chemicals contains the epoxy resin formulations.

Polymers

Eras of civilization have frequently been named for materials discovered and subsequently used extensively by humans (e.g., the Stone Age, the Bronze Age, etc). Toward that end, the 20th century might appropriately be labeled as the Plastics Age or, somewhat more broadly, the Polymer

Age. Whereas biological polymers (e.g., DNA, proteins) have been present since the origins of life, the development of synthetic polymers and the realization that long-chain molecules could exist are relatively recent milestones on the materials calendar. Indeed, it was only in the 1920s that the concept of a polymer or macromolecule came to be grudgingly accepted, slowly displacing the colloid theory that suggested that macromolecules were nothing more than aggregates of small molecules. The subsequent two decades witnessed the rapid development of many types of synthetic fibers, plastics, and elastomers, including Dacron, nylon, and Spandex. Also, guiding principles for the successful linking of monomers into high-molecular-weight polymers were established. The remainder of the 20th century saw explosive growth in new polymers and their applications. Today, polymers are ubiquitous, from commodity plastics such as plumbing pipe, food wrap, and trash bags, to highly specialized and tailored materials for use in aircraft components, personal computers, and health care.

A key reason for the broad acceptance and appeal of polymers is that they are the most versatile of all materials, spanning the range of mechanical properties from soft gels and rubbers to extremely strong fibers. This versatility is a direct consequence of the great latitude conferred by design of the repeat-unit structure and control of chain length (or molecular weight) and chain architecture. While most polymers are based on carbon backbones, inorganic backbones (e.g., silicones) add yet another dimension to the tailoring of material properties. A key aim of this article is therefore to introduce basic ideas of how polymer structure affects properties. A short discussion of the implications of these ideas in the design of polymeric biomaterials will follow. Several examples of such materials are discussed in some detail in additional entries of the encyclopedia. This article concludes with a summary of opportunities and challenges in polymer science for the next two decades and beyond.

Introduction to Structure/Property Relationships

Small Organic Molecules

The properties of small organic molecules (melting point, for example) depend intimately on the structure of the molecule and principally on four factors: 1) size; 2) symmetry; 3) internal flexibility (presence or lack of bond rotations); and 4) strength of intermolecular forces. A consideration of how these dictate properties of small molecules is instructive in understanding structure/property relations in polymers. Melting (or fusion) is a first-order thermodynamic transition, and the melting temperature, T_m, depends on the ratio of the enthalpy and entropy of fusion:

$$T_m = \Delta H_m / \Delta S_m$$

Melting points typically increase with molecular size, as evidenced by tracking the properties $(gas \rightarrow liquid \rightarrow grease \rightarrow wax \rightarrow resin)$ of n-alkanes as a function of carbon number. Highly flexible molecules that are reasonably symmetric crystallize readily because the conformation(s) required in the crystal (for example, the extended zig-zag conformation for n-alkanes) are easily adopted. Such materials have rather low melting points even at high molecular weights, since considerable entropy (translational and rotational) is gained on melting and thus ΔS_m is large. Branching can lower melting points by reducing the ability of a molecule to pack tightly into a crystal, thus lowering the enthalpy of fusion. Moreover, branching introduces asymmetry that can inhibit crystallization when the material is rapidly cooled, the result being an amorphous solid or glass. In cases where branching is extensive and does not lead to lack of symmetry, melting points can actually increase vs. linear counterparts due to loss of conformational flexibility and hence lower entropies of fusion. Rigid molecules such as anthracene tend to have high melting points due to a low entropy of fusion, as only translational entropy is primarily gained on melting. Similarly, molecules with strong intermolecular forces (e.g., hydrogen bonds) have high melting points, usually not due to the need to overcome these forces on melting but rather as the result of their persistence in the molten state affording low entropies of fusion. For this reason, benzoic acid has a relatively high melting point for its molecular weight. These same four factors applied to polymer repeat units, along with chain orientation and extent of crosslinking, are the primary determinants of polymer properties.

Glass Transition Temperature

A glass is appropriately defined as material having the structural properties of a liquid, namely disorder, and the mechanical properties of a solid; it has a glass transition temperature, T_g, above which the material can flow. The idea of a melting point is easily appreciated but T_g is perhaps less familiar and thus deserves a brief discussion here. Cooling of a molten substance can lead to crystallization at T_m and is accompanied by a rapid drop in volume as shown in figure (except for odd substances such as water). The crystalline substance will continue to drop in volume, albeit slowly, as determined by its coefficient of expansion. However, under certain circumstances it is possible to bypass T_m and form a supercooled liquid that eventually forms a glass. This process, known as vitrification, can occur if cooling is rapid, the viscosity of the melt is very high, and/or the molecular structure of the substance is either asymmetric or large and difficult to pack into a crystal. At T_g, molecular motion effectively ceases. However, unlike Tm, the value of T_g depends upon the cooling rate, with rapid cooling "freezing-in" higher energy molecular conformations and organizations. Therefore, it is common to see T_g reported as ranges, and it is important to specify cooling or heating rates in measurements of T_g.

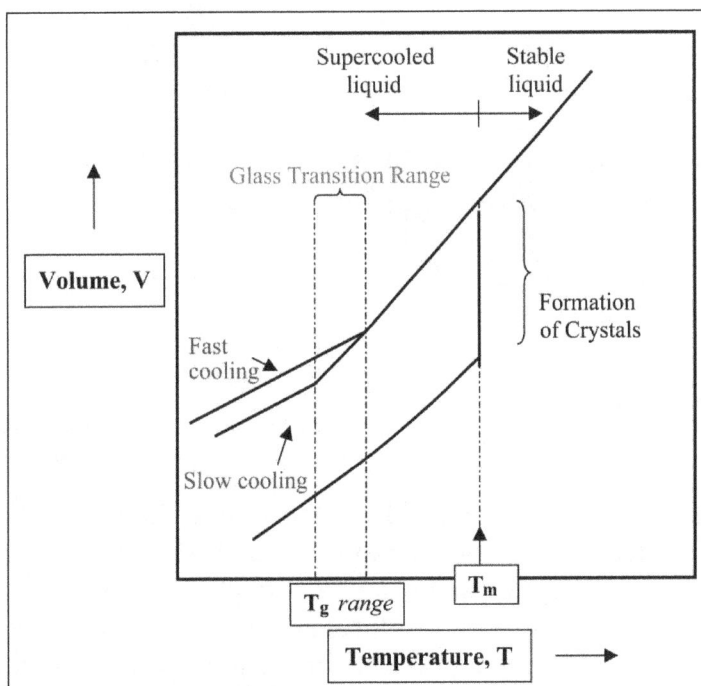

Plot of specific volume vs. temperature, contrasting the formation of a glass vs. crystals from the melt.

Polyethylene and Polymers from Mono-Substituted Ethylenes

Consider first molecular weight, using the n-alkane series as an example with high molecular weight and linear or high-density polyethylene (HDPE) as the upper limit.

Polyethylene: $-(CH_2CH_2)_n$

Several important points derive from inspection of samples of polyethylene and some lower-molecular weight solid n-alkanes such as paraffin wax (roughly 30–40 carbons). The wax, a highly crystalline substance, is easily broken into small pieces with a fingernail, whereas the polymer resists such damage and is seen to be a very tough flexible material. Molten wax has a low viscosity in contrast to polyethylene. More significantly, and unlike the wax, fibers can be drawn from the molten polyethylene very easily. The reason is that the molecules of polyethylene are long enough to physically entangle with one another, thus supporting the ability to form fibers from the melt. The entanglement molecular weight is roughly the length of a molecule at which the properties of polymers begin to manifest themselves. It is dependent upon the chemical structure of the polymer, but most entanglement molecular weights are in the range of 2000 to 20,000 g/mol.

As expected, the viscosity of a molten polymer will rise dramatically with molecular weight since chain entanglements will inhibit the ability of molecules to move past one another. Many other properties such as melting point and tensile strength do not continue to increase with molecular weight but reach an asymptote. (However, the tensile strength can be increased substantially by chain orientation). This is because, at long chain lengths, polymer chains behave as a collection of shorter independent segments. In the case of polyethylene, the asymptotic limit for Tm is about 140 °C. This relatively low temperature is the result of high chain flexibility and thus a high ΔS_m.

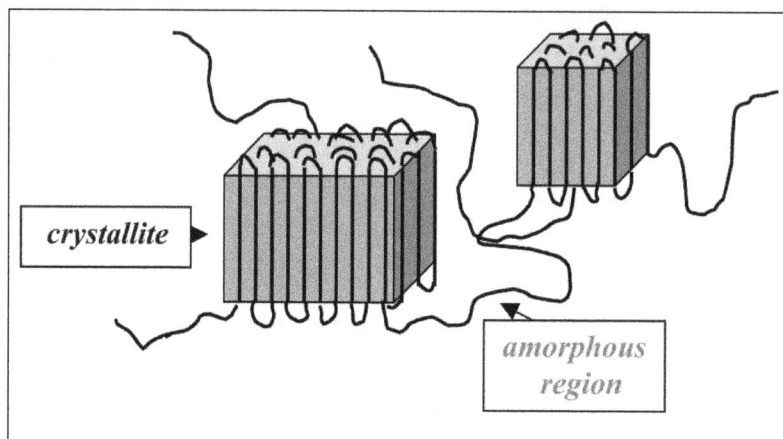

Morphology of a typical semicrystalline polymer such as liner polyethylene. It is kinetically easier for the chains to fold on themselves during crystallization rather than form long extended structures.

Polymers such as linear polyethylene crystallize easily due to the highly symmetric repeat unit and high chain flexibility. However, polymers rarely crystallize completely. Crystallization can begin at multiple points along a long polymer chain, eventually trapping segments of disordered or amorphous chains between crystallites. Therefore, materials such as HDPE are semicrystalline, with a T_m for the crystalline component and a T_g for the amorphous regions, with the degree of crystallinity being about 70–80%. The T_g for polyethylene is approximately -50 to -90 °C, and hence this polymer at room temperature is a composite of crystals and liquidlike amorphous regions. The toughness of polyethylene at room temperature is due to the ability of the amorphous regions above their T_g to absorb mechanical impact. Semicrystalline polymers are typically translucent due to light-scattering by the crystallites. Their solubilities can effectively be nil until temperatures rise to close to T_m, and this is why HDPE containers can be used to store gasoline even though a mixture of alkanes might be expected to dissolve the polymer readily at room temperature because of similar molecular structures.

Mechanical extension of a semi-crystalline polymer typically leads to "necking-down" in the drawn region as chains are pulled out of crystallites and oriented in the stretch direction. Extensive orientation in gel-spun, ultrahigh-molecular-weight HDPE affords fibers the tensile strengths of which exceed that of steel on a weight basis.

Molecular symmetry can affect both the melting point and the degree of crystallinity. Low-density polyethylene (LDPE) has occasional short alkyl branches that serve to disrupt chain packing and thus lower the degree of crystallinity to about 50% and the T_m to 105–110 °C vs. about 140 °C for the linear material. LDPE is softer (and tougher) than HDPE due to its lower crystallinity.

The effect of symmetry is particularly pronounced for polymers derived from mono-substituted ethylenes such as polypropylene.

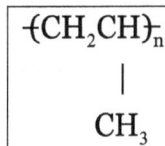

Polypropylene:

$$-(CH_2CH)_n-$$
$$|$$
$$CH_3$$

This repeat unit as drawn is misleading in that there are two possible configurations at the methine carbon (−CH−) and the arrangement of these along a chain leads to three possible isomers (here R=methyl):

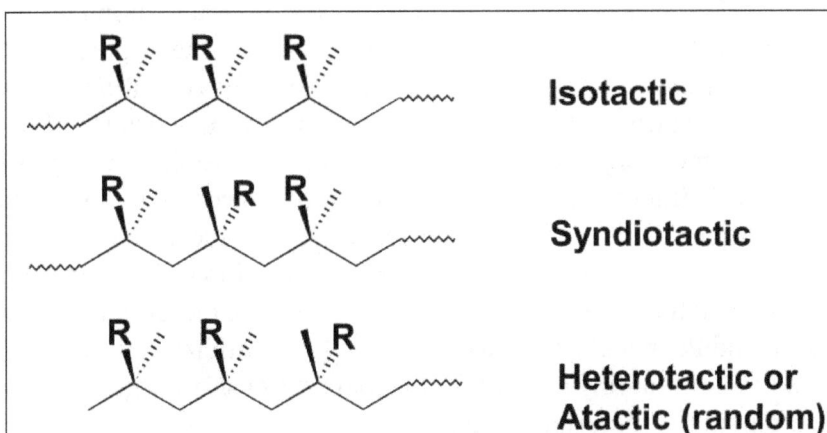

Isomers of $-(CH_2- CH(R))-n$.

After polyethylene, polypropylene is the second most important polymer in terms of annual world production, and it is only the isotactic form of polypropylene that is produced. This material has a higher T_m (165–170 °C) compared with HDPE, the principal reason being that the methyl groups reduce the conformational flexibility of the chain in the melt, affording a lower ΔS_m. Its T_g is about -10 °C due again to restricted bond rotations, and hence, compared to polyethylene, it is not flexible at lower temperatures. Interestingly, the laboratory that discovered catalysts to produce HDPE many decades ago also produced isotactic polypropylene, but the material was ignored since it was not thought that a backbone with pendant methyl groups would have a higher melting point than polyethylene. As pointed out earlier in our discussion about small molecules, branching can lower T_m but that is principally if there is no regular placement of branches and no stereochemical control at the branch points. (This assumes that there are no strong intermolecular forces acting to depress the entropy of the melt). Isotactic polypropylene has a very regular structure and its conformational flexibility dictates its thermal properties. The higher T_m makes this material useful in a variety of medical applications (e.g., plastic syringes and tubing) because it can be sterilized in an autoclave without coming too close to its T_m.

It is interesting to compare the structures of linear polyethylene and isotactic polypropylene in the crystalline regions. In polyethylene, the chains adopt a zig-zag conformation as is the case in crystals of small-molecule alkanes. In polypropylene, by contrast, the pendant methyl groups force the chains to be helical. An analogy exists with the secondary structure of polypeptides; for example, poly(glycine) prefers the â – sheet structure whereas polypeptides with bulky side chains adopt an α – helix.

Polystyrene is another important material derived from a mono-substituted ethylene where R is a benzene ring. Commercial polystyrene is exclusively the atactic form and, because this isomer is irregular, the polymer does not crystallize and therefore is transparent. This is an attractive property and accounts for the widespread use of polystyrene in petri dishes and related lab supplies. Its T_g is about 100 °C due to restricted chain flexibility by virtue of the rather bulky phenyl substituents, and hence the material is rather brittle at room temperature. Like T_m, the T_g of a polymer reaches an asymptote with molecular weight, the 100°C value is the limiting Tg for atactic polystyrene.

Poly(vinyl chloride) or PVC, where R=Cl, is another important atactic polymer. Besides its widespread use in plumbing pipe and connectors, it is a familiar lab staple in the form of flexible Tygon tubing. Tygon is PVC containing a plasticizer such as 2-diethylhexyl phthalate (DEHP). PVC has a T_g of about 85 °C, significantly higher than that of polypropylene because of the slightly larger size of the pendant Cl but more significantly due to stronger intermolecular forces stemming from the polar C–Cl bond. At T_g, intermolecular forces need to be overcome to allow chain segments to move relative to one another. Addition of a plasticizer, which is typically a low-volatility liquid with good solubility in the polymer, serves to disrupt intermolecular forces and hence depress the T_g. In Tygon, the T_g is below room temperature. It is interesting that the material keeps its physical shape rather than forming a puddle of high-viscosity polymer plus plasticizer, and the explanation is that PVC contains a small amount of tiny crystallites (perhaps from short segments of syndiotactic repeat units) that act to hold the material together.

Another means of lowering T_g and thus increasing material flexibility is to copolymerize vinyl chloride with monomers such as vinyl acetate. At least some readers will recall long-playing phonograph records which were frequently made from such copolymers.

Polyesters

Polyesters represent an important class of materials with broad applications as biomaterials. Poly(ethylene tereph thalate), PET, is a ubiquitous material known as Mylar in film form and Dacron in fiber form. Woven Dacron mesh is used for large-diameter vascular grafts. PET, with its rather regular structure, is a semi-crystalline polymer with a T_m of about 265 °C and a T_g near 85 °C.

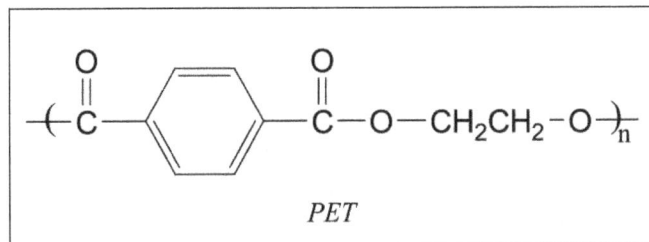

PET

Polycarbonates are formally polyesters of carbonic acid, the most popular being the material with the trade names Lexan or Merlon. The repeat unit, which is symmetric, is more complex and less flexible than that of PET, and thus polycarbonate will typically form a glass (T_g ca. 150 °C) on cooling; hence the polymer is transparent.

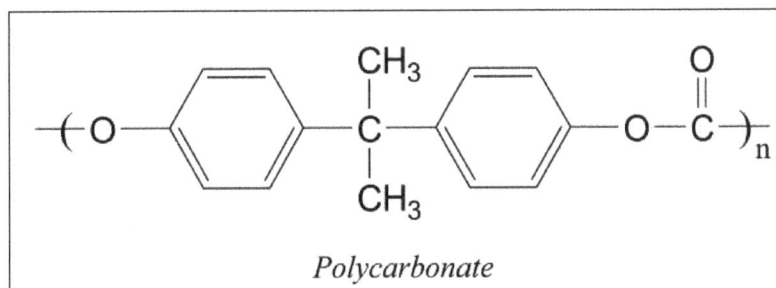

Polycarbonate

The polymer is rigid at room temperature, and it is tempting to think that all glassy polymers below their T_g might be brittle. Polycarbonate is remarkably resistant to impact vs., for example, glassy polystyrene, as evidenced by the properties of compact discs (polycarbonate) and their holders (polystyrene). Polycarbonate has sub-T_g short-range molecular motions associated with the flexible carbonate group that serve to absorb significant impact energy. Such energy-absorbing mechanisms are not available in polystyrene. Similar sub-T_g motions account for the flexibility of PET fibers and films at room temperature.

Both PET and polycarbonate hydrolyze very slowly and are not useful as biodegradable materials. However, simple aliphatic polyesters such as poly(glycolic acid) and poly(lactic acid) are well-known biodegradable polyesters. (They are sometimes referred to as polyglycolide and polylactide, respectively, because they are frequently prepared from dimers of the acids.)

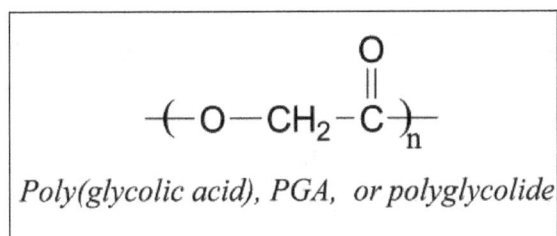

Poly(glycolic acid), PGA, or polyglycolide

$$+O-CH-\overset{\overset{\displaystyle O}{\|}}{C}+_n$$
$$\underset{CH_3}{}$$

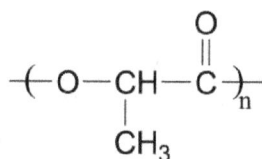

Poly(lactic acid), PLA, or polylactide

PGA is semicrystalline (typically ca. 50% crystallinity) with a T_m of 225–23 °C and a T_g of about 35–40 °C. The small $-CH_2-$ group allows close packing in the crystal and strong dipole interactions among esters, and only rather unusual solvents such as fluorinated alcohols can dissolve the polymer. PLA, like polypropylene, has two possible configurations at the methine carbon. Since lactic acid is made naturally as the L-isomer, the pure (isotactic) polymer derived from it is called poly(L-lactic acid) or PLLA. This material is also semicrystalline (35–40%), but with a lower T_m (ca. 175 °C) than PGA, presumably due to methyl groups acting to partially shield dipole-dipole interactions. The T_g of PLLA is around 60 °C v 40 °C for PGA, consistent with the idea that the methyl group increases rotational barriers along backbone bonds. Copolymers of the D and L isomers of lactic acid are readily made and these are amorphous as expected from their atactic structure. Also, copolymers of lactic and glycolic acids (PLGA) are of great interest because degradation rates can be controlled for a variety of applications in tissue engineering and drug delivery.

Dilution of the polar ester group by methylene units reduces intermolecular forces and thus lowers both T_g and T_m compared with PGA. An example is poly(e-caprolactone), a semicrystalline polymer that has a T_m of 60 °C and a T_g of about -60 °C.

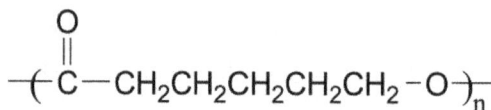

$$+\overset{\overset{\displaystyle O}{\|}}{C}-CH_2CH_2CH_2CH_2CH_2-O+_n$$

Poly(ε-caprolactone)

Cross-linking and Orientation

The molecular origins of structure/property relationships discussed above apply to both small molecules and polymers. However, two additional parameters affect polymers in unique ways. The first is cross-linking, which involves primary bonds (chemical cross-linking) or secondary bonds (physical cross-linking) and affords what are termed polymer networks. Light cross-linking of low T_g polymers gives materials known as elastomers, and these have the typical properties of a rubber band. Crosslinking raises T_g, as chain motions are now restricted by network junction points. Heavy cross-linking can give very hard, brittle materials, as T_g will rise above room temperature and molecular motions will be too restricted to contribute significantly to impact energy absorption (e.g., toughness). A biopolymer example is keratin, the major constituent of hair and fingernails. Keratin in hair is lightly cross-linked via disulfides from coupling of neighboring cystine amino acids, and the ability to introduce a "permanent wave" takes advantage of the reversibility of disulfide formation and cleavage via redox chemistry. Fingernails have a much higher disulfide crosslink density, and hence they are harder. (Tortoise shells are even more highly cross-linked).

Polymer networks can be prepared by covalently linking chains into what is effectively a single molecule using high-energy radiation such as gamma rays, light, or a wide variety of chemical reactions (e.g., disulfide formation). It is also possible to exploit strong intermolecular interactions to afford materials with the properties of a polymer network, yet retaining the ability to reprocess the material upon heating. A classic example is the segmented polyurethanes, which contain a low-T_g block and are cross-linked via thermally labile hydrogen-bonded domains. These have the "snap" of a rubber band but are reprocessable, and are termed thermoplastic elastomers.

New Synthesis Strategies

Polymer chemists continue to explore and optimize polymerization reactions as well as discover new reaction schemes and catalysts that afford a tight molecular weight distribution (MWD), afford more precise control over backbone composition and repeat unit stereochemistry, and allow for the creation of novel chain architectures.

Particularly interesting advances have been made in the area of "living" polymerization, a term for a system where propagating chain ends remain active throughout the duration of the reaction. For example, it has long been thought that free radical polymerization would not be amenable to the characteristics of a living polymerization (e.g., a linear dependence of molecular weight with monomer conversion) due to the inherent high reactivity of radicals. However, it has been discovered over the last decade that it is indeed possible to achieve living polymerization character provided that the number of active radical species is very low at any given time. This can be done by reversibly capping active chain ends using stable nitroxide radicals, halides via redox chemistry with selected transition metals (called atom-transfer radical polymerization, or ATRP), or reversible addition-fragmentation polymerization (RAFT). On another front, it was believed about 15 years ago that polyethylene and polypropylene were mature technologies, although that has changed dramatically with the advent of new transition metal catalysts that, for example, afford fine control of the extent of branching (and hence the degree of crystallinity) of polymers of ethylene and small amounts of α-olefins. The new catalysts are also producing old polymers with a high control of stereospecificity. One recent example is syndiotactic polystyrene.

Finally, polymerization reactions are being developed that yield polymers having interesting chain architectures, such as block and graft structures. Chain architecture is important in that compositionally different regions of a polymer chain can contribute different properties. For example, styrene-butadiene-styrene triblock polymer will self-assemble upon cooling from the melt into various thermodynamically stable morphologies depending on the relative lengths of, and thus volume fractions of, the blocks. At about 20 vol% polystyrene, the material is composed of a low-T_g polybutadiene continuous phase and small (ca. 20-nm diameter) spherical domains of atactic polystyrene (T_g of 100 °C). The domains are interconnected by virtue of the block polymer chain structure, and hence this material behaves as an elastomer with cross-linking via the glassy polystyrene domains. Such a material is a thermoplastic elastomer, because it can flow if heated above the T_g of polystyrene.

Whereas block and graft copolymers have been known for some time, new polymerization methods such as ATRP are expanding the number of examples of these systems.

More complex architectures, such as multiarm stars and dendrimers, are being targeted as polymerization chemistry becomes more versatile. Star polymers (multiarm structures emanating from a central core) are of interest principally for the large numbers of chain ends. For example, a 4-arm star polymer will typically have a lower T_g and a lower viscosity than a linear polymer of a similar total molecular weight. Dendrimers are a special class of star-like polymer that have regularly placed branched repeat units. Construction of such a polymer can begin with a core to which a monomer capable of branching is attached (divergent approach), or conversely by making highly branched arms and then connecting these to a core as a final step (convergent approach). Dendrimers typically have poor physical properties in the solid state due to the absence of chain entanglements. However, their interiors have open volume in which to bind drugs and an exterior with a very high concentration of functional groups due to the high degree of branching. These characteristics make dendrimers of great interest in medical applications, for example, as vehicles for delivery of genes to cells.

Challenges and Opportunities: The Next 20 Years

Polymer science and engineering will continue to be a key discipline for the design and development of new classes of materials with new or improved applications in medicine.

Modeling will be a Routine First Step

The creation of new polymers will begin with a broad evaluation of properties using high-level modeling and simulation to determine critical parameters (isomeric structure, molecular weight, degree of chain orientation) that influence a property of interest. In particular, modeling will be employed to predict complex organization of functional low-molar-mass molecules and polymers, as is beginning to be done for the difficult problem of predicting protein folding motifs. Modeling will also extend to synthesis routes to define the best approach, as well as to processing. Much guiding information will be in hand prior to doing "wet chemistry."

Synthesis and Processing will become Seamless

Synthesis and processing are typically separate operations. This will change over the next two decades as opportunities to carry out simultaneous synthesis and processing emerge. This idea is not new (e.g., reaction injection molding or chemical vapor deposition), but it will become more widespread. Of particular interest will be polymers and small molecules that self-assemble into ordered molecular structures (e.g., liquid crystals) or morphological structures (e.g., block copolymers). Combinatorial synthesis of polymers will become more routine, and the emergence of combinatorial processing approaches to rapidly identify conditions for fabricating polymers to achieve maximum properties is anticipated.

Polymer Synthesis will Boast Sequence Control

Nature has evolved machinery to precisely control the sequence of amino acid additions and hence the primary structure of polypeptides. However, conventional polymerization techniques have limited opportunities for the control the sequence of addition of two or more monomers. Block copolymers are possible due to successive addition of monomer charges to active chain ends, and alternating copolymers can be obtained under special circumstances. However, there currently is no viable means to prepare, for example, vinyl-type copolymers with sequence control (e.g., poly[(monomerA)$_{1-}$ (monomerB)$_2$]$_n$). This is in stark contrast to peptide synthesis on ribosomes within biological cells, which employ a template to code for specific amino acids that are enzymatically linked. A major opportunity and challenge presents itself for the sequence-controlled polymerization of a wide variety of monomers using systems that mimic the functions of ribosomes. Template directed polymerizations are energetically uphill due to the decrease in entropy resulting from monomer ordering, and in living systems the requisite energy is supplied by ATP. Electrochemical polymer synthesis is an attractive option in that properly patterned electrodes may simultaneously serve as solid templates for synthesis and as energy sources. Other lessons from biology, such as self-assembly and development of hierarchical structures, will continue to be borrowed and built upon.

The Definition of Structure in Polymers will Rival that of Low Molar Mass Materials

Polymers are by their very nature complex materials, typically having a distribution of chain lengths, isomer content, degrees of orientation, and fractional crystallinity. Thus, structure in polymeric materials is frequently ill-defined compared with small molecules. This difference will gradually disappear over the next two decades as long-chain molecules are synthesized with greater compositional and structural precision. The implications will be significant in that very precise structure/property relationships will be possible, and maximization of properties through the integration of modeling, synthesis, and processing will be realized.

The Impact on Nanotechnology

Many common examples exist of nanoscale (one dimension of < 100 nm) materials, including block copolymer films and collagen fibers that function as scaffolds for tissues and organs. However, there will be an increasing push to exploit the properties of individual molecules, or very small aggregates of molecules, for the next generation of multifunctional biomaterials. For example, carbon nanotubes and collections of only a few organic molecules are being studied as components of diodes, transistors, and memory elements. The ability to create well-defined organic polymer structures and manipulate these to form complex and functional arrangements will drive a revolution in information storage and processing, sensing, and communications. This effort is predicted to impact biomaterials and biomedical devices in many ways, including the design of surfaces with specific and predictable protein adsorption characteristics, and stable and reliable interfaces between cells and electronic materials, both inorganic and organic.

Hybrid Materials Systems will be Ubiquitous

Opportunities will continue to arise for the use of polymers as components in hybrid materials systems along with metals, ceramics, and/or electronic materials. Familiar composite material (e.g., graphite-reinforced epoxy) will see increased uses, but significant potential is seen for new materials combinations such as organic electroactive materials and silicon for hybrid electronic and optical devices in, for example, biosensors.

Properties of Polymers

A comparison of the properties of polyethylene (both LDPE & HDPE) with the natural polymers rubber and cellulose is instructive. As noted above, synthetic HDPE macromolecules have masses ranging from 105 to 106 amu (LDPE molecules are more than a hundred times smaller). Rubber and cellulose molecules have similar mass ranges, but fewer monomer units because of the monomer's larger size. The physical properties of these three polymeric substances differ from each other, and of course from their monomers.

- HDPE is a rigid translucent solid which softens on heating above 100 °C, and can be fashioned into various forms including films. It is not as easily stretched and deformed as is LDPE. HDPE is insoluble in water and most organic solvents, although some swelling may occur on immersion in the latter. HDPE is an excellent electrical insulator.

- LDPE is a soft translucent solid which deforms badly above 75 °C. Films made from LDPE stretch easily and are commonly used for wrapping. LDPE is insoluble in water, but softens and swells on exposure to hydrocarbon solvents. Both LDPE and HDPE become brittle at very low temperatures (below -80 °C). Ethylene, the common monomer for these polymers, is a low boiling (-104 °C) gas.

- Natural (latex) rubber is an opaque, soft, easily deformable solid that becomes sticky when heated (above 60 °C), and brittle when cooled below -50 °C. It swells to more than double its size in nonpolar organic solvents like toluene, eventually dissolving, but is impermeable to water. The C5H8 monomer isoprene is a volatile liquid (b.p. 34 °C).

- Pure cellulose, in the form of cotton, is a soft flexible fiber, essentially unchanged by variations in temperature ranging from -70 to 80 °C. Cotton absorbs water readily, but is unaffected by immersion in toluene or most other organic solvents. Cellulose fibers may be bent and twisted, but do not stretch much before breaking. The monomer of cellulose is the $C_6H_{12}O_6$ aldohexose D-glucose. Glucose is a water soluble solid melting below 150 °C.

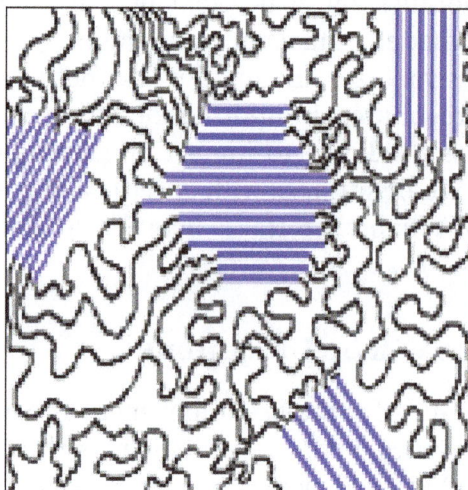

To account for the differences noted here we need to consider the nature of the aggregate macromolecular structure, or morphology, of each substance. Because polymer molecules are so large, they generally pack together in a non-uniform fashion, with ordered or crystalline-like regions mixed together with disordered or amorphous domains. In some cases the entire solid may be amorphous, composed entirely of coiled and tangled macromolecular chains. Crystallinity occurs when linear polymer chains are structurally oriented in a uniform three-dimensional matrix. In the diagram on the right, crystalline domains are colored blue. Increased crystallinity is associated with an increase in rigidity, tensile strength and opacity (due to light scattering). Amorphous polymers are usually less rigid, weaker and more easily deformed. They are often transparent.

Three factors that influence the degree of crystallinity are:

- Chain length.

- Chain branching.

- Interchain bonding.

The importance of the first two factors is nicely illustrated by the differences between LDPE and HDPE. As noted earlier, HDPE is composed of very long unbranched hydrocarbon chains. These pack together easily in crystalline domains that alternate with amorphous segments, and the resulting material, while relatively strong and stiff, retains a degree of flexibility. In contrast, LDPE is composed of smaller and more highly branched chains which do not easily adopt crystalline structures. This material is therefore softer, weaker, less dense and more easily deformed than HDPE. As a rule, mechanical properties such as ductility, tensile strength, and hardness rise and eventually level off with increasing chain length.

The nature of cellulose supports the above analysis and demonstrates the importance of the third factor (iii). To begin with, cellulose chains easily adopt a stable rod-like conformation. These molecules align themselves side by side into fibers that are stabilized by inter-chain hydrogen bonding between the three hydroxyl groups on each monomer unit. Consequently, crystallinity is high and the cellulose molecules do not move or slip relative to each other. The high concentration of hydroxyl groups also accounts for the facile absorption of water that is characteristic of cotton.

Natural rubber is a completely amorphous polymer. Unfortunately, the potentially useful properties of raw latex rubber are limited by temperature dependence; however, these properties can be modified by chemical change. The cis-double bonds in the hydrocarbon chain provide planar segments that stiffen, but do not straighten the chain. If these rigid segments are completely removed by hydrogenation (H2 & Pt catalyst), the chains lose all constrainment, and the product is a low melting paraffin-like semisolid of little value. If instead, the chains of rubber molecules are slightly cross-linked by sulfur atoms, a process called vulcanization which was discovered by Charles Goodyear in 1839, the desirable elastomeric properties of rubber are substantially improved. At 2 to 3% crosslinking a useful soft rubber, that no longer suffers stickiness and brittleness problems on heating and cooling, is obtained. At 25 to 35% crosslinking a rigid hard rubber product is formed. The following illustration shows a cross-linked section of amorphous rubber. The more highly-ordered chains in the stretched conformation are entropically unstable and return to their original coiled state when allowed to relax.

Unstretched Rubber

Stretched Rubber

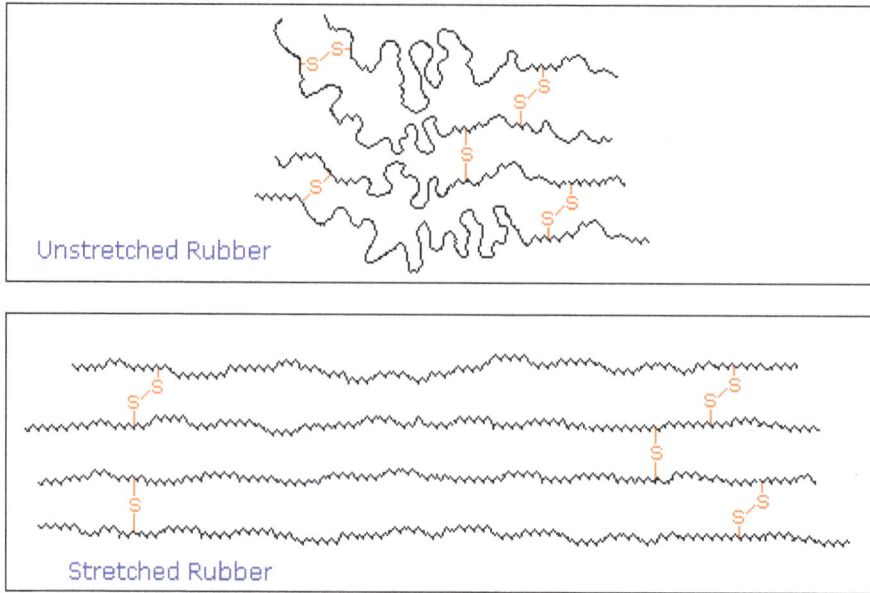

Classification of Polymers

Since Polymers are numerous in number with different behaviours and can be naturally found or synthetically created, they can be classified in various ways. The following below are some basic ways in which we classify polymers.

Classification based on Source

The first classification of polymers is based on their source of origin.

Natural Polymers

The easiest way to classify polymers is their source of origin. Natural polymers are polymers which occur in nature and are existing in natural sources like plants and animals. Some common examples are Proteins (which are found in humans and animals alike), Cellulose and Starch (which are found in plants) or Rubber (which we harvest from the latex of a tropical plant).

Synthetic Polymers

Synthetic polymers are polymers which humans can artificially create/synthesize in a lab. These are commercially produced by industries for human necessities. Some commonly produced polymers which we use day to day are Polyethylene (a mass-produced plastic which we use in packaging) or Nylon Fibers (commonly used in our clothes, fishing nets etc).

Semi-synthetic Polymers

Semi-synthetic polymers are polymers obtained by making modification in natural polymers artificially in a lab. These polymers formed by chemical reaction (in a controlled environment) and are of commercial importance. Example: Vulcanized Rubber (Sulphur is used in cross bonding the polymer chains found in natural rubber) Cellulose acetate (rayon) etc.

Classification based on Structure of Polymers

Classification of polymers based on their structure can be of three types:

- Linear polymers: These polymers are similar in structure to a long straight chain which identical links connected to each other. The monomers in these are linked together to form a long chain. These polymers have high melting points and are of higher density. A common example of this is PVC (Poly-vinyl chloride). This polymer is largely used for making electric cables and pipes.

- Branch chain polymers: As the title describes, the structure of these polymers is like branches originating at random points from a single linear chain. Monomers join together to form a long straight chain with some branched chains of different lengths. As a result of these branches, the polymers are not closely packed together. They are of low density having low melting points. Low-density polyethene (LDPE) used in plastic bags and general purpose containers is a common example.

- Crosslinked or Network polymers: In this type of polymers, monomers are linked together to form a three-dimensional network. The monomers contain strong covalent bonds as they are composed of bi-functional and tri-functional in nature. These polymers are brittle and hard. Ex:- Bakelite (used in electrical insulators), Melamine etc.

Based on Mode of Polymerization

Polymerization is the process by which monomer molecules are reacted together in a chemical reaction to form a polymer chain (or three-dimensional networks). Based on the type of polymerization, polymers can be classified as.

Addition Polymers

These types of polymers are formed by the repeated addition of monomer molecules. The polymer is formed by polymerization of monomers with double or triple bonds (unsaturated compounds). Note, in this process, there is no elimination of small molecules like water or alcohol etc (no by-product of the process). Addition polymers always have their empirical formulas same as their monomers. Example: ethene $n(CH_2=CH_2)$ to polyethene $-(CH_2-CH_2)n-$.

Condensation Polymers

These polymers are formed by the combination of monomers, with the elimination of small molecules like water, alcohol etc. The monomers in these types of condensation reactions are bi-functional or tri-functional in nature. A common example is the polymerization of Hexamethylenediamine and adipic acid. to give Nylon – 66, where molecules of water are eliminated in the process.

Classification Based on Molecular Forces

Intramolecular forces are the forces that hold atoms together within a molecule. In Polymers, strong covalent bonds join atoms to each other in individual polymer molecules. Intermolecular forces (between the molecules) attract polymer molecules towards each other.

The properties exhibited by solid materials like polymers depend largely on the strength of the forces between these molecules. Using this, Polymers can be classified into 4 types.

Elastomers

Elastomers are rubber-like solid polymers that are elastic in nature. When we say elastic, we basically mean that the polymer can be easily stretched by applying a little force.

The most common example of this can be seen in rubber bands (or hair bands). Applying a little stress elongates the band. The polymer chains are held by the weakest intermolecular forces, hence allowing the polymer to be stretched. But as you notice removing that stress also results in the rubber band taking up its original form. This happens as we introduce crosslinks between the polymer chains which help it in retracting to its original position, and taking its original form. Our car tyres are made of Vulcanized rubber. This is when we introduce sulphur to cross bond the polymer chains.

Thermoplastics

Thermoplastic polymers are long-chain polymers in which inter-molecules forces (Van der Waal's forces) hold the polymer chains together. These polymers when heated are softened (thick fluid like) and hardened when they are allowed to cool down, forming a hard mass. They do not contain any cross bond and can easily be shaped by heating and using moulds. A common example is Polystyrene or PVC (which is used in making pipes).

Thermosetting

Thermosetting plastics are polymers which are semi-fluid in nature with low molecular masses. When heated, they start cross-linking between polymer chains, hence becoming hard and infusible. They form a three-dimensional structure on the application of heat. This reaction is irreversible in nature. The most common example of a thermosetting polymer is that of Bakelite, which is used in making electrical insulation.

Fibres

In the classification of polymers, these are a class of polymers which are a thread like in nature, and can easily be woven. They have strong inter-molecules forces between the chains giving them less elasticity and high tensile strength. The intermolecular forces may be hydrogen bonds or dipole-dipole interaction. Fibres have sharp and high melting points. A common example is that of Nylon-66, which is used in carpets and apparels.

The above was the general ways to classify polymers. Another category of polymers is that of Biopolymers. Biopolymers are polymers which are obtained from living organisms. They are biodegradable and have a very well defined structure. Various biomolecules like carbohydrates and proteins are a part of the category.

Polymerizations

A process called polymerization is used to join single molecules (monomers) together to form a chain of many, the above example shows Vinyl Chloride bonded into Poly Vinyl Chloride. The left portion of the image shows the vinyl chloride while on the right hand side these single molecules of vinyl chloride have bonded together to form poly vinyl chloride, "poly" meaning many. Polymers are often formed from crude oil however more recently polymers have been created from corn starch and vegetable fats to form Bioplastics.

Polymerization is the chemical process of monomers joining together to form polymers, often it takes many thousands of monomers to make a single polymer. Two types of polymerization reactions are listed below:

- Addition polymerization where monomers addon to each other with the addition of a catalyst, these are usually alkenes such as ethene and propene. Alkenes can act as monomers because they have a double bond.

- Condensation polymerization: This is when monomers join or polymerise with a byproduct such as water, carbon dioxide or ammonia. This usually requires two different types of monomers that join alternately.

Polymers can have different structures and these can affect their properties, some examples are shown below.

Linear Chain

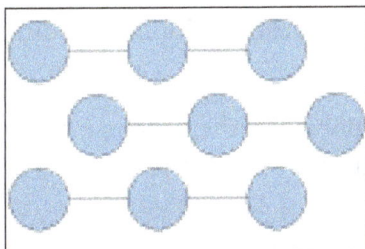

The chains when heated can flow easily this is a good example of the chain of a thermoplastic. Properties often associated with this type of chain are a weak material, ductile with a low density and melting point.

Branched Chain

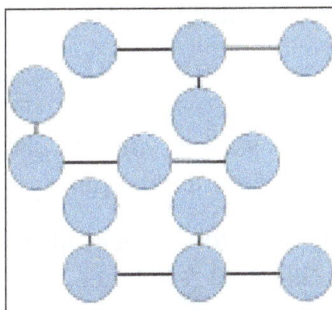

Branched chains will not flow as easily when heated they will have a higher melting point than linear chains and typically have the properties of a slightly less ductile material (stronger and stiffer).

Cross Linked

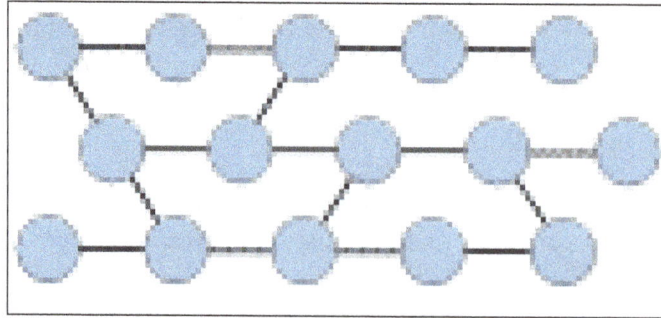

The Cross linked bonding retains its shape when heated and is a good example of the bonding found in thermo set plastics. These also have a higher melting point than linear and branched bonds, they are usually a harder material but also more brittle.

The *two main types of polymers* are thermo sets and thermo plastics. Thermo plastics can be reshaped after heating whereas thermo sets cannot. Thermo plastics have the benefit of being recyclable and generally cheaper and easier to process than thermo set plastics.

Applications where polymers may be used:

Acrylic Acrylic is a polymer called poly(methyl methacrylate) and is often used as an alternative to glass, examples of these include small fish tanks/aquariums visors in goggles and crash helmets covers for skylights and as it has good weatherproofing qualities it is commonly used for outdoor signage.

High Density Polythene (HDPE) Babies baths, kitchen equipment, children's toys, fabric filaments. Polythene is cheap and acid resisting. A strong polymer which softens at aroiund 120 degrees celcius.

Low Density Polythene (LDPE) Used in plastic bags, film and packaging. LDPE is a more flexible polymer than HDPE. This softens at around 85 degrees celcius.

Poly Vinyl Chloride

- Industrial and domestic piping, artificial leather (e.g. shoes), protective clothing.
- PVC is reasonably hard and hardwearing it is also fairly cheap.

Poly Ether Ketone

- Used in aggressive environments such as nuclear plants, oil and geo thermal wells, high pressure steam valves, aircraft and car engine parts.
- Peek has a high melting point, very good tensile strength, has a low coefficient of friction and is light weight with good chemical resistance.

Types of Polymerization

Polymers are huge chains or sometimes even 3D structures of repeating units known as monomers. The monomer is the basic unit of a polymer. These monomers can bond to each other on each side, potentially forever.

So this reaction of combining these monomers to form long chains or three-dimensional networks is known as polymerization. Broadly polymerization can be classified into two categories:

- Step-Growth or Condensation Polymerization.

- Chain-Growth or Addition Polymerization.

Addition Polymerization

As the name suggests addition polymers form when an addition reaction occurs. The repeating monomers form a linear or branch structure depending on the type of monomer. During addition polymerization, the monomers rearrange themselves to form a new structure. But there is no loss of an atom or a molecule. Again there are four types of addition polymerizations which are:

- Free Radical Polymerization: Here the addition polymer forms by addition of atoms with a free electron in its valence shells. These are known as free radicals. They join in a successive chain during free radical polymerization.

- Cationic polymerization: A polymerization where a cation is formed causing a chain reaction. It results in forming a long chain of repeating monomers.

- Anionic Vinyl Polymerization: Involves the polymerization of particularly vinyl polymers with a strong electronegative group to form a chain reaction.

- Coordination Polymerization: This method was invented by two scientists Ziegler and Natta who won a Nobel Prize for their work. They developed a catalyst which let us control the free radical polymerization. It produces a polymer which has more density and strength.

Condensation Polymerization

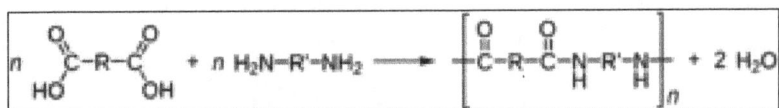

Condensation polymers form from the step growth polymerization. Here when molecules of monomers react to form a bond they replace certain molecules. These molecules are the by-product of the reaction. In most cases, this by-product is a water molecule.

The type of polymers that result from a condensation polymerization depends on the monomers. If the monomer has only one reactive group, the polymers that form have low molecular weight. When monomers have two reactive end groups we get linear polymers. And monomers with higher than two reactive groups results in a polymer with a three-dimensional network.

Polyester and nylon are two common condensation polymers. Even Proteins and Carbohydrates are a result of condensation polymerization.

Addition Polymerization vs. Condensation Polymerization

Let us do a comparative analysis of the two types of polymerization to understand them better:

- In addition polymerization monomers only join at the active site of the chain. But in condensation polymerization, any molecule can react with any other.

- In addition polymerization, there are three distinct steps. Initiation, propagation and finally termination. In condensation polymerization, there is no termination step. The end groups remain reactive through the entire process.

- Addition polymerization results in homo-chain polymers whereas condensation polymerization results in hetro-chain polymers.

- The most significant difference is that in addition polymers there is no loss of atom. But in condensation reaction, there is a loss of a molecule of water, ammonia etc as a by-product.

Polymerization Mechanisms

Addition Polymerization

During the process of polymeric chain growth if there is only addition of monomers without the loss of any small molecule and the polymer is an exact multiple of used monomer unit is called addition polymerization. Example of addition polymers are PE PVC, PP etc.

Mechanism of Addition Polymerization

Addition polymerization proceeds through two types of mechanism:

- Free radical mechanism.

- Ionic mechanism. A) Ionic mechanism are two type:
  - Cationic mechanism.
  - Anionic mechanism.

Every polymerization reaction proceeds through three steps:

Chain initiation: This is the first step of polymerization reaction where initiator activate the monomer unit resulting in the formation of active center in monomer unit. As,

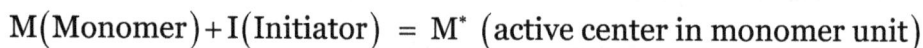

$$M(Monomer) + I(Initiator) = M^* \ (active \ center \ in \ monomer \ unit)$$

I (Free radical, heat, pressure, catalyst etc).

Chain propagation: This is the second step of polymerization reaction where second third fourth monomer units are added with activated monomer unit consecutively resulting in the formation of large chain unit.

As,

$$
\begin{aligned}
M^*(\text{active center in monomer}) + M(\text{Monomer}) &= M - M^* \\
M - M^* + M(\text{Monomer}) &= M - M - M^* \\
M - M - M^* + M(\text{Monomer}) = M - M - M - M^* & \\
-------------------- \quad &----- \\
-------------------- \quad &----- \\
&= (M)_n - M^*
\end{aligned}
$$

Chain termination: This is the third and final step of polymerization reaction where termination of active center is removed by the following possible way:

- Recombination of two smaller activated monomers.

- Recombination of two larger activated polymeric chain.

- Recombination of one smaller activated monomer with a larger activated monomer.

- Disproportion(Simultaneous oxidation and reduction of two activated monomers).

Free Radical Mechanism

It is the way of polymerization where the initiator of the reaction is a free radical. This initiator free radical is produced by many ways but one of the important ways is decomposition of the compound by sun light, heat, catalyst etc. As example peroxide, hydroperoxide, peracid is very useful initiator in organic polymerization reaction.

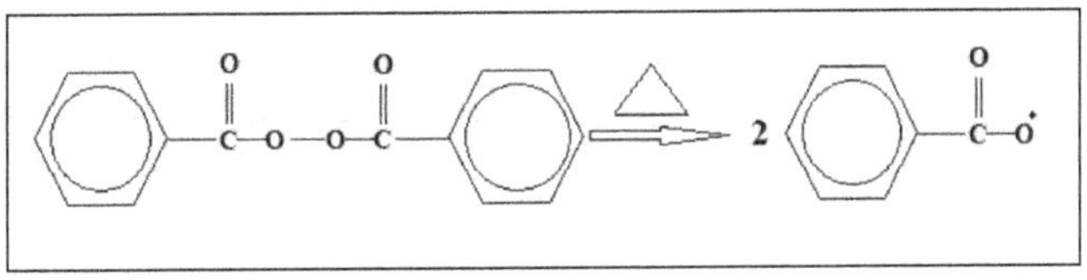

Benzoyl peroxide.

Decomposition of benzoyl peroxide in presence of sun light or heat form free radical as initiator-Now, the following three common steps of free radical polymerization reaction are:

Chain initiation: This is the first step of polymerization reaction where initiator is added to the monomer unit resulting in the formation of free radical center in monomer unit.

As,

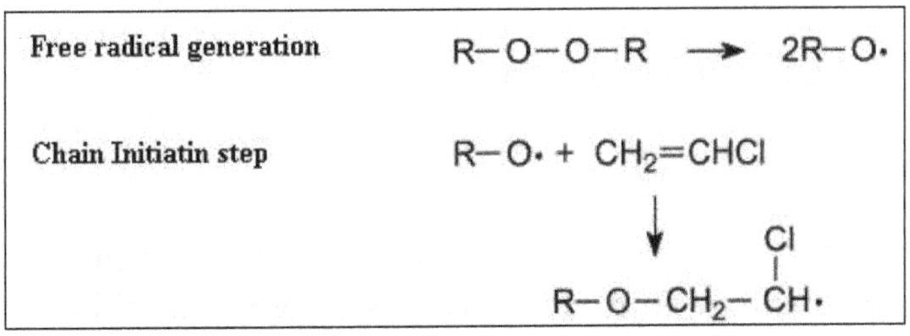

Free radical chain initiation of vinyl chloride.

Chain propagation: This is the second rapid step of polymerization reaction where second third forth monomer units are added with generated free radical monomer unit consecutively resulting in the formation of large chain radical.

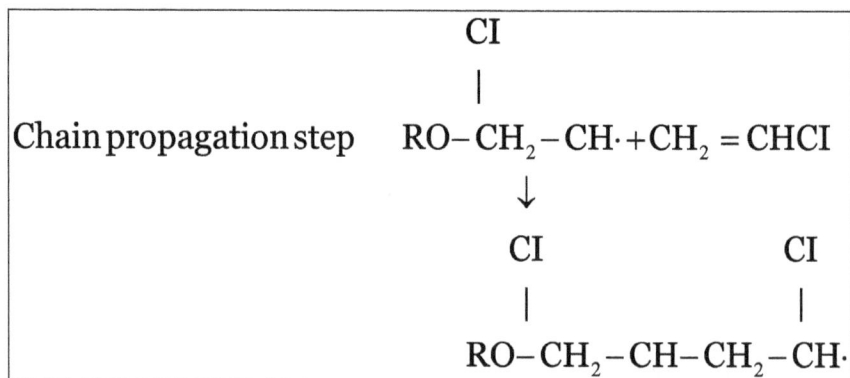

$$
\begin{array}{l}
\text{Chain propagation step}\quad \underset{\displaystyle \overset{\displaystyle |}{\text{Cl}}}{\text{RO}-\text{CH}_2-\text{CH}\cdot}+\text{CH}_2=\text{CHCl}\\[2em]
\qquad\qquad\qquad\qquad\qquad\downarrow\\[1em]
\qquad\qquad\qquad \underset{\text{Cl}}{\text{RO}-\text{CH}_2-\text{CH}}-\underset{\text{Cl}}{\text{CH}_2-\text{CH}\cdot}
\end{array}
$$

Free radical chain prpagation of vinyl chloride.

Chain termination: This is the third and final step of polymerization reaction where termination of free radical after long propagation is takes place by the following possible way:

- Recombination of two smaller free radicals:

- Recombination of two larger polymeric free radical chain.

- Recombination of one smaller free radical monomer with a larger free radical.

- Disproportion: (Simultaneous oxidation and reduction of activated long chain free radical) In this process any one free radical centered carbon atom abstracted hydrogen free radical (reduction) from other (oxidation) free radical and thereby chain termination takes place.

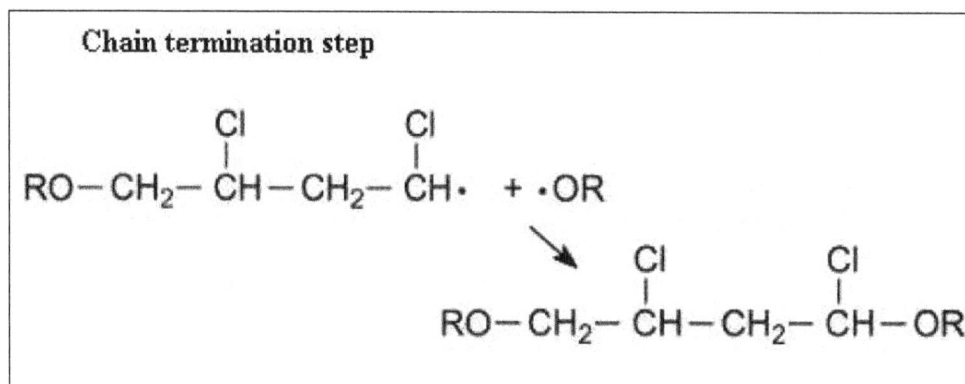

Chain termination step

$$
\begin{array}{l}
\text{RO}-\text{CH}_2-\underset{\overset{|}{\text{Cl}}}{\text{CH}}-\text{CH}_2-\underset{\overset{|}{\text{Cl}}}{\text{CH}}\cdot \quad +\cdot\text{OR}\\[2em]
\qquad\qquad\qquad\searrow\\
\qquad\qquad \text{RO}-\text{CH}_2-\underset{\overset{|}{\text{Cl}}}{\text{CH}}-\text{CH}_2-\underset{\overset{|}{\text{Cl}}}{\text{CH}}-\text{OR}
\end{array}
$$

Free radical chain termination of vinyl chloride.

Other important way of chain termination are:

Inhibitor Reaction

The process of termination is done by inhibitor, a compound which inhibit the growth of polymerization reaction by termination. This termination is done when inhibitor combine with generated free radical and give comparatively stable product. Oxygen, Nitrobenzene, Dinitrobenzene and phenolic materials are act as inhibitor in many organic polymerization reaction. Example of polymers formed by free radical mechanism are polyethylene, polypropylene, styrene, styrene butadiene etc.

Ionic polymerization: It is the way of polymerization where the initiator of the reaction is an ion. This initiator ion is formed from specific Lewis acid or base produce cation(carbonium ion) or anion(carbanion)on the monomer unit. Depending up on this cation and or anion formation ionic polymerization is devided into two parts.

Cationic Mechanism

It is the way of polymerization where the initiator of the reaction is formed by the combination of Lewis acid($AlCl_3$, $FeCl_3$, BF_3) used as catalyst with a co catalyst like water.

1. Chain initiation: This is the first step of cationic polymerization reaction where initiator is added to the monomer unit resulting in the formation of cation center(carbinium ion) in monomer unit.

As,

Free radical generation	$AlCl_3 + H_2O \longrightarrow \overline{AlCl_3}(OH) + H^+$
	$\underbrace{\qquad}_{A^-}$
Chain Initiatin step	$A^-\ H^+\quad +\ CH_2{=}CHCl$

$$CH_3-\overset{\overset{\displaystyle Cl}{\vert}}{C}H^+A^-$$

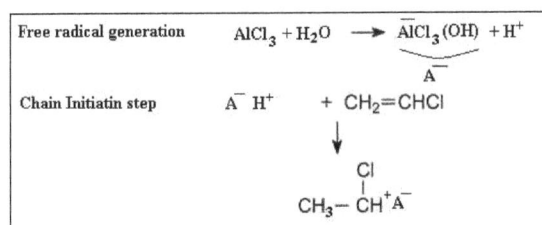

Cationic chain initiation of vinyl chloride.

2. Chain propagation: This is the second rapid step of cationic polymerization reaction where second third forth monomer units are added with generated cation center(carbinium ion) in monomer unit consecutively resulting in the formation of large chain cation.

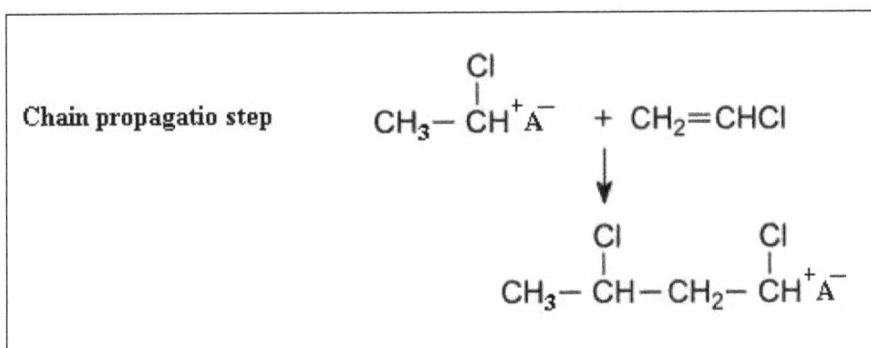

Chain propagatio step	$CH_3-\overset{\overset{\displaystyle Cl}{\vert}}{C}H^+A^- \quad + \quad CH_2{=}CHCl$

$$CH_3-\overset{\overset{\displaystyle Cl}{\vert}}{C}H-CH_2-\overset{\overset{\displaystyle Cl}{\vert}}{C}H^+A^-$$

Cationic chain propagation of vinyl chloride.

3. Chain termination: This is the third and final step of polymerization reaction where termination of cation after long propagation is takes place by, a) Proton release: It is the way of long chain termination of cationic polymerization by release of proton from carbonium ion chain to the counter ion attached consequently form unsaturated compound. b) combination with any part of initiator: It is the way of long chain termination of cationic polymerization by the formation of covalent bond by the combination of carbonium ion with negative part of initiator.

Chain termination step

$$CH_3-\underset{\underset{Cl}{|}}{CH}-CH_2-\underset{\underset{Cl}{|}}{CH}^+A^- + \overline{Al}Cl_3(OH) + H^+ \downarrow$$

$$CH_3-\underset{\underset{Cl}{|}}{CH}-CH_2-\underset{\underset{Cl}{|}}{CH}-OH$$

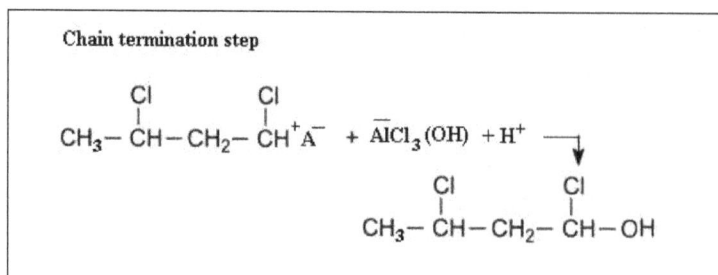

Cationic chain termination of vinyl chloride.

Example of polymers formed by cationic mechanism are Vinyl ethers, Styrene Isobutylene etc.

Anionic Mechanism

It is the way of polymerization where the initiator of the reaction is formed by the combination of any Lewis base (n-BuLi, Alkyl-Na) alkali meal amide etc.) used as catalyst.

Chain initiation: This is the first step of anionic polymerization reaction where initiator is added to the monomer unit resulting in the formation of anion center(carbanion) in monomer unit.

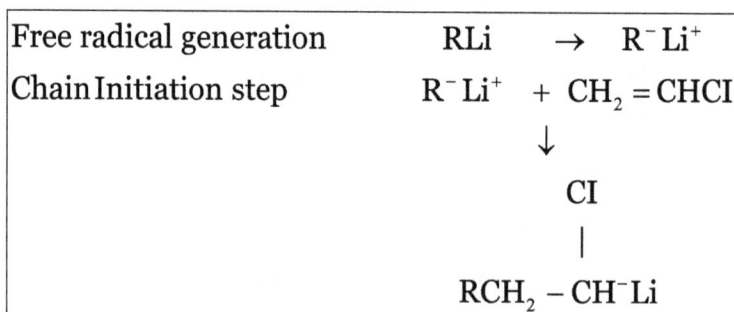

Free radical generation	$RLi \rightarrow R^-Li^+$
Chain Initiation step	$R^-Li^+ + CH_2=CHCl$
	$\downarrow$
	Cl
	$|$
	RCH_2-CH^-Li

Anionic chain initiation of vinyl chloride.

Chain propagation: This is the second rapid step of anionic polymerization reaction where second third forth monomer units are added with generated anion center(carbanion) in monomer unit consecutively resulting in the formation of large chain anion.

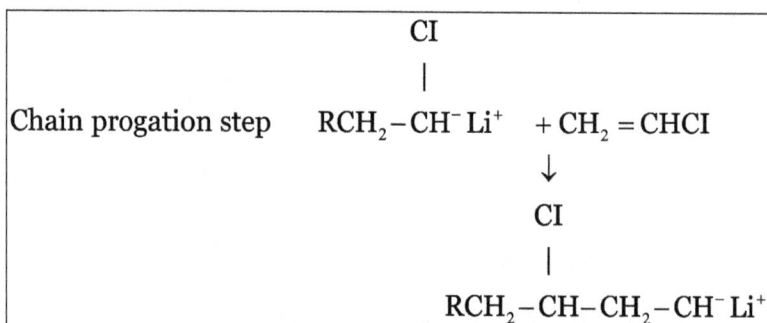

	Cl
	$|$
Chain progation step	$RCH_2-CH^-Li^+ + CH_2=CHCl$
	$\downarrow$
	Cl
	$|$
	$RCH_2-CH-CH_2-CH^-Li^+$

Anionic chain propagation of vinyl chloride.

Chain termination: This is the third and final step of polymerization reaction where termination of anion after long propagation is takes place by the combination with any foreign compound(- such as CH_3OH, H_2O, CO_2) added from outside. Without the addition of foreign compound the termination of anionic polymerization is not possible.

Chain termination step

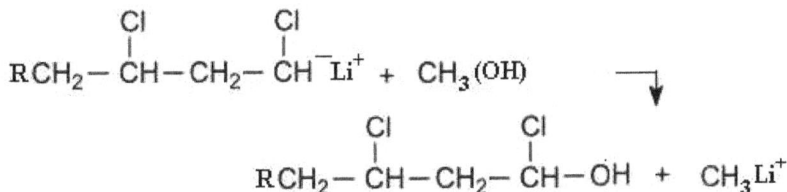

$$RCH_2-\overset{\overset{\displaystyle Cl}{|}}{CH}-CH_2-\overset{\overset{\displaystyle Cl}{|}}{CH}^-Li^+ + CH_3(OH) \longrightarrow$$

$$RCH_2-\overset{\overset{\displaystyle Cl}{|}}{CH}-CH_2-\overset{\overset{\displaystyle Cl}{|}}{CH}-OH + CH_3Li^+$$

Anionic chain termination of vinyl chloride.

Electron withdrawing group like -CN, -COOH, -CHO stabilized the carbanion and hence boost up the anionic polymerization reaction. Example of polymers formed by anionic mechanism are Styrene, Isoprene, butadiene etc.

Condensation Polymerization

During the process of polymeric chain growth if there is reaction takes place between different functional group of single monomer or different monomers with the loss or elimination of compound like HCl , H_2o or other compound and the polymer is not an exact multiple of used monomer unit is called condensation polymer polymerization. Example of condensation polymer are PF UF, MF, Nylon PET etc.

Coordination Polymerization

Organometallic compound has a great role in many chemical reaction as well as polymerization reaction also. When Organometallic compound act as initiator in Polymerization reaction is called coordination polymerization reaction. Zieglar and Natta first prepared polymer by coordination polymerization halide of Ti V, Zr, Cr, Wand Mo.The coordination polymerization can be done by the combination of halide of different transition metal like $TiCl_2$, $TiCl_3$, $TiCl_4$ etc. and organometallic compound such as trialkyl(Trimethyl, Triethyl) aluminium. The most common and useful catalyst is the combination of triethyl aluminium with titanium trichloride. These catalysts are heterogeneous catalyst. The monomer catalyst complex involves coordination between a carbon atom of a monomer and a metal atom of the catalyst. In this polymerization the triethyl aluminium act as electron donor and and metal halide acts as an electron acceptor. The different catalysts for coordination polymerization reaction. Mechanism of coordination polymerization:

Chain initiation: This is the first step of coordination polymerization reaction where complex catalyst is added to the monomer unit resulting in the formation of metal-carbon bond in monomer unit.

Chain propagation: This is the second rapid step of coordination polymerization reaction where second third forth monomer units are added with generated metal-carbon bond in monomer unit, consecutively resulting in the formation of large chain metal-carbon bond.

Chain termination: This is the third and final step of coordination polymerization reaction where termination of large chain metal-carbon is terminated by the transfer of hydrogen atom from monomer or from active hydrogen containing compound.

Importance of Coordination Polymerization:

Normal free radical polymerization involves atactic polymer whereas coordination polymerization isotactic polymer. That means stereospecific polymer can be prepared by this coordination polymerization. Using this process(Zeiglar-Natta catalyst) of polymerization isoprene produces only stereospecific Cis-1,4-polyisoprene

Coordination polymerization always involves mainly linear polymer hence highly crystalline in nature.

Addition Polymers

An addition polymer is a polymer formed by chain addition reactions between monomers that contain a double bond. Molecules of ethene can polymerize with each other under the right conditions to form the polymer called polyethylene.

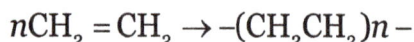

$$n\text{CH}_2 = \text{CH}_2 \rightarrow -(\text{CH}_2\text{CH}_2)n-$$

The letter n stands for the number of monomers that are joined in repeated fashion to make the polymer and can have a value in the hundreds or even thousands.

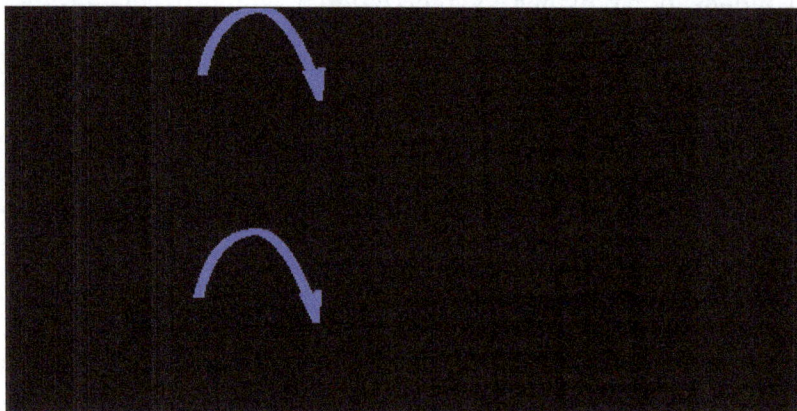

Polyethylene synthesis.

The reactions above show the basic steps to form an addition polymer:

- Initiation - A free radical initiator (X^*) attacks the carbon-carbon double bond (first step above). The initiator can be something like hydrogen peroxide. This material can easily split to form two species with a free electron attached to each: $\text{H}-\text{O}-\text{O}-\text{H} \rightarrow 2\text{H}-\text{O}$. This free radical attacks a carbon-carbon double bond. One of the pi electrons forms a single bond with the initiator while the other pi electron forms a new free radical on the carbon atom.

- Propagation - The new free radical compound interacts with another alkane, continuing the process of chain growth.

- Termination occurs whenever two free radicals come in contact with one another (not shown). The two free electrons form a covalent bond and the free radical on each molecule no longer exists.

Polyethylene can have different properties depending on the length of the polymer chains and on how efficiently they pack together. Some common products made from different forms of polyethylene include plastic bottles, plastic bags, and harder plastic objects such as milk crates.

Several other kinds of unsaturated monomers can be polymerized and are components in common household products. Polypropylene is stiffer than polyethylene and is in plastic utensils and some other types of containers.

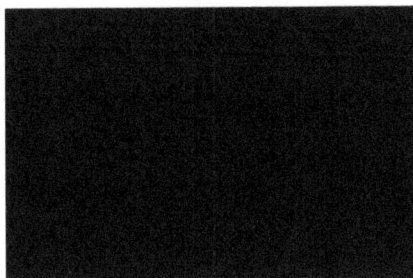

Polypropylene structure.

Polystyrene is used in insulation and in molded items such as coffee cups.

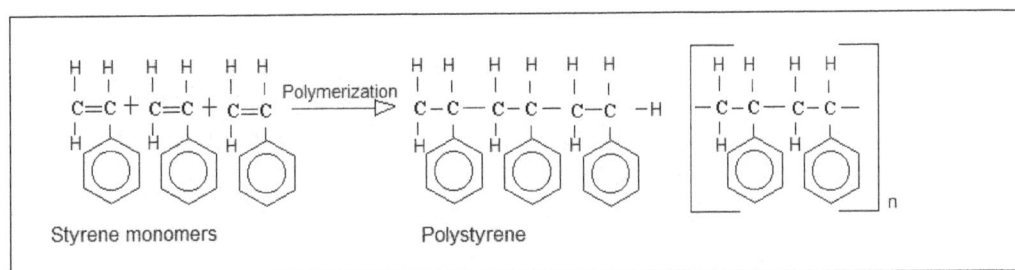

Polystyrene synthesis and structure.

Polyvinyl chloride (PVC) is extensively used for plumbing pipes.

Polyvinyl chloride.

Polyisoprene is a polymer of isoprene and is better known as rubber. It is produced naturally by rubber trees, but several variants have been developed which demonstrate improvements on the properties of natural rubber.

Polyisoprene.

Heterogeneous Catalysis

When air flows past a body moving at high Mach numbers, a bow shock forms in front of the body. If the flight Mach number is high enough, air will dissociate and form species as O, N, NO; if the Mach number corresponds to energies of order about 1 eV (about 11,000 K), ionization also takes place, and O^+, NO^+, N^+ and e^- will appear.

Past the bow shock the air flow entering the shock layer will eventually reach the body surface, which (for structural reasons) must be kept at surface (or: wall) temperatures T_w of order 300 K - 2000 K at most. At these temperatures the greatest part of the species above will tend to recombine. Recombination will occur throughout the shock layer, since the gas temperature drops going from the shock toward the wall, and a (usually) substantial percentage of atoms that have escaped gas-phase recombination will recombine on the surface of the body.

Solid surfaces may be thought as a medium where bonds present inside the bulk of the material have been severed by the fabrication process. Besides, the complex solid structure has many more degrees of freedom than molecules in the gas. It is intuitive that gas species will naturally tend to form bonds with the surface that may facilitate recombination. When this occurs, the surface is said to be catalytic.

If surface recombination occurs, the formation energy of the recombining species will initially stay within the new species formed. For instance, when two O atoms recombine, initially the total formation energy of the two O atoms will tend to stay within the O_2 formed. But then the O_2 just formed would be translationally and roto-vibrationally excited so much that it would again dissociate, unless part of its energy is deposited on the surface.

Since formation energies of O, N or ionized species are large, even if the percentage deposited is small, the surface may be considerably heated by catalytic recombination. This heat flux is additional to molecular conduction and that due to sensible enthalpy transported by diffusing individual species (convective fluxes at the surface are zero, if the surface velocity is assumed zero, as is the case for non-permeable or non-porous walls). Convincing estimates of the heat flux contribution due to catalysis yield as much as 40% of the total heat flux inside the stagnation region.

Since the inception of the manned space program, there has been therefore much interest in preventing recombination over the Thermal Protection System (TPS). In essence, one would like to coat TPS with materials as little catalytic as possible. Old work on recombination (dating back to the '50s and '60s, and performed at room temperature) indicated that glass is among the best material, a fact well known to combustion researchers investigating explosion limits. Follow-on research and testing have shown that even glasses tends to increase their "catalyticity" with temperature, and by a large factor. Much work has been performed to predict this increase, since the actual composition of dissociated air during a re-entry trajectory is difficult to measure, and some sort of modeling is needed to preliminarily predict heat fluxes.

To bypass this problem, back in the early '70s, an extreme assumption became popular, i.e., the surface was assumed to be "fully catalytic", meaning that all species would recombine and form O_2 and N_2 only. In this case the heat flux contribution due to catalysis is maximum, which at first sight sounds like a conservative way of designing TPS. However, this assumption results also in overestimating Boundary Layer temperatures, therefore underestimating densities, and eventually wrong C_p predictions. Ever since, there has been an acute awareness of the need to understand finite-rate recombination.

What follows wants to be a primer of what is known to-date in this area; it is hoped that the material covered will be used as a starting point toward better understanding and modeling of surface catalysis in hypersonics.

Heterogeneous Catalysis

The word catalysis was first used by Berzelius in 1835: "Catalysts are substances that by their mere presence evoke chemical reactions that would not otherwise take place". Wilhelm Ostwald defined a catalyst as: "a substance that changes the velocity of a chemical reaction without itself appearing in the end products". In accord with that, the Chinese word used for a catalyst is "tsoo mei" which also mean "marriage broker".

This brief historical introduction brings to the real meaning of catalyst. A catalyst participating to a reaction does not appear in the stoichiometric equation of the reaction. His effect, for an assigned reaction, is to increase the reaction velocities for an assigned temperature or to decrease the temperature at which the reaction achieves a given rate. For simple reactions, without thermodynamically unstable intermediate product, this means that the catalyst action is to increase both the reaction velocities (forward k_f and backward k_b) in such a manner that the ratio k_f/k_b does not change. For this reason equilibrium obtained with a catalyst is the same as that ultimately arrived at when no catalyst is present. It is important to notice that the catalyst cannot initiate a reaction which is thermodynamically impossible. Its action is always subject to the laws of thermodynamics. The "speed-up" obtained using a catalyst is due to a reduction of the activation energy E necessary for a specific reaction. For example, we consider a solid catalyst: according to "transition state theory" a higher reaction velocity means a higher velocity constant:

$$k_v = \frac{kT}{h}e^{-\Delta G^{\ddagger}=RT},$$

where k is the Boltzmann constant, h is the Plank constant, T is the temperature, R is the gas constant and $\Delta G^{\ddagger}$ is the free energy for the activated state. For a fixed temperature, an higher k_v means a lower free energy $\Delta G^{\ddagger}$ for the catalysed reaction:

$$\Delta G^{\ddagger} = \Delta H^{\ddagger} - T\Delta S^{\ddagger},$$

and for the catalysed reaction $\Delta S^{\ddagger}$ is lower than the same quantity for the non catalysed reaction because the particle bonded on the surface of the catalyst loses some degree of freedom. This means that, to have a lower $\Delta G^{\ddagger}$ with respect to the non-catalysed reaction, the $\Delta H^{\ddagger}$ must be lower. From transition state theory we know that $\Delta H^{\ddagger} = E_a$ provided there is no change in the number of molecules in the reaction, which explains the activation energy reduction.

It is possible to specify different kinds of catalysis depending on the chemical phase of catalyst and reactants:

- Homogeneous Catalysis: The catalyst is in the same phase as the reactants and no phase boundary exist.

- Heterogeneous Catalysis: There is a phase boundary separating the catalyst from the reactants.

- Enzymatic Catalysis: Neither a homogeneous nor heterogeneous system.

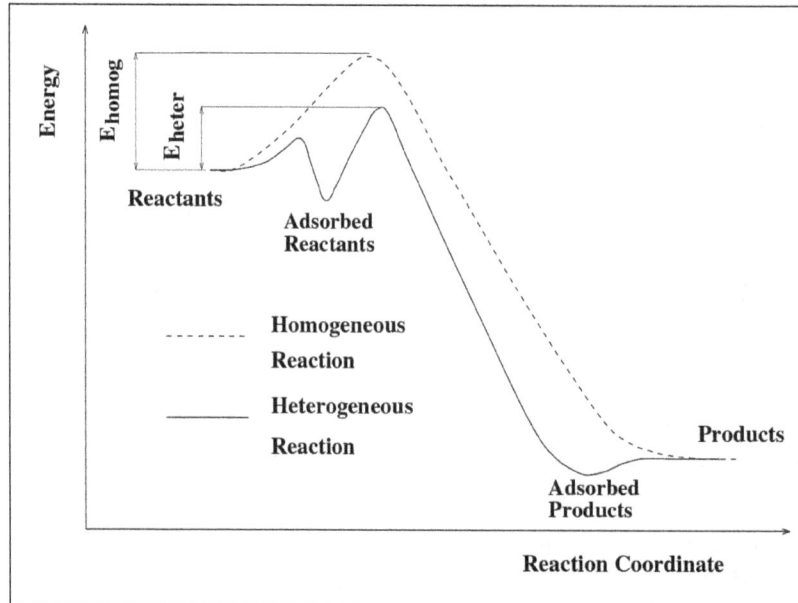

Energy reaction path for heterogeneous and homogeneous reaction.

A gas flow past a solid wall belongs to the second case: the reactants are the gas chemical species and the wall is the catalyst.

Modelling Heterogeneous Catalysis

A scheme representing the elementary steps involved in heterogeneous catalysis can be drawn as follow:

- Diffusion of reactants to the surface.

- Adsorption of reactants at surface.

- Chemical reactions on the surface.

- Desorption of products from the surface.

- Diffusion of products away from the surface.

Each one of these steps has a different velocity and the slower one is the process-determining rate. Steps 1 and 5 are usually fast; exceptions exist when the catalyst is very efficient. The limit for the catalytic reaction rate is the quantity of reactants that goes to the wall, and the surface reaction rates become independent of the kinetics properties of the surface. This is the case of "diffusion controlled" catalysis.

Coatings for space vehicles TPS are studied to have very low catalytic efficiency, to lower, as much as possible, the surface heat flux due to recombination effects. Therefore, for these applications, the rate determining step for the heterogeneous catalysis process has to be found among steps.

Adsorption

Adsorption is the formation of a bond between the atom, or the molecule, and the solid surface. We speak of physisorption (physical adsorption) when the bond between surface and gas particles is due to van der Waals forces (forces between inert atoms and molecules). The forces involved are electrostatic forces for molecules with permanent dipole moments, induced polar attractions for readily polarizable molecules, dispersion forces for non-polar atoms and molecules (forces due to the fluctuation of electron density). The strength of these bonds depends upon the physical properties of the adsorbed gas, and is connected to the boiling point (or condensation) of the gas. Physisorption does not depend much on the chemical nature of the solid. The particle-surface bond energy is low (10-50 kJ/mole) and the bond is important only at low temperatures (~100-300 K): as the temperature rises the gas is removed more or less completely. In heterogeneous catalysis, the importance of this kind of adsorption lies in the fact that it can be a precursor step for chemisorption.

Chemisorption is associated to the solid surface properties. The surface of a solid can have properties markedly different from those of the bulk solid because on the surface there are unsaturated bonds. In fact, an atom on the surface is not in the same condition of a "bulk" atom, because it does not have its full complement of neighbors. This is true either for a covalent solid (e.g. SiO_2, SiC) as for an ionic solid (e.g. NaCl). A simple way to understand this situation is to think about an ideal fracture of a crystal: the fracture sides are new surfaces where atoms which were before inside the bulk are now surface atoms with "dangling" bonds.

When a free gas atom is near to a solid surface there is an attractive interaction between it and the surface atoms: the atom is attracted to the surface and it can find in one of the "dangling" bonds a site, i.e. a physical location corresponding to a potential well (low energy state) where it remains trapped. The bond formed is a true chemical bond, usually covalent: there is an interpenetration of particle shells with electron sharing and the bond energy is high (40-800 kJ/mole). Chemisorption occurs until the unsaturated surface valences are filled. This explains why, if multilayer physisorption is possible, no more than one layer can be adsorbed via chemisorption although new species can form at surface (e.g. due to oxidation in metals). Physisorption and chemisorption are spontaneous $\left(\Delta H < 0\right)$ but the latter can have substantial activation energy E_A. This means that it starts at temperatures higher than those relative to physisorption.

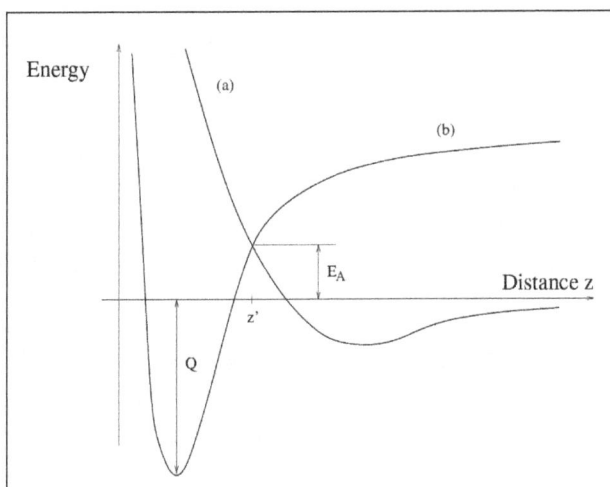

Potential energy curves for adsorption: (a) physisorption of a molecule; (b) chemisorption of two atoms.

Two kinds of chemisorption are possible: the first one is called associative or non-dissociative and it occurs to atoms (O, N, H,) and to molecules (CO, OH) without molecular bond breaking. A very simple schematic representation of associative chemisorption is:

$$A + * \rightarrow A^*,$$

where the "*" represent the site and the A* the adsorbed atom (adatom). Assuming a uniform surface with uniformly distributed sites and assuming sites activity do not be function of the fraction of surface already covered, θ, the adsorption rate can be expressed as:

$$v_{ads} = s_0 v_c \left(1-\theta\right),$$

where s_0 is the initial sticking coefficient, representing the probability of a molecule to stick on a bare surface, and v_c is the collision rate:

$$v_c = \frac{n}{4}\left(\frac{8kT}{\pi m}\right),$$

where n is the gas particle density and m the particle mass.

The second kind of chemisorption is called dissociative and it may occur to diatomic molecules like O_2, N_2, H_2 and CO. In this case chemisorption occurs with breaking of molecular bonds. Each atom is adsorbed from one site:

$$A_2 + 2* \rightarrow 2A^*$$

For this reaction the rate of adsorption reads:

$$v_{ads} = s_0 v_c \left(1-\theta\right)^2 \frac{z}{2},$$

where z represents the number of neighbors sites. In fact, in this case we assume that the atoms coming from the broken molecule can adsorb only on adjacent sites.

Dissociative adsorption may occur also following another path:

$$A_2 + * \rightarrow A^* + A$$

In this case the striking molecule dissociates, an atom is adsorbed and the other goes back to the gas. Usually this process is far less probable than the previous one. In fact the energy content of the system adatom plus atom is greater than the energy associated to the system of two adatoms.

Before describing the reaction step, we need to spend a few words about the initial sticking coefficient expression. This important quantity, assuming an uniform sites activity over the surface, depends on T. For atoms impinging on a surface a general expression is:

$$s_0 = Pe^{-E/(kT)}$$

where P is expected to be <1. s_0 decreases with temperature and this behavior can be understood with simple considerations: with increasing temperature gas atoms impinging on the surface have higher kinetic energy and thus higher probability to "rebound" after the collision with a site without sticking. Notice that for impinging molecules this trend can be the reverse (s_0 increases with T): in fact, if there is an activation energy associated to dissociative adsorption, at higher temperature it becomes easier to overcome this energy barrier.

Reaction

The kind of reactions interesting to hypersonics applications are recombinations of atoms forming O_2 and N_2 and reactions producing NO, CO, and CO_2.

In heterogeneous catalysis at least one of the reacting particles must be first chemisorbed in order to react catalytically on a solid surface. Therefore, if physisorption does not play a direct role, chemisorption is a fundamental step to "prepare" atoms and molecules to react. This mechanism furnishes a reaction path with an activation energy lower than the one of the correspondent homogeneous reaction.

The first reaction mechanisms we consider is the recombination between an atom from the gas and an adsorbed one: the free gas atom strikes the adatom and reacts with it. The result is a recombined molecule. This is called the Eley-Rideal mechanism (E-R) and a simple scheme of it is:

$$A + B^* \rightarrow AB^*,$$

where the symbols A, B mean an atom. After the recombination the molecule AB* is still adsorbed. The rate of this reaction depends by the partial pressure of the gas phase atoms and by the surface coverage. On this basis for the step rate we can write:

$$v_{ER} = k [A] [B^*]$$

The second mechanism is the recombination between two adatoms: this is called the Langmuir-Hinshelwood mechanism (L-H) and a schematic representation is:

$$A^* + B^* \rightarrow AB^* + *$$

Obviously this kind of reaction is possible if adatoms diffuse over the surface. For adatoms a way to move it is to "hop" from one potential well to another overcoming a potential barrier E_m that is lower than the desorption energy E_d; in fact particles during this movement do not leave the surface completely. In literature it can be found that $E_m \sim 0.1 - 0.2 \, E_d$. This value depends on the surface coverage extent (it decreases when θ increases), on the surface defects and on the crystallographic orientation of the surface. The presence of the energy barrier E_m implies that the L-H mechanism is an activated process which becomes efficient at higher temperature with respect to the E-R mechanism. This is evident in catalysis involving metal surfaces were the recombination coefficient drops sharply when the L-H mechanism starts.

Desorption

After a E-R or a L-H recombination, there is not chemical bond between surface and molecule, which is free to leave the wall. This phenomenon is called desorption. It may happen that desorption is not "immediate" and the molecule may fall in a physisorption well before the leaving. The importance of this phenomenon lies in the fact that recombination energy can be left, in part or totally, on the wall when the molecule leaves the wall. The characteristic time of desorption is strictly linked to the amount of energy left on the wall by the molecular internal energy relaxation process. This means that the molecule can leave the surface in an excited state and "quench" later in the gas near the wall, or they can leave the surface in thermal equilibrium with it. To take in to account this phenomenon a chemical energy accomodation (CEA) coefficient β is defined as the fraction of the equilibrium dissociation energy delivered to the catalyst surface per recombination event:

$$\beta = \frac{\dot{q}}{D_{AB}\dot{\Delta n}/2},$$

where $\dot{q}$ is the effective energy flux to the wall, $\dot{\Delta n}$ is the flux of recombined atoms and DAB is the AB molecule dissociation energy.

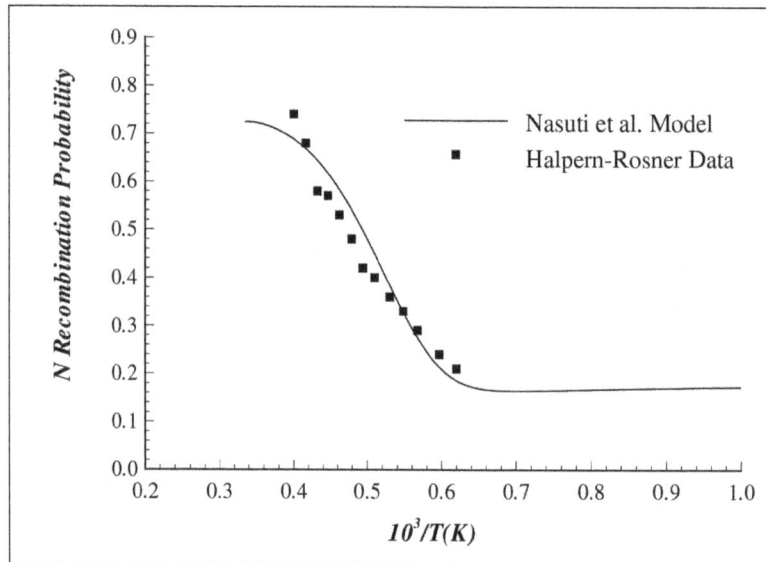

Nitrogen recombination over tungsten: $\tilde{a}_N$ vs inverse temperature.

Experimental values of β show a dependence on the interaction between surface and recombined molecules. If the bond between recombined molecule and surface is "strong" there is enough time to have a total accommodation ($\beta = 1$). On the contrary, if the bond is "weak" the molecules desorb rapidly and the energy left on the wall is smaller ($\beta < 1$).

Large savings in TPS appear, in principle, possible if could be deter- mined reliably and accurately, and if it could be "designed" a priori based on material properties and recombining atoms type. This is a frontier of surface catalysis still in need of exploration.

Thermal Desorption

Besides the mechanisms cited, we have to consider also thermal desorption. In fact, adsorbed atoms vibrate with the frequency kT/h; at sufficiently high temperatures this vibrations can extract the atom from the potential well leading to desorption,

$$A^* \rightarrow A + *.$$

Following transition state theory, the rate of desorption reads:

$$v_{des} = N_a \frac{k_T}{h} e^{\Delta S/k} e^{-\Delta H/(kT)},$$

where ΔS and ΔH are respectively entropy and enthalpy differences between two states (adsorbate-gas). Usually ΔH is supposed to be of the same order of magnitude of the adsorption energy. Due to the high values of the latter, thermal desorption becomes important at high temperatures (e.g. T > 2000 for N over W).

Rate Determining Step

For this reason we will speak about low and high temperatures without any precise specifications. In any case, what we mean for "low" are temperatures not far for ambient, and for "high" temperatures over 1000 K.

At low temperatures the surface coverage θ is near to unity; this because the atoms mobility is very low and thermal desorption is not efficient. In these conditions the E-R recombination mechanism is expected to be more effective with respect to L-H. In fact, there is a very high probability, for an incoming atom, to strike an adsorbed one. In these conditions the rate of adsorption is expected to be larger than the rate of surface atoms depletion:

$$v_{ads} \gg v_{reac}, v_{des}$$

When the temperature rises, the coverage reduces both due to thermal and L-H recombination desorption. High temperature enhances adatom mobility and consequently their probability of recombination. Besides, the L-H mechanism is very efficient because it removes two adatoms at once. The E-R mechanism become less effective due to the low probability of a collision between a gas atom and an adatom, since surface coverage is low and adatoms are very mobile and di cult "targets" for striking gas atoms. Under these conditions the rate limiting step can be adsorption:

$$v_{des}, v_{reac} \gg v_{ads}$$

Besides defining a rate limiting steps, some further comment. We can observe that the molecules formed via a E-R mechanism with high probability are, more likely to leave the surface in an excited state. They would then take with them part of the recombination energy. This is not the case of L-H recombined molecules, because they would have enough time to leave all the energy excess energy to the surface, and would desorb in thermal equilibrium. Therefore the β factor would be expected to be small at low temperatures and to increase toward unity with increasing temperature.

Surface Kinetics for Reentry Flows

For reentry applications, one of the problems to solve is to predict the catalytic activity of TPS surfaces. The US Shuttle TPS coatings are made of borosilicate Reacting Cured Glass (RCG: 92%

SiO_2, 5% B_2O_3, 3% SiB_4), a surface that can be thought as a pure SiO_2 surface. In fact, there is experimental evidence that a RCG surface may be more similar to SiO_2 than to the borosilicate bulk. This is the reason why the existing models simulating RCG catalytic activity, built by starting from the gas/surface physics seen above, are based on SiO_2 structure. These models can be used to set realistic catalytic boundary conditions in CFD codes.

Before going through these models we recall here some definitions starting from the Molecular Recombination Coefficient:

$$\gamma = \frac{\text{flux of atoms recombining at surface}}{\text{flux of atoms impinging the surface}}$$

γ represents also the recombination probability for an atom impinging the surface. The Thermal Recombination Coefficient is defined as:

$$\gamma' = \beta\gamma$$

This coefficient is a measure of the energy effectively left on the surface by heterogeneous chemistry. In fact values of β can be lower than 0.1, meaning that only a small portion of recombination energy is left on the wall. If recombination rates (or probabilities) are measured via calorimetry, telling γ and γ' apart is a di cult task.

Finally we recall the expression for the surface catalytic recombination rate, usually named catalitycity:

$$K_w = \gamma\sqrt{\frac{kT_w}{2\pi m}},$$

where the subscript w stands for "wall" and m is the atom mass.

Heterogeneous Catalysis Models

A model for recombination of Oxygen atoms on SiO_2 has been presented by Jumper and Seward the surface adsorption sites are the Si atoms of the SiO_2 crystal lattice. They assume that an O atom of the SiO_2 surface can desorb, either thermally or after a recombination with a gas atom (E-R mechanism), leaving available a Si atom that becomes a site where an oxygen atom can adsorb forming a double bond. Therefore the number of adsorption sites available on the surface is equal to the number of Si atoms on the surface.

To obtain the recombination coefficient, Jumper and Seward write the equation for the time rate variation of the surface concentration of oxygen atoms (n_s):

$$\frac{dn_s}{dt} = \dot{n}_{ads} - \dot{n}_{tdes} - \dot{n}_{rec},$$

where $\dot{n}_{ads}$ represents the rate at which O adsorbs at surface, $\dot{n}_{tdes}$ the rate at which the adatoms desorb without recombination, and where $\dot{n}_{rec}$ is the rate at which atoms leave the surface after recombination with a gas atom following the E-R mechanism. Jumper and Seward, assuming steady state $\left(\dfrac{dn_s}{dt} = 0\right)$, calculate the surface coverage and the recombination coefficient that are functions of wall temperature and O partial pressure. The initial sticking coefficients are function of wall temperature too. This model was extended to RCG in together with Newman and Kitchen, proposed a similar model for N recombination over SiO_2- based TPS also. These models are the first that can simulate heterogeneous catalysis in Hypersonics taking into account the local flow and wall characteristics.

A model for air recombination over SiO_2-based materials has been presented by Nasuti et al.. In this model the recombination coefficients are also obtained by writing a surface balance for incoming and leaving particles based on. For steady state conditions the flux of atoms adsorbing on the surface has to be balanced by the flux of atoms desorbing due to thermal desorption, E-R recombination and L-H recombination (in the latter case two atoms desorb at one time). Therefore simultaneous E-R and L-H recombination mechanisms are accounted for. The surface considered is SiO_2 and adsorption sites are specified as potential energy wells. This model calculates the recombination coefficients d for N_2, O_2 and also N O surface formation. Therefore it produces 3 independent recombination coefficients functions of wall temperature and of O and N surface partial pressures. Constant values are assumed for the initial sticking coefficients. This model depends on knowledge of basic surface physics, and in principle may be extended to any kind of gas species and any surface.

The last model recently proposed is by Deutschmann et al. In this case, again for a SiO_2 surface, a more conventional kinetic approach is used. Rate constants for adsorption, thermal desorption surface reactions, including E-R and L-H mechanisms, and desorption after recombination are calculated. The rate constants for reaction and desorption are in Arrhenius- like form. Recombination coefficients (for N_2 and O_2 only) depend on the wall temperature and on adsorbed species concentration. As in Nasuti et al., constant values for the initial sticking coefficients are assumed.

Whereas in the other two models the desorption step is always assumed to be collapsed in the reaction step, Deutschmann et al. assume this to be true for the E-R mechanism only. In fact, after L-H recombination, the molecule is assumed still adsorbed and a rate for the desorption step is calculated (a very low activation energy is actually used: 20 kJ/mol).

The crucial point of these models is the incorporation of physiochemical quantities, such as, for example, atoms adsorption energies, sticking coefficients, number of sites for unit area and others. There is still a large range of uncertainty in the determination of these quantities and in each model more or less complicated expressions for steric factor are used to cover these "holes". In any case the e ort to produce such a models, in our opinion, has to be appreciated. In fact, these tools are valid under several flow conditions, and wall temperatures. They can cover a wide range of right and wind tunnel flow conditions with a larger flexibility with respect to experimental fits obtained for a small range of temperature. For example, a spacecraft with a complex geometry re-entering our atmosphere, undergoes different flow conditions near to the surface at different locations and in this case using a single fit could be too restrictive.

Table: Main parameters for recombination coefficient models: initial sticking coefficient so surface site density N_s (sites $= m^2$); steric factor S; desorption energy E_d (kJ $=$ mole).

Authors	S_o	N_s	S	E_d
Jumper-Seward (SiO$_2$)	$0.05 \cdot e^{0.002T}$	$5.00 \cdot 10^{18}$	$2.24 \cdot 10^{-5} \cdot e^{0.00908T}$	339
Jumper-Seward (RCG)	$1.00 \cdot e^{0.002T}$	$2.00 \cdot 10^{18}$	$2.00 \cdot 10^{-4} \cdot e^{0.003T}$	339
Nasuti et al.	0.05	$4.50 \cdot 10^{18}$	0.1	250
Deutschmann et al.	0.10	$1.39 \cdot 10^{19}$	–	200

Oxygen recombination coefficient calculated using the three models presented above are shown in figure. These results, obtained for an O partial pressure = 400 P a, are compared with the experimental data of Greaves and Linnett (SiO$_2$) and of Kolodziej and Stewart (RCG). All the three models represent qualitatively well the high temperature behavior by reproducing the O rollover for T ~ 1600 K. In all three this second order effect has been found to be due mainly to thermal desorption, the most important process at those temperatures. Moving toward the low temperature range, the Deutschmann et al. model underpredicts γ of order of magnitude showing its intentional high-temperature-addressed design. The other two models represent qualitatively and quantitatively well the recombination coefficient trend. An evident difference between these two models is their different curvatures. These factors are one critical to such a models, because by necessity, they try to include physics not otherwise modeled by elementary steps. A way to clarify this point may be to have more experimental data for γ and γ' over a wide range of temperature going from ambient to the material melting limit.

Heterogeneous Catalysis Applications to Hypersonic Flows

It is already some time that the CFD community is moving towards the inclusions of this kind of models in complex thermochemical nonequilibrium flow solvers, but often "rude" boundary conditions, such as "fully catalytic wall" or "non-catalytic wall", are still used.

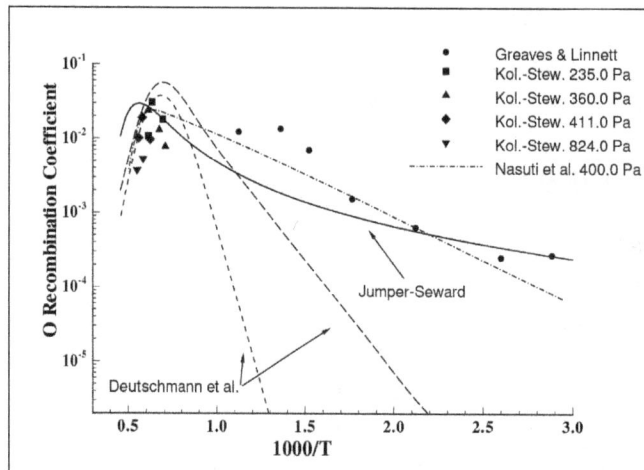

γ_o for TPS coating materials: comparison among several models.

The only justification for them is that they are simple to understand and to implement in a code. Clearly this is not acceptable when real conditions are very far from these extremes some results showing the large differences due to the use of one or another condition are reported. A brief summary of the different wall boundary conditions used in CFD for heterogeneous catalysis and for a

5 species air is:

- Non catalytic wall:

$$\gamma_i = 0 \quad i = O, N$$

The wall is completely indifferent to the flow chemistry.

- Fully catalytic wall:

$$\gamma_i = 1 \quad i = O, N$$

In this case the wall is a "perfect catalyst" and catalysis is "diffusion controlled". Since each atom hitting the wall recombines, the catalytic efficiency depends by the velocity of diffusion of atomic species trough the boundary layer.

- Fully Equilibrium wall:

$$Y_i = \bar{Y}_i = O_2, N_2, NO, O, N$$

where $\bar{Y}_i$ are the equilibrium species mass fractions for the local values of temperature and pressure. This is a very strong condition. It forces the flux of atoms to the wall (usually is $Y_O = Y_N = 0$) imposing very strong mass fractions gradients at the wall. Without any doubt these boundary conditions are stressing over the limit the near wall chemistry.

- Finite rate catalysis:

$$0 < \gamma_i < 1 \quad i = O, N$$

In this case a measure of real interface physics is introduced in the CFD flow solver. This means attributing the same importance to homogeneous and heterogeneous chemistry.

Some Applications

Solving hypersonic flows, the additional computational effort required by Finite Rate Catalysis (FRC) models is balanced by the increase in quality of the results. This concept can be clarified by a simple example: we consider a flow past a 40 cm long blunt cone (scaled down ELECTRE reentry capsule geometry). The inflow conditions are those of the HEG shock tunnel test chamber for $p_0 = 500\,Pa, T_0 = 9500\,K$ stagnation conditions. The body surface is assumed to be at 1500 K. The flow is solved with the TINA code and two different Boundary Conditions (BCs): (a) finite rate catalysis; (b) fully catalytic wall. The results of these calculations are shown in figure. From the stagnation line results, shown in figure we see that, besides the expected results for the wall mass fractions, i.e. $Y_{N_2}^{(b)} > Y_{N_2}^{(a)}, Y_{O_2}^{(b)} > Y_{O_2}^{(a)}$, the wall NO mass fraction in the fully catalytic case is higher than in the finite rate catalysis case, notwithstanding the fact that the FRC model used predicts also catalytic NO formation. The reason for this result is that in case (b) the strong production of N_2 and O_2 $\left(K_{wO_2}^{(b)} \simeq K_{wN_2}^{(b)} \right)$ and the diffusion of these species away from the surface due to the strong species gradients, drive the gas phase mechanism:

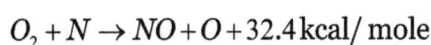

$$O_2 + N \rightarrow NO + O + 32.4\,kcal/mole$$

$$N_2 + O + 76.4 \text{ kcal/mole} \rightarrow NO + N$$

This mechanism pumps NO in the flow and stores energy in the gas phase; it is very effective slightly away from the surface and it quenches near to the wall, where $Y_O, Y_N \rightarrow 0$.

For case (a) instead this mechanism is not effective: in the gas phase a smaller quantity of N O is produced by the reaction $N_2 + O + 76.4 \text{ kcal/mole} \rightarrow NO + N$, whereas in the zone close to the wall, due to the surface formation of O_2 and to the presence of N atoms, reaction above produces N O and releases energy. This latter phenomenon is similar to what was reported by Tirsky in, where an exchange mechanism "$O_2 + N \rightarrow NO + O + 32.4 \text{ kcal/mole}$ + reverse of $N_2 + O + 76.4 \text{ kcal/mole} \rightarrow NO + N$" is presented as very effective in the gas close to the wall when $K_{wO_2} > K_{wN_2}$. Notice that in the case(a) shown here, in the zone closer to the wall, there is not any sensible contribution from either reaction $N_2 + O + 76.4 \text{ kcal/mole} \rightarrow NO + N$ direct or reverse.

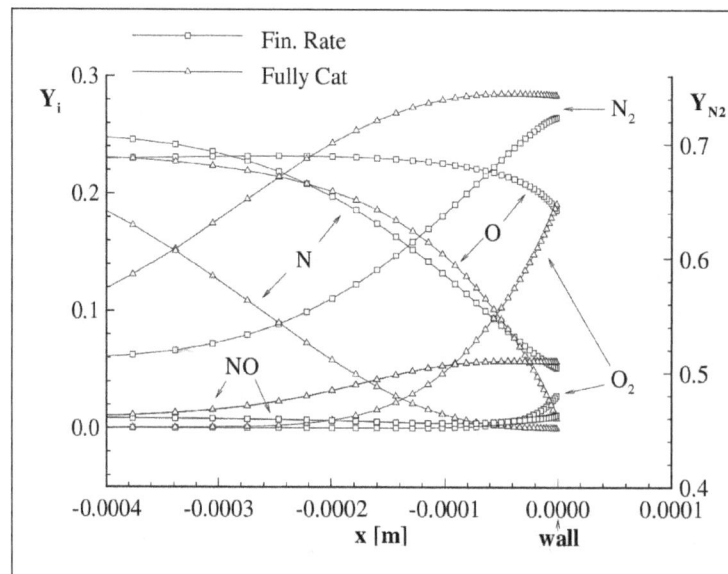

Comparison between two solutions with different catalytic boundary conditions at T_w = 1500 K: stagnation line mass fractions.

The effects of this on the surface heat flux can be seen in figure where the ratio q_i / q_{tot} for i = "tr (translational), vib (vibrational), df (diffusive)" are plotted along the body. For case (b) the strong recombination leads to $q_{df} / q_{tot} > 0.4$ all along the body, whereas for case (a) this ratio is < 0.3. Still for case (a), the energy released by reaction $O_2 + N \rightarrow NO + O + 32.4 \text{ kcal/mole}$ near to the surface yields a quite larger q_{tr} / q_{tot} with respect to case (b). This result is also partly due to the desorption of a larger amount of molecules than in case (b). In fact, assuming that the recombined molecules leave the wall in thermal equilibrium ($T = T_V = T_w$), the BL gas is cooled by this "cold" flow of particles. Therefore, due to the larger recombination, in case (b) the wall ∇T and ∇T_V are smaller than in case (a).

Another important issue arises when we compare flow conditions for flight and wind tunnels simulations. In figure we show the results of two calculations performed for the same shape (i.e. sphere plus cone) and obeying to the binary scaling rule $(\rho_\infty L = 0.0006)$.

For the two cases the amount of dissociated species reaching the wall is quite different, as shown by figure This leads to different amounts of recombined species. In fact the atoms depletion term can be expressed as:

$$\dot{w}_a = K_{wa}\rho_w Y_{wa}$$

where Y_{wa} is the atoms mass fraction at the wall.

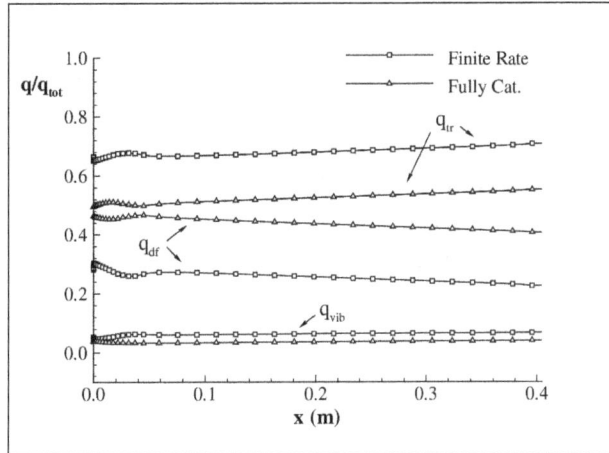

Comparison between solutions with different catalytic boundary conditions at $T_w = 1500$ K: different contributions as a percent of the total wall heat ux.

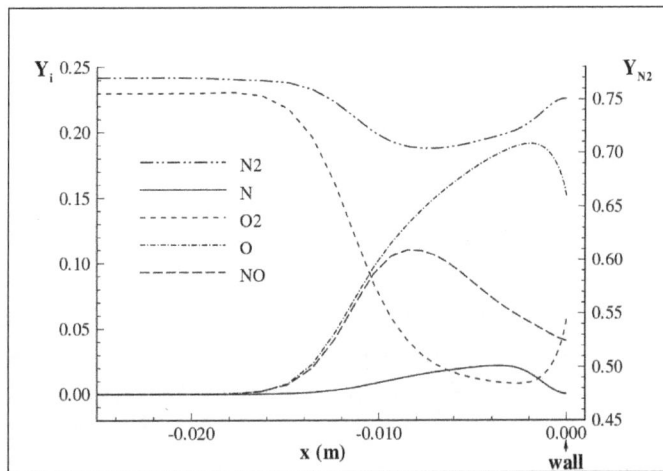

Stagnation line mass fractions for shock tunnel numerical simulation:
$Ma = 9.7,\ \rho_\infty = .002017\,kg/m^3,\ L = .33\,m,\ T_w = 1000\,K.$

Therefore the catalytic activity has different effects as it is shown by diffusive contribution to the total surface heat flux in figure. In fact, along the body the ratio $q\,df/q_{tot}$ is higher than 0.15 for the wind tunnel numerical simulation whereas, apart from the zone near to the stagnation point, it is lower than 0.1 for the "in flight" case. Near the stagnation zone this difference reduces, and on the stagnation point the two contributions are very close to each other. For the "in flight" numerical simulation we see a rapid decrease of the catalytic activity from the stagnation point to $x/L \simeq 0.16$.

In cases similar to the latter presented here, the necessary "binary scaling" principle is insufficient to guarantee reactive flow similarity then, to do that, the Damköheler numbers related to heterogeneous chemistry (i.e., inside the BCs) must be the same of flight conditions. This further condition yields that, to reproduce "real flight conditions" in a wind tunnel, a coating material different from that applied to the flying object must be used.

A consequence of what presented above is that numerical simulations with realistic catalytic boundary conditions may help in the design of ground testing models. In fact the choice of the model skin has to be done on the basis of "scaling factors" for the heterogeneous chemistry. Therefore the main question is how important is the influence of catalysis on the quantities measured in a test; if this importance is high, the similarity parameters (Damköheler numbers), associated with catalytic activity must be set identical to those in flight. This conclusion should be supported by two suggested actions: (a) an experimental study of the catalytic behavior of several materials, from oxides to metals, resulting in a database containing data for a wide range of temperatures; (b) starting from these data, the definition of more accurate and more effective models for the catalytic activity of a set of materials of interest for hypersonic flight. Finally, the right coupling among numerical simulations, ground testing and in flight measurements may lead to more reliable predictions.

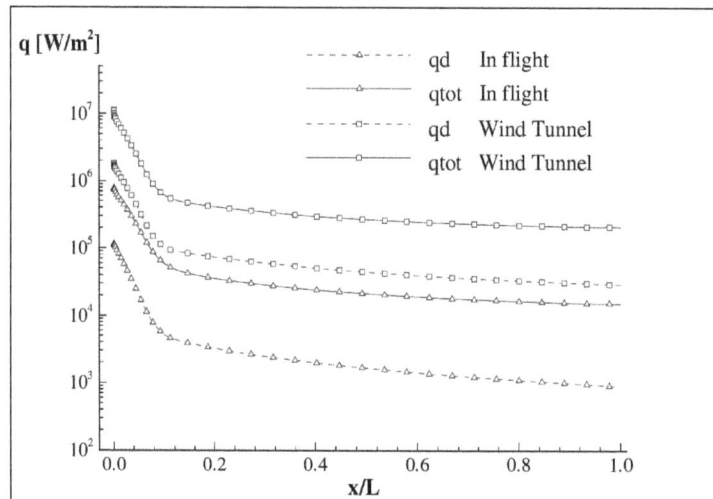

Wall heat fluxes for shock tunnel and "in flight" numerical simulations.

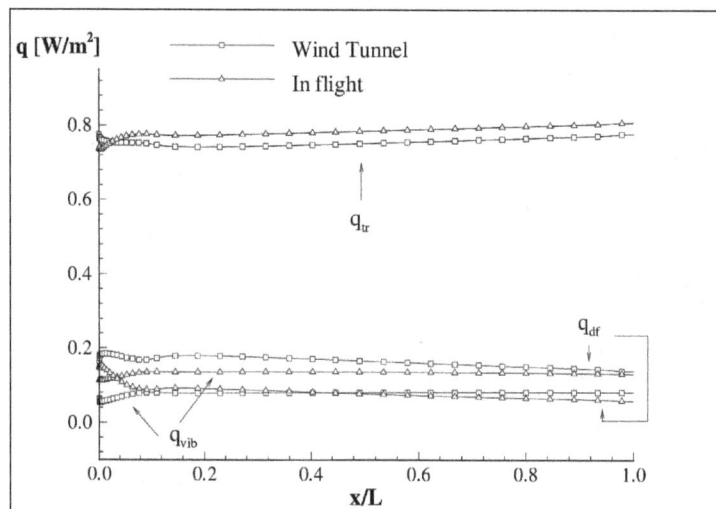

Comparison between shock tunnel and "in flight" numerical simulations:
different contributions as a percent of the total wall heat ux.

Homogeneous Catalysis

Catalysts are compounds that, when added to chemical reactions, reduce the activation energy and increase the reaction rate. The amount of a catalyst does not change during a reaction, as it is not consumed as part of the reaction process. Catalysts lower the energy required to reach the transition state of the reaction, allowing more molecular interactions to achieve that state. However, catalysts do not affect the degree to which a reaction progresses. In other words, though catalysts affect reaction kinetics, the equilibrium state remains unaffected.

Catalysts can be classified into two types: homogeneous and heterogeneous. Homogeneous catalysts are those which exist in the same phase (gas or liquid) as the reactants, while heterogeneous catalysts are not in the same phase as the reactants. Typically, heterogeneous catalysis involves the use of solid catalysts placed in a liquid reaction mixture.

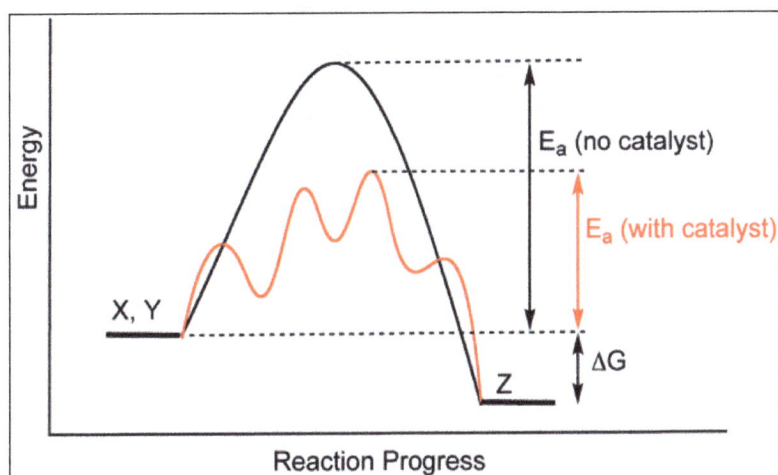

Examples of Homogeneous Catalysts

Acid catalysis, organometallic catalysis, and enzymatic catalysis are examples of homogeneous catalysis. Most often, homogeneous catalysis involves the introduction of an aqueous phase catalyst into an aqueous solution of reactants. In such cases, acids and bases are often very effective catalysts, as they can speed up reactions by affecting bond polarization.

An advantage of homogeneous catalysis is that the catalyst mixes into the reaction mixture, allowing a very high degree of interaction between catalyst and reactant molecules. However, unlike with heterogeneous catalysis, the homogeneous catalyst is often irrecoverable after the reaction has run to completion.

Homogeneous catalysts are used in variety of industrial applications, as they allow for an increase in reaction rate without an increase in temperature.

Interactive: Catalysis

The model contains reactants which will form the reaction: $A_2 + B_2 \rightarrow 2\,AB$. In this case the model has been set so the activation energy is high. Try running the reaction with and without a catalyst to see the effect catalysts have on chemical reactions. 1. Run the model to observe what happens without a catalyst. 2. Pause the model. 3. Add a few (3 – 4) catalyst atoms to the container by clicking the button. 4. Run the model again, and observe how the catalyst affects the reaction.

Step Growth Polymerization

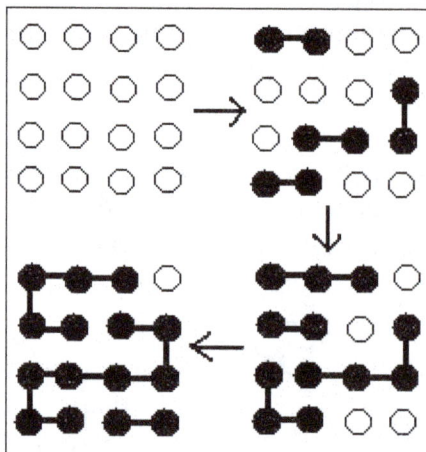

A generic representation of a step-growth polymerization. (Single white dots represent monomers and black chains represent oligomers and polymers).

Step-growth polymerization refers to a type of polymerization mechanism in which bi-functional or multifunctional monomers react to form first dimers, then trimers, longer oligomers and eventually long chain polymers. Many naturally occurring and some synthetic polymers are produced by step-growth polymerization, e.g. polyesters, polyamides, polyurethanes, etc. Due to the nature of the polymerization mechanism, a high extent of reaction is required to achieve high molecular weight. The easiest way to visualize the mechanism of a step-growth polymerization is a group of people reaching out to hold their hands to form a human chain—each person has two hands (= reactive sites). There also is the possibility to have more than two reactive sites on a monomer: In this case branched polymers production take place.

IUPAC deprecates the term step-growth polymerization and recommends use of the terms polyaddition, when the propagation steps are addition reactions and no molecules are evolved during these steps, and polycondensation when the propagation steps are condensation reactions and molecules are evolved during these steps.

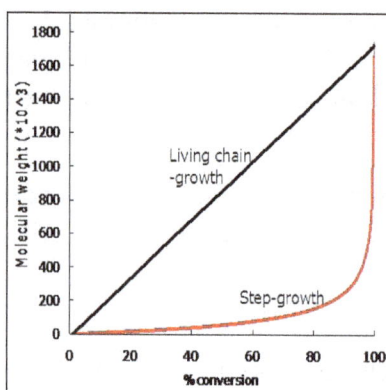

Comparison of Molecular weight vs conversion plot between step-growth and living chain-growth polymerization.

Most natural polymers being employed at early stage of human society are of condensation type. The synthesis of first truly synthetic polymeric material, Bakelite, was announced by Leo Baekeland in 1907, through a typical step-growth polymerization fashion of phenol and formaldehyde. The pioneer of synthetic polymer science, Wallace Carothers, developed a new means of making polyesters through step-growth polymerization in 1930s as a research group leader at DuPont. It was the first reaction designed and carried out with the specific purpose of creating high molecular weight polymer molecules, as well as the first polymerization reaction whose results had been predicted beforehand by scientific theory. Carothers developed a series of mathematic equations to describe the behavior of step-growth polymerization systems which are still known as the Carothers equations today. Collaborating with Paul Flory, a physical chemist, they developed theories that describe more mathematical aspects of step-growth polymerization including kinetics, stoichiometry, and molecular weight distribution etc. Carothers is also well known for his invention of Nylon.

Condensation Polymerization

"Step growth polymerization" and condensation polymerization are two different concepts, not always identical. In fact polyurethane polymerizes with addition polymerization (because its polymerization produces no small molecules), but its reaction mechanism corresponds to a step-growth polymerization.

The distinction between "addition polymerization" and "condensation polymerization" was introduced by Wallace Hume Carothers in 1929, and refers to the type of products, respectively:

- A polymer only (addition).

- A polymer and a molecule with a low molecular weight (condensation).

The distinction between "step-growth polymerization" and "chain-growth polymerization" was introduced by Paul Flory in 1953, and refers to the reaction mechanisms, respectively:

- By Functional Groups (Step-Growth Polymerization).

- By Free-radical or ion (chain-growth polymerization).

Differences from Chain-Growth Polymerization

This technique is usually compared with chain-growth polymerization to show its characteristics.

Step-growth polymerization	Chain-growth polymerization
Growth throughout matrix.	Growth by addition of monomer only at one end or both ends of chain.
Rapid loss of monomer early in the reaction.	Some monomer remains even at long reaction times.
Similar steps repeated throughout reaction process.	Different steps operate at different stages of mechanism (i.e. Initiation, propagation, termination, and chain transfer).
Average molecular weight increases slowly at low conversion and high extents of reaction are required to obtain high chain length.	Molar mass of backbone chain increases rapidly at early stage and remains approximately the same throughout the polymerization.
Ends remain active (no termination).	Chains not active after termination.
No initiator necessary.	Initiator required.

Classes of Step-growth Polymers

Examples of monomer systems that undergo step-growth polymerization.
The reactive functional groups are highlighted.

Classes of step-growth polymers are:

- Polyester has high glass transition temperature T_g and high melting point T_m, good mechanical properties to about 175 °C, good resistance to solvent and chemicals. It can exist as fibers and films. The former is used in garments, felts, tire cords, etc. The latter appears in magnetic recording tape and high grade films.

- Polyamide (nylon) has good balance of properties: high strength, good elasticity and abrasion resistance, good toughness, favorable solvent resistance. The applications of polyamide include: rope, belting, fiber cloths, thread, substitute for metal in bearings, jackets on electrical wire.

- Polyurethane can exist as elastomers with good abrasion resistance, hardness, good resistance to grease and good elasticity, as fibers with excellent rebound, as coatings with good resistance to solvent attack and abrasion and as foams with good strength, good rebound and high impact strength.

- Polyurea shows high T_g, fair resistance to greases, oils, and solvents. It can be used in truck bed liners, bridge coating, caulk and decorative designs.

- Polysiloxane are available in a wide range of physical states—from liquids to greases, waxes, resins, and rubbers. Uses of this material are as antifoam and release agents, gaskets, seals, cable and wire insulation, hot liquids and gas conduits, etc.

- Polycarbonates are transparent, self-extinguishing materials. They possess properties like crystalline thermoplasticity, high impact strength, good thermal and oxidative stability. They can be used in machinery, auto-industry, and medical applications. For example, the cockpit canopy of F-22 Raptor is made of high optical quality polycarbonate.

- Polysulfides have outstanding oil and solvent resistance, good gas impermeability, good resistance to aging and ozone. However, it smells bad, and it shows low tensile strength as well as poor heat resistance. It can be used in gasoline hoses, gaskets and places that require solvent resistance and gas resistance.

- Polyether shows good thermoplastic behavior, water solubility, generally good mechanical properties, moderate strength and stiffness. It is applied in sizing for cotton and synthetic fibers, stabilizers for adhesives, binders, and film formers in pharmaceuticals.

- Phenol formaldehyde resin (Bakelite) have good heat resistance, dimensional stability as well as good resistance to most solvents. It also shows good dielectric properties. This material is typically used in molding applications, electrical, radio, televisions and automotive parts where their good dielectric properties are of use. Some other uses include: impregnating paper, varnishes, decorative laminates for wall coverings.

- Poly-Triazole polymers are produced from monomers which bear both an alkyne and azide functional group. The monomer units are linked to each other by the a 1,2,3-triazole group; which is produced by the 1,3-Dipolar cycloaddition, also called the Azide-alkyne Huisgen cycloaddition. These polymers can take on the form of a strong resin, or a gel. With oligopeptide monomers containing a terminal alkyne and terminal azide the resulting clicked peptide polymer will be biodegradable due to action of endopeptidases on the oligopeptide unit.

Branched Polymers

A monomer with functionality of 3 or more will introduce branching in a polymer and will ultimately form a cross-linked macrostructure or network even at low fractional conversion. The point at which a tree-like topology transits to a network is known as the gel point because it is signalled by an abrupt change in viscosity. One of the earliest so-called thermosets is known as bakelite. It is not always water that is released in step-growth polymerization: in acyclic diene metathesis or ADMET dienes polymerize with loss of ethylene.

Kinetics

The kinetics and rates of step-growth polymerization can be described using a polyesterification mechanism. The simple esterification is an acid-catalyzed process in which protonation of the acid is followed by interaction with the alcohol to produce an ester and water. However, there are a few assumptions needed with this kinetic model. The first assumption is water (or any other condensation product) is efficiently removed. Secondly, the functional group reactivities are independent of chain length. Finally, it is assumed that each step only involves one alcohol and one acid.

$$\frac{1}{1-p^{n-1}} = 1 + (n-1)kt[\text{COOH}]^{n-1}$$

This is a general rate law degree of polymerization for polyesterification where n= reaction order.

Self-Catalyzed Polyesterification

If no acid catalyst is added, the reaction will still proceed because the acid can act as its own catalyst. The rate of condensation at any time t can then be derived from the rate of disappearance of -COOH groups:

$$rate = \frac{-d[\text{COOH}]}{dt} = k[\text{COOH}]^2[\text{OH}]$$

The second-order [\ce{COOH}] term arises from its use as a catalyst, and k is the rate constant. For a system with equivalent quantities of acid and glycol, the functional group concentration can be written simply as:

$$rate = \frac{-d[\text{COOH}]}{dt} = k[\text{COOH}]^3$$

After integration and substitution from Carothers equation, the final form is the following:

$$\frac{1}{(1-p)^2} = 2kt[\text{COOH}]^2 + 1 = X_n^2$$

For a self-catalyzed system, the number average degree of polymerization (Xn) grows proportionally with $\sqrt{t}$.

External Catalyzed Polyesterification

The uncatalyzed reaction is rather slow, and a high X_n is not readily attained. In the presence of a catalyst, there is an acceleration of the rate, and the kinetic expression is altered to,

$$\frac{-d[\text{COOH}]}{dt} = k[\text{COOH}][\text{OH}]$$

which is kinetically first order in each functional group. Hence,

$$\frac{-d[\text{COOH}]}{dt} = k[\text{COOH}]^2$$

and integration gives finally,

$$\frac{1}{1-p} = 1 + [\text{COOH}]kt = X_n$$

For an externally catalyzed system, the number average degree of polymerization grows proportionally with t.

Molecular Weight Distribution in Linear Polymerization

The product of a polymerization is a mixture of polymer molecules of different molecular weights. For theoretical and practical reasons it is of interest to discuss the distribution of molecular weights in a polymerization. The molecular weight distribution (MWD) had been derived by Flory by a statistical approach based on the concept of equal reactivity of functional groups.

Probability

Step-growth polymerization is a random process so we can use statistics to calculate the probability of finding a chain with x-structural units ("x-mer") as a function of time or conversion,

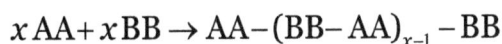

$$x\,AA + x\,BB \rightarrow AA - (BB - AA)_{x-1} - BB$$

$$x\,AB \rightarrow A - (B - A)_{x-1} - B$$

Probability that an 'A' functional group has reacted,

$$p^{x-1}$$

Probability of finding an 'A' unreacted,

$$(1 - p)$$

Combining the above two equations leads to,

$$P_x = (1 - p)p^{x-1}$$

Where P_x is the probability of finding a chain that is x-units long and has an unreacted 'A'. As x increases the probability decreases.

Number Fraction Distribution

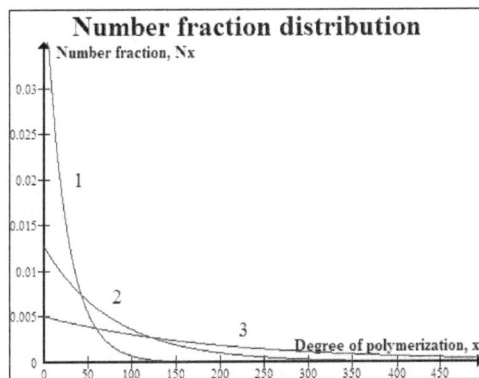

Number-fraction distribution curve for linear polymerization. Plot 1, p=0.9600; plot 2, p=0.9875; plot 3, p=0.9950.

The number fraction distribution is the fraction of x-mers in any system and equals the probability of finding it in solution.

$$\frac{N_x}{N} = (1-p)p^{x-1}$$

Where N is the total number of polymer molecules present in the reaction.

Weight Fraction Distribution

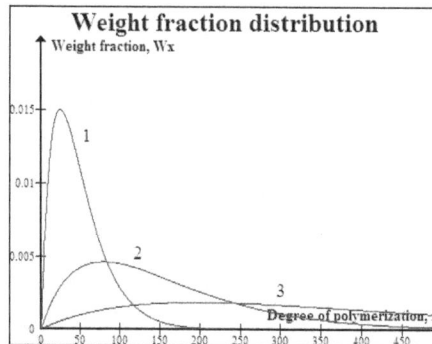

Weight fraction distribution plot for linear polymerization.
Plot 1, p=0.9600; plot 2, p=0.9875; plot 3, p=0.9950.

The weight fraction distribution is the fraction of x-mers in a system and the probability of finding them in terms of mass fraction:

$$\frac{W_x}{W_o} = \frac{xN_xM_o}{N_oM_o} = \frac{xN_x}{N_o} = x\frac{N_x}{N}\frac{N}{N_o}$$

- M_o is the molar mass of the repeat unit,
- N_o is the initial number of monomer molecules,
- and N is the number of unreacted functional groups.

Substituting from the Carothers equation,

$$X_n = \frac{1}{1-p} = \frac{N_o}{N}$$

We can now obtain:

$$\frac{W_x}{W_o} = x(1-p)^2 p^{x-1}$$

PDI

The polydispersity index (PDI), is a measure of the distribution of molecular mass in a given polymer sample.

$$PDI = \frac{M_w}{M_n}$$

However, for step-growth polymerization the Carothers equation can be used to substitute and rearrange this formula into the following:

$$PDI = 1 + p$$

Therefore, in step-growth when p=1, then the PDI=2.

Molecular Weight Control in Linear Polymerization

Need for Stoichiometric Control

There are two important aspects with regard to the control of molecular weight in polymerization. In the synthesis of polymers, one is usually interested in obtaining a product of very specific molecular weight, since the properties of the polymer will usually be highly dependent on molecular weight. Molecular weights higher or lower than the desired weight are equally undesirable. Since the degree of polymerization is a function of reaction time, the desired molecular weight can be obtained by quenching the reaction at the appropriate time. However, the polymer obtained in this manner is unstable in that it leads to changes in molecular weight because the ends of the polymer molecule contain functional groups that can react further with each other.

This situation is avoided by adjusting the concentrations of the two monomers so that they are slightly nonstoichiometric. One of the reactants is present in slight excess. The polymerization then proceeds to a point at which one reactant is completely used up and all the chain ends possess the same functional group of the group that is in excess. Further polymerization is not possible, and the polymer is stable to subsequent molecular weight changes.

Another method of achieving the desired molecular weight is by addition of a small amount of monofunctional monomer, a monomer with only one functional group. The monofunctional monomer, often referred to as a chain stopper, controls and limits the polymerization of bifunctional monomers because the growing polymer yields chain ends devoid of functional groups and therefore incapable of further reaction.

Quantitative Aspects

To properly control the polymer molecular weight, the stoichiometric imbalance of the bifunctional monomer or the monofunctional monomer must be precisely adjusted. If the nonstoichiometric imbalance is too large, the polymer molecular weight will be too low. It is important to understand the quantitative effect of the stoichiometric imbalance of reactants on the molecular weight. Also, this is necessary in order to know the quantitative effect of any reactive impurities that may be present in the reaction mixture either initially or that are formed by undesirable side reactions. Impurities with A or B functional groups may drastically lower the polymer molecular weight unless their presence is quantitatively taken into account.

More usefully, a precisely controlled stoichiometric imbalance of the reactants in the mixture can provide the desired result. For example, an excess of diamine over an acid chloride would eventually produce a polyamide with two amine end groups incapable of further growth when the acid chloride was totally consumed. This can be expressed in an extension of the Carothers equation as,

$$X_n = \frac{(1+r)}{(1+r-2rp)}$$

where r is the ratio of the number of molecules of the reactants,

$$r = \frac{N_{AA}}{N_{BB}} \text{ were } N_{BB} \text{ is the molecule in excess.}$$

The equation above can also be used for a monofunctional additive which is the following,

$$r = \frac{N_{AA}}{(N_{BB} + 2N_B)}$$

where N_B is the number of monofunction molecules added. The coefficient of 2 in front of N_B is require since one B molecule has the same quantitative effect as one excess B-B molecule.

Multi-chain Polymerization

A monomer with functionality 3 has 3 functional groups which participate in the polymerization. This will introduce branching in a polymer and may ultimately form a cross-linked macrostructure. The point at which this three-dimensional 3D network is formed is known as the gel point, signaled by an abrupt change in viscosity.

A more general functionality factor f_{av} is defined for multi-chain polymerization, as the average number of functional groups present per monomer unit. For a system containing N_0 molecules initially and equivalent numbers of two function groups A and B, the total number of functional groups is $N_0 f_{av}$,

$$f_{av} = \frac{\sum N_i \cdot f_i}{\sum N_i}$$

And the modified Carothers equation is:

$$x_n = \frac{2}{2 - pf_{av}}, \text{ where p equals to } \frac{2(N_0 - N)}{N_0 \cdot f_{av}}$$

Advances in Step-growth Polymers

The driving force in designing new polymers is the prospect of replacing other materials of construction, especially metals, by using lightweight and heat-resistant polymers. The advantages of lightweight polymers include: high strength, solvent and chemical resistance, contributing to a variety of potential uses, such as electrical and engine parts on automotive and aircraft components, coatings on cookware, coating and circuit boards for electronic and microelectronic devices, etc. Polymer chains based on aromatic rings are desirable due to high bond strengths and rigid polymer chains. High molecular weight and crosslinking are desirable for the same reason. Strong dipole-dipole, hydrogen bond interactions and crystallinity also improve heat resistance. To obtain desired mechanical strength, sufficiently high molecular weights are necessary, however, decreased solubility is a problem. One approach to solve this problem is to introduce of some flexibilizing linkages, such as isopropylidene, C=O, and SO$_2$ into the rigid polymer chain by using an appropriate monomer or comonomer. Another approach involves the synthesis of reactive telechelic oligomers containing functional end groups capable of reacting with each other, polymerization of the oligomer gives higher molecular weight, referred to as chain extension.

Aromatic Polyether

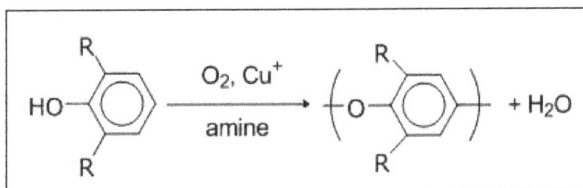

The oxidative coupling polymerization of many 2,6-disubstituted phenols using a catalytic complex of a cuprous salt and amine form aromatic polyethers, commercially referred to as poly(p-phenylene oxide) or PPO. Neat PPO has little commercial uses due to its high melt viscosity. Its available products are blends of PPO with high-impact polystyrene (HIPS).

Polyethersulfone

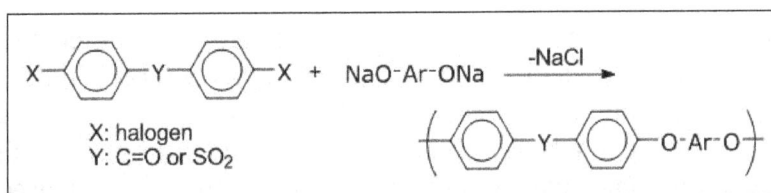

Polyethersulfone (PES) is also referred to as polyetherketone, polysulfone. It is synthesized by nucleophilic aromatic substitution between aromatic dihalides and bisphenolate salts. Polyethersulfones are partially crystalline, highly resistant to a wide range of aqueous and organic environment. They are rated for continuous service at temperatures of 240-280 °C. The polyketones are finding applications in areas like automotive, aerospace, electrical-electronic cable insulation.

Aromatic Polysulfides

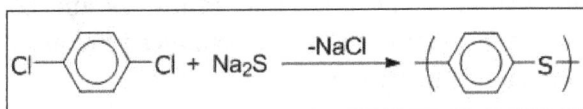

Poly(p-phenylene sulfide) (PPS) is synthesized by the reaction of sodium sulfide with p-dichlorobenzene in a polar solvent such as 1-methyl-2-pyrrolidinone (NMP). It is inherently flame-resistant and stable toward organic and aqueous conditions; however, it is somewhat susceptible to oxidants. Applications of PPS include automotive, microwave oven component, coating for cookware when blend with fluorocarbon polymers and protective coatings for valves, pipes, electromotive cells, etc.

Aromatic Polyimide

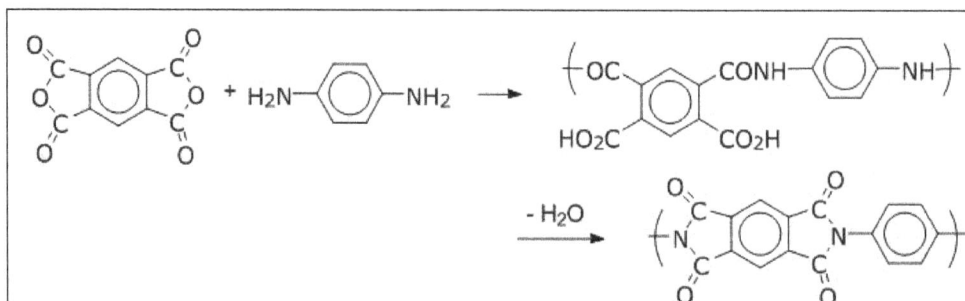

Aromatic polyimides are synthesized by the reaction of dianhydrides with diamines, for example, pyromellitic anhydride with *p*-phenylenediamine. It can also be accomplished using diisocyanates in place of diamines. Solubility considerations sometimes suggest use of the half acid-half ester of the dianhydride, instead of the dianhydride itself. Polymerization is accomplished by a two-stage process due to the insolubility of polyimides. The first stage forms a soluble and fusible high-molecular-weight poly(amic acid) in a polar aprotic solvent such as NMP or N,N-dimethylacetamide. The poly(amic aicd) can then be processed into the desired physical form of the final plymer product (e.g., film, fiber, laminate, coating) which is insoluble and infusible.

Telechelic Oligomer Approach

Telechelic oligomer approach applies the usual polymerization manner except that one includes a monofunctional reactant to stop reaction at the oligomer stage, generally in the 50-3000 molecular weight. The monofunctional reactant not only limits polymerization but end-caps the oligomer with functional groups capable of subsequent reaction to achieve curing of the oligomer. Functional groups like alkyne, norbornene, maleimide, nitrite, and cyanate have been used for this purpose. Maleimide and norbornene end-capped oligomers can be cured by heating. Alkyne, nitrile, and cyanate end-capped oligomers can undergo cyclotrimerization yielding aromatic structures.

Dendritic Polymers

Dendritic polymers are belonging to a special class of macromolecules. They are called "Dendrimers". Similar to linear polymers, they composed of a large number of monomer units that were chemically linked together. Due to their unique physical and chemical properties, dendrimers have wide ranges of potential applications. These include adhesives and coatings, chemical sensors, medical diagnostics, drug-delivery systems, high-performance polymers, catalysts, building blocks of supermolecules, separation agents and many more.

The Origin of Dendrimers

The name dendrimer is derived from Greek words *dendron* meaning "tree" and *meros* meaning "part". A major difference between linear polymers and dendrimers is that a linear polymer consists of long chains of molecules, like coils, crisscrossing each other. A dendrimer consists of molecular chains that branched out from a common center, and there is no entanglement between each dendrimer molecules. The first synthesis of these macromolecules is credited to Fritz Vogtle and coworkers in 1978[1]. It consisted of Michael Addition of acrylonitrile to primary amine groups. Each successive step involved reductions of the nitrile groups followed by additions of acrylonitrile.

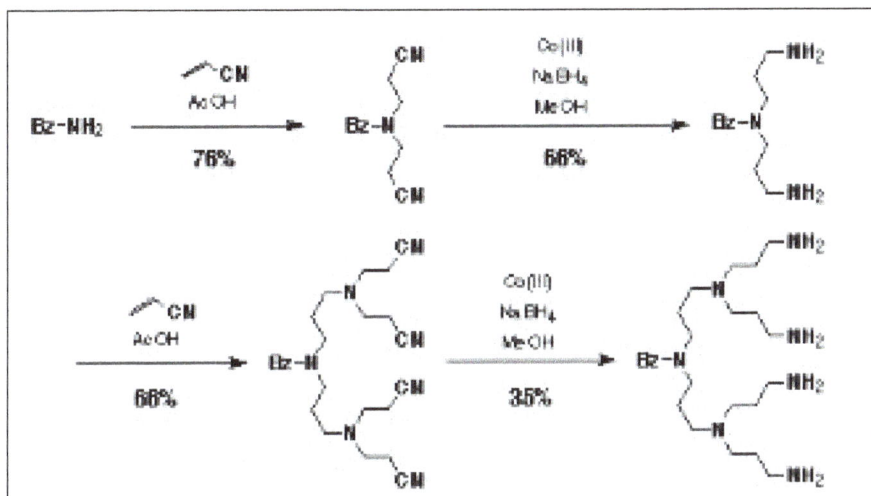

"Cascade" synthesis of poly(propylenimine).

The synthesis of dendrimers is very difficult and expensive. The industry sector was not keen to invest in dendrimer research. Until recently, key-pioneering works by Tomalia and Frechet solved many of the difficulties. Dendrimer research is still at its infant stage.

The Structures of a Dendrimer Molecule

Dendrimers have a globular configuration with monomer units branching out from a center cord. The structure is highly defined and organized.

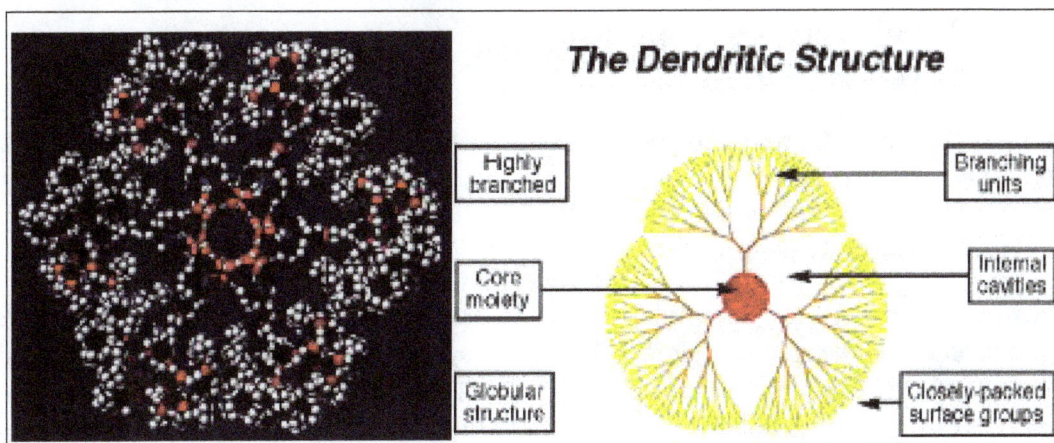

The number of branches increases exponentially extending from core to the periphery. The branching would come to a stop when the steric hindrance stopped any further growth. There are three distinct architectural components[2]. The multi-functionalized core (initiator core) forms the heart of the molecules; all branches emanate from this core.

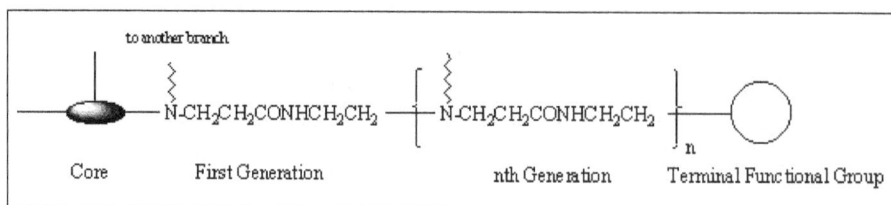

Components of a dendrimer, poly (amidoamine).

The monomers that attach to the core form the first branches (Tomalia called them the First Generation). On the successive generations, two monomers will attach to the ends of the monomers in the previous generation. At the terminating generation, a terminal functional group is added to the tail of the monomer. Most of the chemical properties of the molecule depend on types of terminal groups. The physical properties of the molecules, such as solubility and viscosity are also affected by the terminal groups.

Some of these dendrimers have diameters that are greater than ten nanometers. The molecular weights range from about 50,000 to 200,000 g/mol. The outer surface area of the molecule increases with the number of generations. There is a significant of void space within the molecule. These voids consist of channels and cavities. These unique geometries give the molecule special properties such as adhesiveness and ability to entrap foreign molecules. The number of terminal groups is easily calculated as follows:

Number of terminal groups $= N_c (N_r)^G$

Where N_c is the number of branches at the core (core multiplicity); N_r is the number of branches on each monomer unit (repeating unit multiplicity); G is the number of generation. The degree of polymerization can be computed using these quantities.

$$N_c \left(\frac{N_r^G - 1}{N_r - 1} \right) \text{ Degree of Polymerization =}$$

$$M_c + N_c \left(M_r \left(\frac{N_r^G - 1}{N_r - 1} \right) \right) + M_t N_r^G \text{ Similarly, the molecular weight is given as:}$$

$M_w =$ where M_c, M_r, and M_t are the molecular weight of the core, the repeating monomer, and the terminal group respectively.

Properties of Dendrimers

Theoretically, dendrimers are monodispersive. All molecules have the exact same molecular weight and structure. Due to minor defect during the synthesizing process, the polydispersity index is about 1.001. Polydispersity of 1.0007 for PAMAM has been reported[5]. The intrinsic viscosity of dendrimers has a peculiar behavior. It increases with increasing molecular weight (number of generations). Contrary to linear polymers, the viscosity will reach a maximum value then starts to decline. It is suggested that the space between the branches is smaller in higher generation dendrimers than lower generation dendrimers.

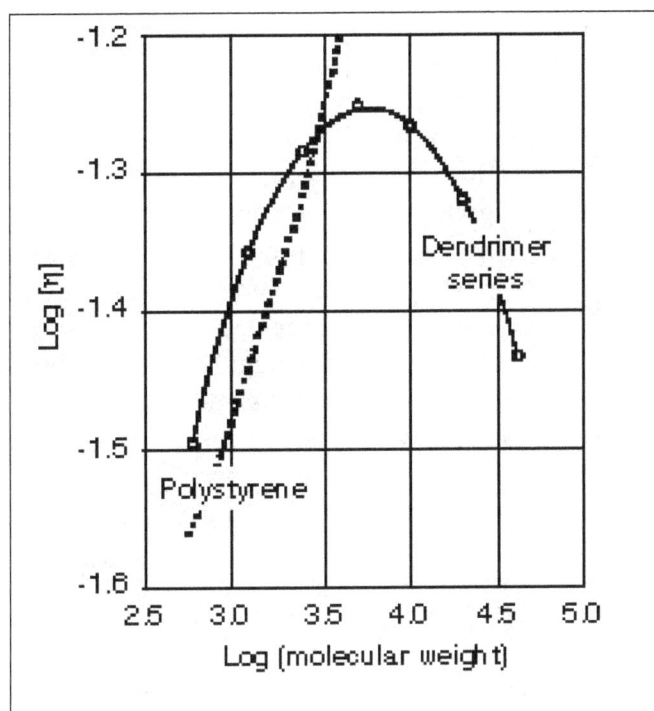

Comparison between the intrinsic viscosities of polystyrene and dendritic polyether.

The decline in viscosity is a consequence of prohibiting the interaction of the outer branches between molecules at a higher generation. The glass transition temperature (T_g) of dendrimers follows similar trend. It reaches to a maximum T_g and levels off at higher molecular weights. This behavior is explained by the absence of entanglement at higher molecular weights.

Synthesizing and Designing Dendrimers

Divergent Growth Method

This type of synthesis involves two steps: The activation of the functional surface groups and the addition of branching monomer units. The reaction starts at the core.

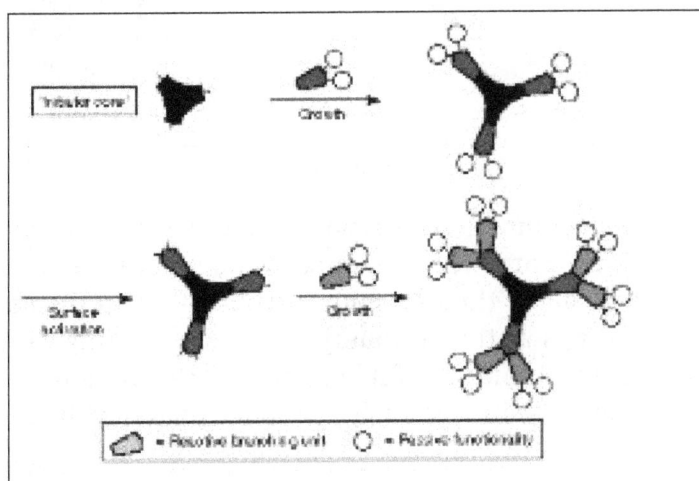

Divergent growth method.

The initiator core contains several reaction sites. The first generation monomer units react with the core readily. Once all reactive sites are taken, the addition stops. Since the end groups on the first generation are protected, addition of monomers to the end chain is impossible. The end groups must be activated before any further addition. The passive functionalities on the end groups are removed by a secondary reaction. Additional monomers are attached to the molecule. Steps are repeated for synthesizing higher generations. Ideally, one can control the structure and molecular weight of the molecule precisely. Divergent growth method is labor intensive and repetitive. In successive generation growth, side reactions and incomplete additions become more apparent. This is attributing to steric hindrance. In each step, the desire products must be purified. The overall yield is considerably small. One of the great advantages of this method is the ability to modify the surface of the dendrimer molecule. By changing the end groups at the outermost generation, the overall chemical and physical properties of the dendrimer can be configured to specific needs. It is due to this versatility that sparked the interest in dendrimer research. Tomalia's PAMAM dendrimers were synthesized using divergent method, starting with an initiator core and expanding to the periphery.

Convergent Growth Method

One of the shortcomings of divergent growth method is that the outermost generation has only one kind of functional group. Convergent growth method would eliminate such weakness. This method was first introduced by Frechet. The reaction starts at the periphery and proceeds to the core. Similar to divergent growth, it involves two steps:

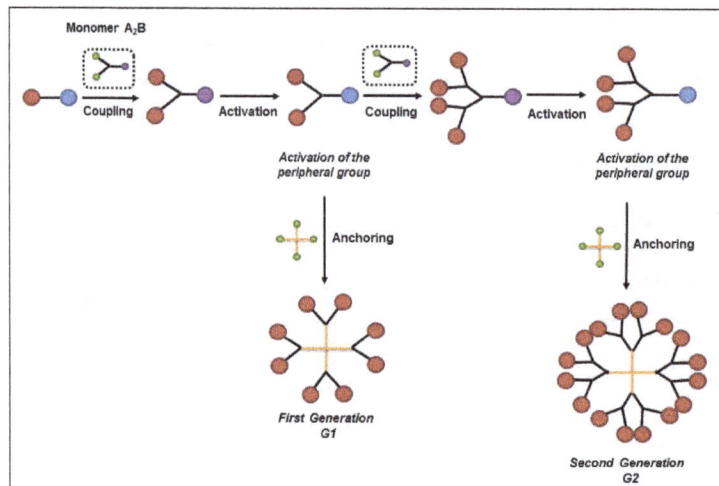

Convergent growth.

the attachment of the outermost functional groups to an inner generation and the attachment of the inner generations to the core. The structural units before the final attachment to the core is called the "wedge". Usually, three to four wedges attach to the core. Each wedge can have different functional groups at the periphery. Thus, the making of unsymmetrical dendrimers is possible. This modification is useful in monolayer formation at the organic-aqueous interface. Half of the dendrimer is submerged in the water phase, while the other half is in the organic phase.

A combination of these two methods can be used to suit for special needs.

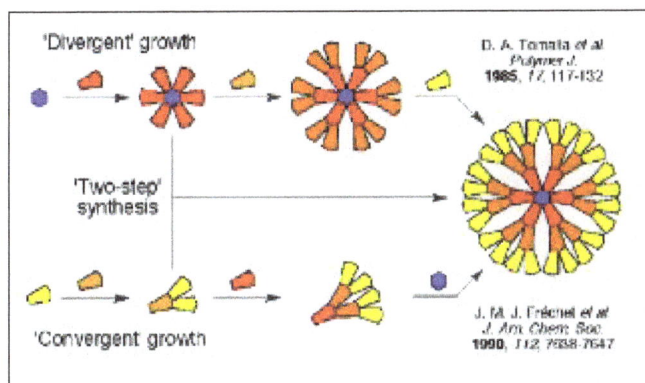

'Divergent' growth

D. A. Tomalia et al
Polymer J.
1985, 17, 117-132

'Two-step'
synthesis

'Convergent growth

J. M. J. Frechet et al
J. Am. Chem. Soc.
1990, 112, 7638-7647

Comparison between divergent and convergent growth.

Potential Applications of Dendrimers

Medicinal Applications

The idea of dendrimer serving as host for foreign molecules was first stumble upon by Meijer. He trapped several small molecules (Bengal Rose dye) in the cavities of water soluble dendrimer molecules with a diameter about 5 nm. The "dendritic box" is a fifth generation poly (propylene imine) consisting of 64 functional groups at the periphery. The trapped foreign molecules cannot diffuse out of the box. Only upon prolong heating, the trapped molecules were able to escape. It is possible to use the "dendritic box" as vehicle for drug delivery. Meijer's group has been designing a box that could be opened enzymatically or photochemically. This unique feature is also explored by Tomalia's group. The capability of hosting small organic molecules in water is the key to transport biological molecules.

Frechets group at Cornell is working on a dendrimer for chemotherapy. A chemotherapeutic drug is weakly bonded to the periphery of a dendrimer. Other functional groups add to the dendrimer to increase water solubility. Once the dendrimer reaches its target, the bond between the drug and the dendrimer is cleaved (enzymatically or photochemically). Since dendrimers are inert and stable, they are nontoxic to human. It was shown that dendrimers could eliminate through the kidneys as urine.

Willich, a German scientist is working on a dendrimer for magnetic resonance angiography and is currently entering clinical trials. The dendrimer is polylysine with gadolinium ion complexes on the end groups. Dendrimer provides multiple bonding sites on the periphery, allowing many MRI contrasting agent complexes to attach to one dendrimer. One dendrimer molecule can host up to twenty-four contrasting agent complexes and hence attaining higher signal-to-noise ratio. The dendrimer prevents any complex from diffusing into untargeted area. Thus, a better contrast MRI picture is obtained.

Several types of PAMAM dendrimers have been tested as gene vectors for gene therapy. Experiments showed that PAMAM dendrimers are effective transfection agents, providing high successful rate of transferring genetic materials into the cell. Dendrimers also tested in boron neutron capture therapy[15]. It is possible to attach boron complexes and antibodies on the surface of the dendrimer. While the antibodies target at the cancer cells, the boron complexes capture neutrons from an external source and release energetic radiation that can kill the cancer cells.

Dendrimer Films

Frechets group at Cornell is working on polyether dendrimers films that can isolate metal ions[3]. His target application is for signal amplification in fiber-optic communication technology. Apparently, his dendrimer coatings on the metal ions are able to prevent interference between ions when they excited by light. Crooks and Well at A&M are exploring the possibilities of using dendrimer films as sensitive interfaces for sensing applications. The dendrimers formed a thin monolayer film onto a gold surface. When it exposed to volatile organic compounds, the film was able to capture the volatile molecules. The contains of the film were then analyzed by a device called surface acoustic wave mass balance. The functionalities on the periphery of the dendrimer molecule could be modified to sense different organic compounds selectivly. Dendrimer films also serve as anti-corrosive coating on metal surfaces. The film is able to trap corrosive agent in the dendritic cavities, preventing any diffusion to the surface of the metal.

Interphases Applications

The convergent method is able to make unsymmetrical dendrimers. This arrangement allows the dendrimers to form monolayers at the gas-liquid interfaces or aqueous-organic interfaces. Amphiphilic dendrimer are useful in forming interfacial liquid membranes for stabilizing aqueous-organic emulsion. One other application is to use dendrimer film to extract chemical compounds between two phases. Dendrimers with carboxylate chain ends can form micelles in water. Their hydrophobic interiors dissolve organic molecules that are insoluble in water. They act like carrier for organic molecules in aqueous phase. This arrangement holds promise for the development of organic chemistry in aqueous medium. Hydrophilic dendrimers with hydrophobic functionalities on the periphery form micelles in organic solvents. These types of dendrimers can extract organic compound from the water phase to the organic phase. Dendrimers films also can use as purifiers. They can selectively permit molecules to diffuse through the interface.

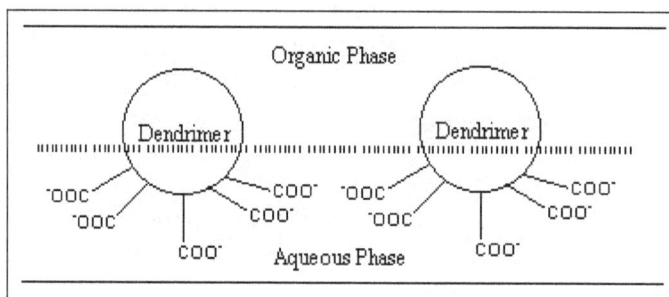

Liquid membrane of amphiphilic dendrimer at the interface between water and an immiscible organic solvent.

Catalysis and Reaction Sites

Catalysis is one of the most promising applications in dendrimer research. Dendrimers have nanoscopic cavities that act like microenvironment for molecular reactions. The cavities provide nanoscale reactor sites for catalysis. There are two possible catalytic sites being investigated, one at the core and the other at the surface respectively. Many attempts have been made on using dendrimers to enhance reaction rate and reaction selectivity. There is a micropolarity around the core, thereby influencing its molecular recognition and catalytic properties. One of the possible schemes for catalytic reaction in water is shown in figure. It is found that dendrimers are very useful in enantiomeric catalysis. Bolm et al had developed a dendrimer catalysis for enantioselective for reduction of benzaldehyd. The dendritic cavities provide a confined environment around the catalytic core and inducing regio and shape selectivity.

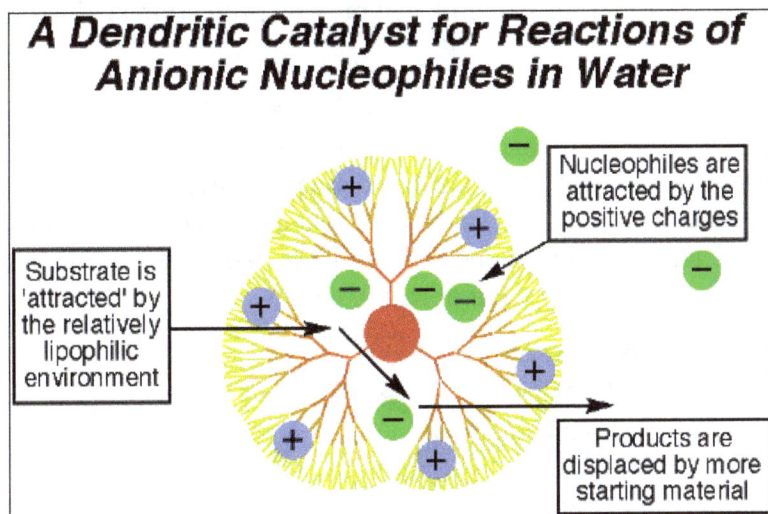

A Dendritic Catalyst for Reactions of Anionic Nucleophiles in Water

Nucleophiles are attracted by the positive charges

Substrate is 'attracted' by the relatively lipophilic environment

Products are displaced by more starting material

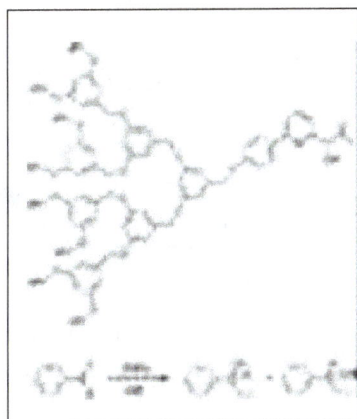

A third generation chiral catalyst as used by Bolm et al. in the diethylzinc addition to benzaldehyde.

The first peripheral catalytic site has been reported by Ford et al. Since there are multiple of catalytic sites concentrated at periphery and the dendrimer provides a good anchorage, the catalytic activity increase significantly comparing to one-dimensional catalytic site. Detty et al has been working on a dendrimer catalytic film that can kill algae and bacteria in sea water.

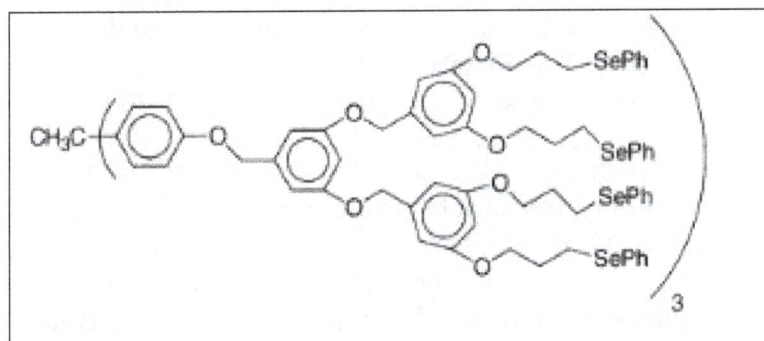

The selenides groups at the periphery catalyze peroxide activation of bromine Cation.

Phenylseleno groups are connected to a third generation dendrimeric polyether molecule. The selenides catalyze the oxidation of bromide ion with hydrogen peroxide to give positive bromine radicals[21]. These radicals are very toxic to algae and bacteria. Its potential application is in ship building industry. The dendrimer film on hulk will provide protection against the accumulation of algae.

Frechets group is investigating a dendrimer that can harvest light. The harvested light energy can transform into chemical energy for reactions, electric current or convert the energy into monochromic light. A laser dye has been placed at the dendritic core. The system acts like an optical amplifier. One other potential application is to use the dendrimer as a host for polymerization. It is possible for polymerization to occur in the cavities of the dendrimer and is well protected, therefore avoiding termination with other polymerization chains.

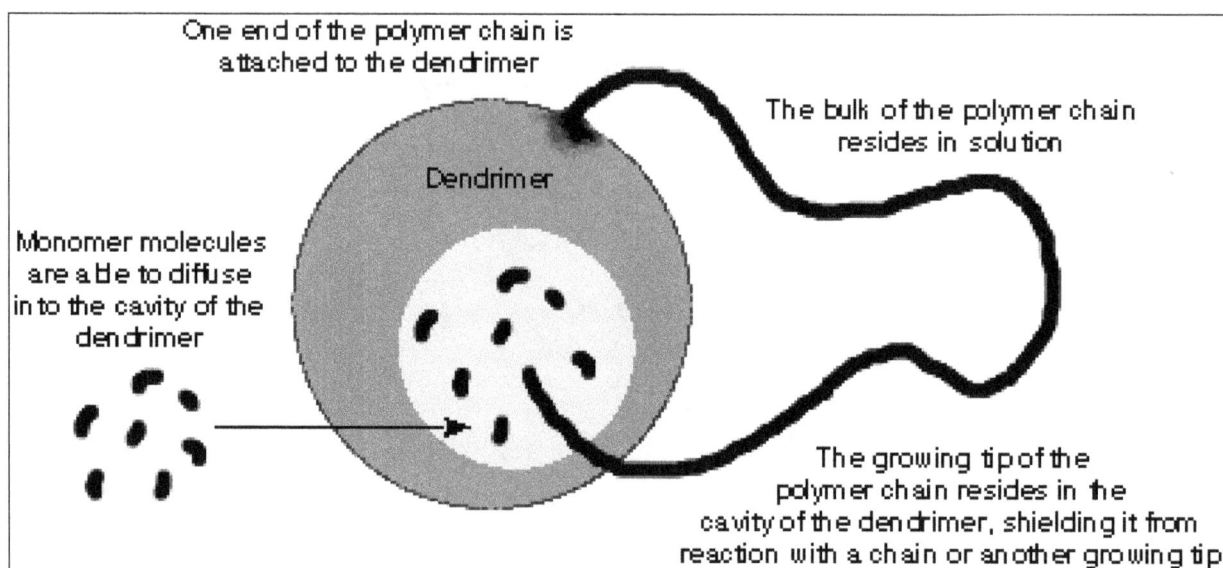

Scheme for control polymerization.

Applications of Polymers

Polymers are a highly diverse class of materials which are available in all fields of engineering from avionics through drug delivery system, bio-sensor devices, Holography, 3D printing, tissue engineering, cosmetics etc. and the improvement and usage of these depends on polymer applications. The applications of polymeric materials and their composites are still increasing rapidly due to their below average cost and ease of manufacture. This in turn fuels further development in research.

Research and development of bioplastics substances for medical, dental and pharmaceutical use have hovered on the front lines for years. Gelatin-based capsules made of animal or vegetable matter, for example, which naturally dissolve in the digestive tract, are in common use. Biodegradable stitches, which do not require manual removal after healing, are regularly used to suture wounds and surgical incisions.

Biopolymers are available as coatings for paper rather than the more common petrochemical coatings. Bioplastics are used for disposable items, such as packaging and crockery. They are also often used for bags, trays, fruit and vegetable containers and in marine sciences .These plastics are also used in non-disposable applications including mobile phone casings, carpet fibers, insulation car interiors, fuel lines, and plastic piping. New electroactive bioplastics are being developed that can be used to carry electric current. Medical implants made of **PLA** (polylactic acid), which dissolve in the body, can save patients a second operation. Compostable mulch films can also be produced from starch polymers and used in agriculture.

- In Aircraft, Aerospace and Sports Equipments.

- Polymers in Biochemistry.

- In Marine Sciences.

- Organic Polymer Flocculants in Water Purification.

- Polymers in Holography.

- 3D Printing Plastics.

- Tissue Engineering and Regenerative Medicine.

- Food Packaging and processing Industry.

References

- Polymers-283090639: researchgate.net, Retrieved 28 May, 2019

- Classification-of-polymers, chemistry-polymers: toppr.com, Retrieved 18 June, 2019

- Schamboeck V, Iedema PD, Kryven I (February 2019). "Dynamic Networks that Drive the Process of Irreversible Step-Growth Polymerization". Scientific Reports. 9 (1): 2276. doi:10.1038/s41598-018-37942-4. PMC 6381213. PMID 30783151

- Types-of-polymerization, chemistry-polymers: toppr.com, Retrieved 14 August, 2019

- Homogeneous-catalysis, introchem-chapter: courses.lumenlearning.com, Retrieved 16 June, 2019

- Dendrimers: courses.sens.buffalo.edu, Retrieved 28 May, 2019

- Applications-of-polymers: polymerchemistry.chemistryconferences.org, Retrieved 15 February, 2019

Nanomaterials and Nanochemistry

Materials that have a unit size between 1 to 1000 nanometers are referred to as nanomaterials. Nanochemistry combines the principles of chemistry and nanoscience to study the synthesis of size, shape, surface and defect properties. This chapter closely examines nanomaterials and nanochemistry to provide an extensive understanding of the subject.

Nanomaterials

Nanomaterials are cornerstones of nanoscience and nanotechnology. Nanostructure science and technology is a broad and interdisciplinary area of research and development activity that has been growing explosively worldwide in the past few years. It has the potential for revolutionizing the ways in which materials and products are created and the range and nature of functionalities that can be accessed. It is already having a significant commercial impact, which will assuredly increase in the future.

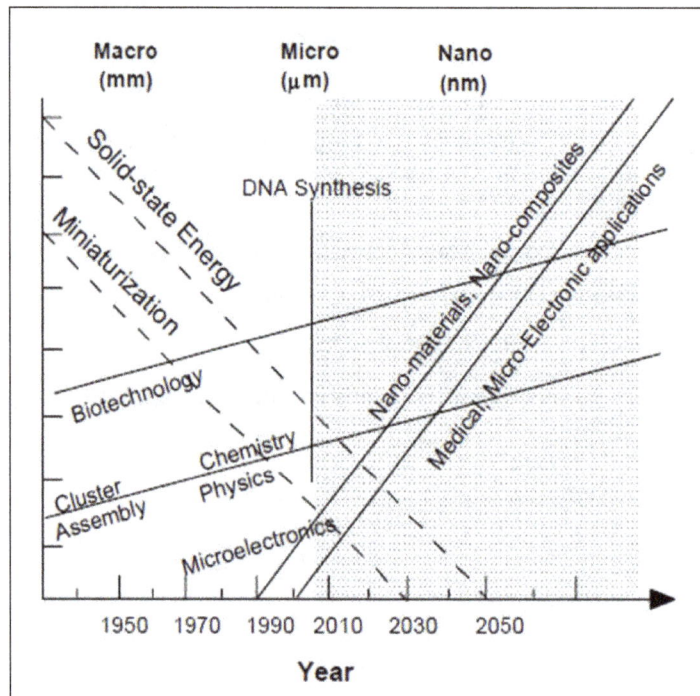

Evolution of science & technology and the future.

Nanoscale materials are defined as a set of substances where at least one dimension is less than approximately 100 nanometers. A nanometer is one millionth of a millimetre - approximately 100,000 times smaller than the diameter of a human hair. Nanomaterials are of interest because at this scale unique optical, magnetic, electrical, and other properties emerge. These emergent properties have the potential for great impacts in electronics, medicine, and other fields.

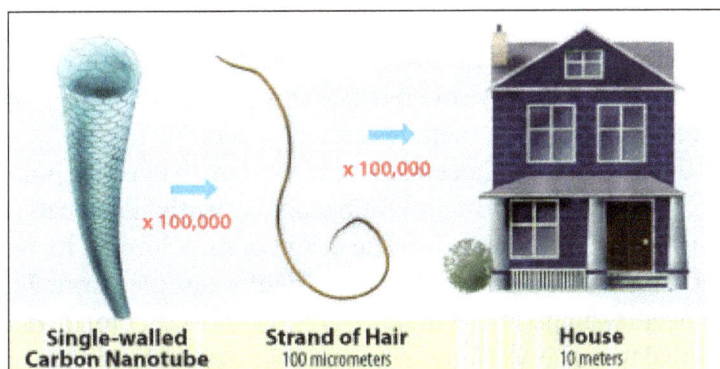

Nanomaterial (For example: Carbon nanotube).

Where are Nanomaterials Found?

Some nanomaterials occur naturally, but of particular interest are engineered nanomaterials (EN), which are designed for, and already being used in many commercial products and processes. They can be found in such things as sunscreens, cosmetics, sporting goods, stainresistant clothing, tires, electronics, as well as many other everyday items, and are used in medicine for purposes of diagnosis, imaging and drug delivery.

Engineered nanomaterials are resources designed at the molecular (nanometre) level to take advantage of their small size and novel properties which are generally not seen in their conventional, bulk counterparts. The two main reasons why materials at the nano scale can have different properties are increased relative surface area and new quantum effects. Nanomaterials have a much greater surface area to volume ratio than their conventional forms, which can lead to greater chemical reactivity and affect their strength. Also at the nano scale, quantum effects can become much more important in determining the materials properties and characteristics, leading to novel optical, electrical and magnetic behaviours.

Nanomaterials are already in commercial use, with some having been available for several years or decades. The range of commercial products available today is very broad, including stainresistant and wrinkle-free textiles, cosmetics, sunscreens, electronics, paints and varnishes. Nanocoatings and nanocomposites are finding uses in diverse consumer products, such as windows, sports equipment, bicycles and automobiles. There are novel UV-blocking coatings on glass bottles which protect beverages from damage by sunlight, and longer-lasting tennis balls using butyl-rubber/nano-clay composites. Nanoscale titanium dioxide, for instance, is finding applications in cosmetics, sun-block creams and self-cleaning windows, and nanoscale silica is being used as filler in a range of products, including cosmetics and dental fillings.

The history of nanomaterials began immediately after the big bang when Nanostructures were formed in the early meteorites. Nature later evolved many other Nanostructures like seashells, skeletons etc. Nanoscaled smoke particles were formed during the use of fire by early humans. The scientific story of nanomaterials however began much later. One of the first scientific report is the colloidal gold particles synthesized by Michael Faraday as early as 1857. Nanostructured catalysts have also been investigated for over 70 years. By the early 1940's, precipitated and fumed silica nanoparticles were being manufactured and sold in USA and Germany as substitutes for ultrafine carbon black for rubber reinforcements.

Nanosized amorphous silica particles have found large-scale applications in many every-day consumer products, ranging from non-diary coffee creamer to automobile tires, optical fibers and catalyst supports. In the 1960s and 1970's metallic nanopowders for magnetic recording tapes were developed. In 1976, for the first time, nanocrystals produced by the now popular inert- gas evaporation technique was published by Granqvist and Buhrman. Recently it has been found that the Maya blue paint is a nanostructured hybrid material. The origin of its color and its resistance to acids and biocorrosion are still not understood but studies of authentic samples from Jaina Island show that the material is made of needle-shaped palygorskite (clay) crystals that form a superlattice with a period of 1.4 nm, with intercalates of amorphous silicate substrate containing inclusions of metal (Mg) nanoparticles. The beautiful tone of the blue color is obtained only when both these nanoparticles and the superlattice are present, as has been shown by the fabrication of synthetic samples.

Today nanophase engineering expands in a rapidly growing number of structural and functional materials, both inorganic and organic, allowing to manipulate mechanical, catalytic, electric, magnetic, optical and electronic functions. The production of nanophase or cluster-assembled materials is usually based upon the creation of separated small clusters which then are fused into a bulk-like material or on their embedding into compact liquid or solid matrix materials. e.g. nanophase silicon, which differs from normal silicon in physical and electronic properties, could be applied to macroscopic semiconductor processes to create new devices. For instance, when ordinary glass is doped with quantized semiconductor "colloids," it becomes a high performance optical medium with potential applications in optical computing.

Classification of Nanomaterials

Nanomaterials have extremely small size which having at least one dimension 100 nm or less. Nanomaterials can be nanoscale in one dimension (eg. surface films), two dimensions (eg. strands or fibres), or three dimensions (eg. particles). They can exist in single, fused, aggregated or agglomerated forms with spherical, tubular, and irregular shapes. Common types of nanomaterials include nanotubes, dendrimers, quantum dots and fullerenes. Nanomaterials have applications in the field of nano technology, and displays different physical chemical characteristics from normal chemicals (i.e., silver nano, carbon nanotube, fullerene, photocatalyst, carbon nano, silica).

According to Siegel, Nanostructured materials are classified as Zero dimensional, one dimensional, two dimensional, three dimensional nanostructures.

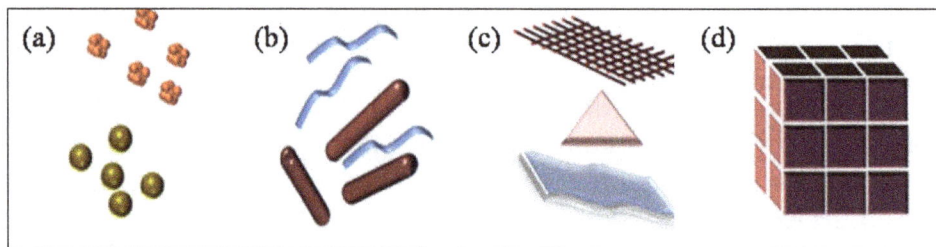

Classification of Nanomaterials (a) 0D spheres and clusters; (b) 1D nanofibers, nanowires, and nanorods; (c) 2D nanofilms, nanoplates, and networks; (d) 3D nanomaterials.

Nanomaterials are materials which are characterized by an ultra fine grain size (< 50 nm) or by a dimensionality limited to 50 nm. Nanomaterials can be created with various modulation dimensionalities as defined by Richard W. Siegel: zero (atomic clusters, filaments and cluster assemblies), one (multilayers), two (ultrafine-grained overlayers or buried layers), and three (nanophase materials consisting of equiaxed nanometer sized grains) as shown in the above figure.

These materials have created a high interest in recent years by virtue of their unusual mechanical, electrical, optical and magnetic properties. Some examples are given below:

- Nanophase ceramics are of particular interest because they are more ductile at elevated temperatures as compared to the coarse-grained ceramics.

- Nanostructured semiconductors are known to show various non-linear optical properties. Semiconductor Q-particles also show quantum confinement effects which may lead to special properties, like the luminescence in silicon powders and silicon germanium quantum dots as infrared optoelectronic devices. Nanostructured semiconductors are used as window layers in solar cells.

- Nanosized metallic powders have been used for the production of gas tight materials, dense parts and porous coatings. Cold welding properties combined with the ductility make them suitable for metal-metal bonding especially in the electronic industry.

- Single nanosized magnetic particles are mono-domains and one expects that also in magnetic nanophase materials the grains correspond with domains, while boundaries on the contrary to disordered walls. Very small particles have special atomic structures with discrete electronic states, which give rise to special properties in addition to the superparamagnetism behaviour. Magnetic nanocomposites have been used for mechanical force transfer (ferrofluids), for high density information storage and magnetic refrigeration.

- Nanostructured metal clusters and colloids of mono- or plurimetallic composition have a special impact in catalytic applications. They may serve as precursors for new type of heterogeneous catalysts (Cortex-catalysts) and have been shown to offer substantial advantages concerning activity, selectivity and lifetime in chemical transformations and electrocatalysis (fuel cells). Enantioselective catalysis was also achieved using chiral modifiers on the surface of nanoscale metal particles.

- Nanostructured metal-oxide thin films are receiving a growing attention for the realization of gas sensors (NO_x, CO, CO_2, CH_4 and aromatic hydrocarbons) with enhanced sensitivity and selectivity. Nanostructured metal-oxide (MnO_2) finds application for rechargeable batteries for cars or consumer goods. Nanocrystalline silicon films for highly transparent contacts in thin film solar cell and nano-structured titanium oxide porous films for its high transmission and significant surface area enhancement leading to strong absorption in dye sensitized solar cells.

- Polymer based composites with a high content of inorganic particles leading to a high dielectric constant are interesting materials for photonic band gap structure.

Examples of Nanomaterials

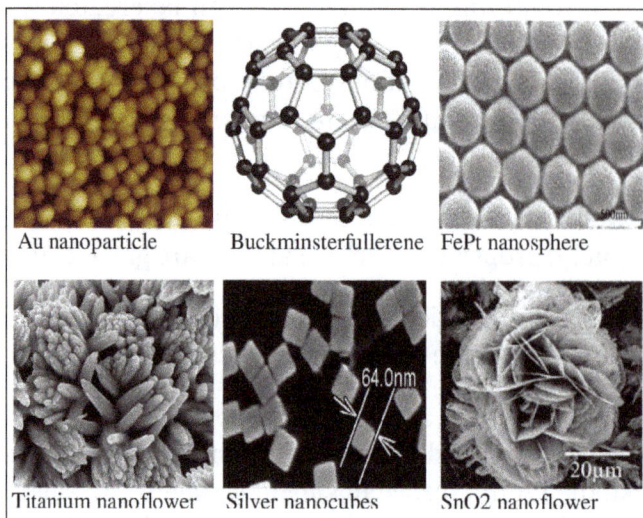

Au nanoparticle   Buckminsterfullerene   FePt nanosphere

Titanium nanoflower   Silver nanocubes   SnO2 nanoflower

Nanomaterial Synthesis and Processing

We are dealing with very fine structures: a nanometer is a billionth of a meter. This indeed allows us to think in both the 'bottom up' or the 'top down' approaches to synthesize nanomaterials, i.e. either to assemble atoms together or to dis-assemble (break, or dissociate) bulk solids into finer pieces until they are constituted of only a few atoms. This domain is a pure example of interdisciplinary work encompassing physics, chemistry, and engineering upto medicine.

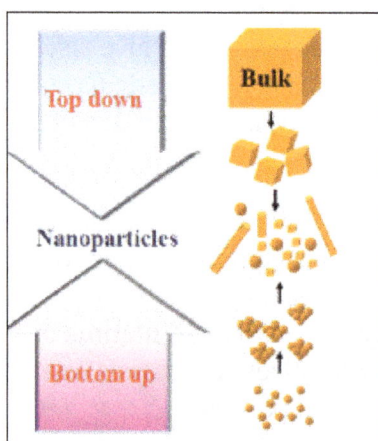

Schematic illustration of the preparative methods of nanoparticles.

Methods for Creating Nanostructures

There are many different ways of creating nanostructures: of course, macromolecules or nanoparticles or buckyballs or nanotubes and so on can be synthesized artificially for certain specific materials. They can also be arranged by methods based on equilibrium or nearequilibrium thermodynamics such as methods of self-organization and self-assembly (sometimes also called bio-mimetic processes). Using these methods, synthesized materials can be arranged into useful shapes so that finally the material can be applied to a certain application.

Mechanical Grinding

Mechanical attrition is a typical example of 'top down' method of synthesis of nanomaterials, where the material is prepared not by cluster assembly but by the structural decomposition of coarser-grained structures as the result of severe plastic deformation. This has become a popular method to make nanocrystalline materials because of its simplicity, the relatively inexpensive equipment needed, and the applicability to essentially the synthesis of all classes of materials. The major advantage often quoted is the possibility for easily scaling up to tonnage quantities of material for various applications. Similarly, the serious problems that are usually cited are:

- Contamination from milling media and/or atmosphere.

- To consolidate the powder product without coarsening the nanocrystalline microstructure.

In fact, the contamination problem is often given as a reason to dismiss the method, at least for some materials. Here we will review the mechanisms presently believed responsible for formation of nanocrystalline structures by mechanical attrition of single phase powders, mechanical alloying of dissimilar powders, and mechanical crystallisation of amorphous materials. The two important problems of contamination and powder consolidation will be briefly considered.

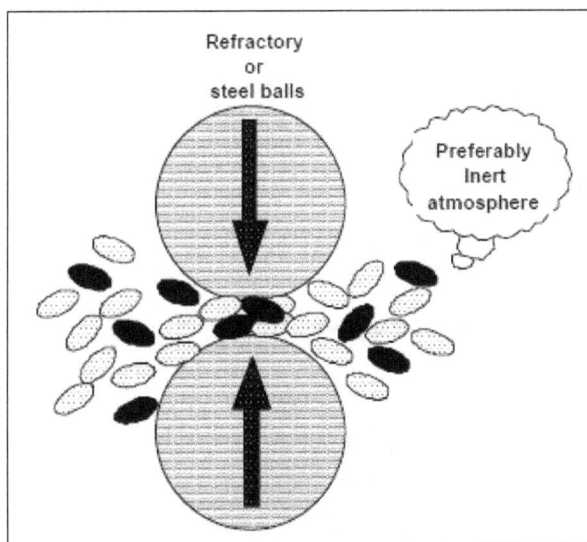

Schematic representation of the principle of mechanical milling.

Mechanical milling is typically achieved using high energy shaker, planetary ball, or tumbler mills. The energy transferred to the powder from refractory or steel balls depends on the rotational (vibrational) speed, size and number of the balls, ratio of the ball to powder mass, the time of milling and the milling atmosphere. Nanoparticles are produced by the shear action during grinding.

Milling in cryogenic liquids can greatly increase the brittleness of the powders influencing the fracture process. As with any process that produces fine particles, an adequate step to prevent oxidation is necessary. Hence this process is very restrictive for the production of non-oxide materials since then it requires that the milling take place in an inert atmosphere and that the powder particles be handled in an appropriate vacuum system or glove box. This method of synthesis is suitable for producing amorphous or nanocrystalline alloy particles, elemental or compound powders. If the mechanical milling imparts sufficient energy to the constituent powders a homogeneous alloy can be formed. Based on the energy of the milling process and thermodynamic properties of the constituents the alloy can be rendered amorphous by this processing.

Wet Chemical Synthesis of Nanomaterials

In principle we can classify the wet chemical synthesis of nanomaterials into two broad groups:

- The top down method: where single crystals are etched in an aqueous solution for producing nanomaterials, For example, the synthesis of porous silicon by electrochemical etching.

- The bottom up method: consisting of sol-gel method, precipitation etc. where materials containing the desired precursors are mixed in a controlled fashion to form a colloidal solution.

Sol-gel Process

The sol-gel process, involves the evolution of inorganic networks through the formation of a colloidal suspension (sol) and gelation of the sol to form a network in a continuous liquid phase (gel). The precursors for synthesizing these colloids consist usually of a metal or metalloid element surrounded by various reactive ligands. The starting material is processed to form a dispersible oxide and forms a sol in contact with water or dilute acid. Removal of the liquid from the sol yields the gel, and the sol/gel transition controls the particle size and shape. Calcination of the gel produces the oxide.

Sol-gel processing refers to the hydrolysis and condensation of alkoxide-based precursors such as $Si(OEt)_4$ (tetraethyl orthosilicate, or TEOS). The reactions involved in the sol-gel chemistry based on the hydrolysis and condensation of metal alkoxides M(OR)z can be described as follows:

$$MOR + H_2O \rightarrow MOH + ROH \, (hydrolysis)$$

$$MOH + ROM \rightarrow M\text{-}O\text{-}M + ROH \, (condensation)$$

Sol-gel method of synthesizing nanomaterials is very popular amongst chemists and is widely employed to prepare oxide materials. The sol-gel process can be characterized by a series of distinct steps.

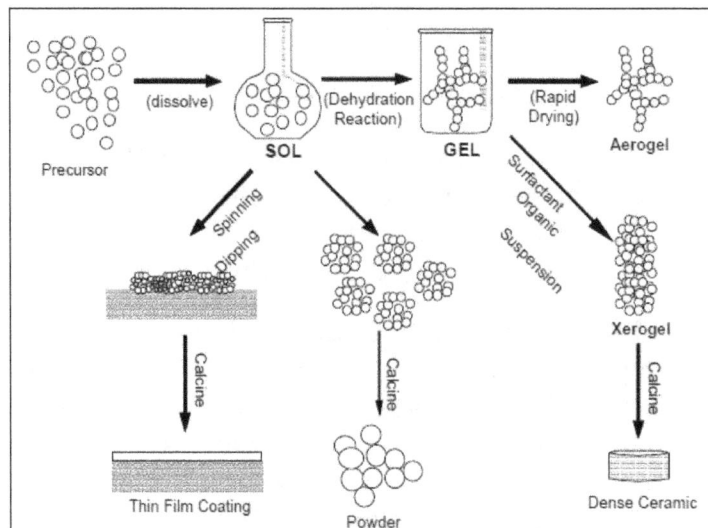

Schematic representation of sol-gel process of synthesis of nanomaterials.

- Formation of different stable solutions of the alkoxide or solvated metal precursor.

- Gelation resulting from the formation of an oxide- or alcohol- bridged network (the gel) by a polycondensation reaction that results in a dramatic increase in the viscocity of the solution.

- Aging of the gel (Syneresis), during which the polycondensation reactions continue until the gel transforms into a solid mass, accompanied by contraction of the gel network and expulsion of solvent from gel pores. Ostwald ripening (also referred to as coarsening, is the phenomenon by which smaller particles are consumed by larger particles during the growth process) and phase transformations may occur concurrently with syneresis. The aging process of gels can exceed 7 days and is critical to the prevention of cracks in gels that have been cast.

- Drying of the gel, when water and other volatile liquids are removed from the gel network. This process is complicated due to fundamental changes in the structure of the gel. The drying process has itself been broken into four distinct steps: (i) the constant rate period, (ii) the critical point, (iii) the falling rate period, (iv) the second falling rate period. If isolated by thermal evaporation, the resulting monolith is termed a xerogel. If the solvent (such as water) is extracted under supercritical or near super critical conditions, the product is an aerogel.

- Dehydration, during which surface- bound M-OH groups are removed, there by stabilizing the gel against rehydration. This is normally achieved by calcining the monolith at temperatures up to 8000C.

- Densification and decomposition of the gels at high temperatures (T>8000C). The pores of the gel network are collapsed, and remaining organic species are volatilized. The typical steps that are involved in sol-gel processing are shown in the schematic diagram below.

The interest in this synthesis method arises due to the possibility of synthesizing nonmetallic inorganic materials like glasses, glass ceramics or ceramic materials at very low temperatures compared to the high temperature process required by melting glass or firing ceramics.

The major difficulties to overcome in developing a successful bottom-up approach is controlling the growth of the particles and then stopping the newly formed particles from agglomerating. Other technical issues are ensuring the reactions are complete so that no unwanted reactant is left on the product and completely removing any growth aids that may have been used in the process. Also production rates of nano powders are very low by this process. The main advantage is one can get monosized nano particles by any bottom up approach.

Gas Phase Synthesis of Nanomaterials

The gas-phase synthesis methods are of increasing interest because they allow elegant way to control process parameters in order to be able to produce size, shape and chemical composition controlled nanostructures. In conventional chemical vapour deposition (CVD) synthesis, gaseous products either are allowed to react homogeneously or heterogeneously depending on a particular application.

- In homogeneous CVD, particles form in the gas phase and diffuse towards a cold surface due to thermophoretic forces, and can either be scrapped of from the cold surface to give nano-powders, or deposited onto a substrate to yield what is called 'particulate films'.

- In heterogeneous CVD, the solid is formed on the substrate surface, which catalyses the reaction and a dense film is formed.

In order to form nanomaterials several modified CVD methods have been developed. Gas phase processes have inherent advantages, some of which are noted here:

- An excellent control of size, shape, crystallinity and chemical composition.

- Highly pure materials can be obtained.

- Multicomonent systems are relatively easy to form.

- Easy control of the reaction mechanisms.

Most of the synthesis routes are based on the production of small clusters that can aggregate to form nano particles (condensation). Condensation occurs only when the vapour is supersaturated and in these processes homogeneous nucleation in the gas phase is utilised to form particles. This can be achieved both by physical and chemical methods.

Furnace

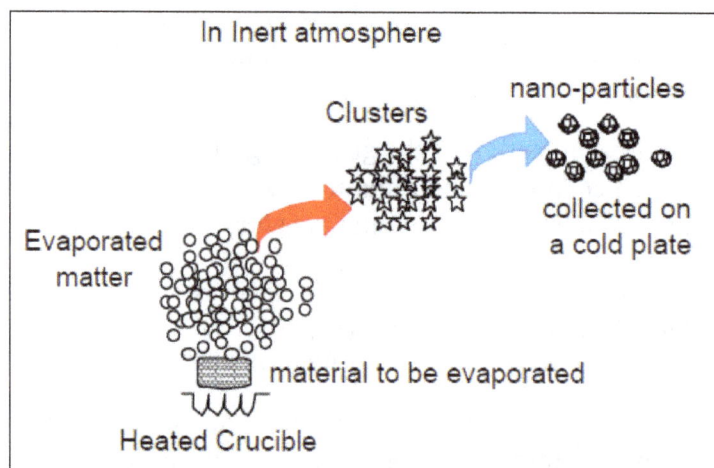

Schematic representation of gas phase process of synthesis
of single phase nanomaterials from a heated crucible.

The simplest fashion to produce nanoparticles is by heating the desired material in a heatresistant crucible containing the desired material. This method is appropriate only for materials that have a high vapour pressure at the heated temperatures that can be as high as 2000°C. Energy is normally introduced into the precursor by arc heating, electron-beam heating or Joule heating. The atoms are evaporated into an atmosphere, which is either inert (e.g. He) or reactive (so as to form a compound). To carry out reactive synthesis, materials with very low vapour pressure have to be fed into the furnace in the form of a suitable precursor such as organometallics, which decompose in the furnace to produce a condensable material. The hot atoms of the evaporated matter lose energy by collision with the atoms of the cold gas and undergo condensation into small clusters via homogeneous nucleation. In case a compound is being synthesized, these precursors react in the gas phase and form a compound with the material that is separately injected in the reaction chamber. The clusters would continue to grow if they remain in the supersaturated region. To control their size, they need to be rapidly removed from the supersaturated environment by a carrier gas. The cluster size and its distribution are controlled by only three parameters:

- The rate of evaporation (energy input).

- The rate of condensation (energy removal).

- The rate of gas flow (cluster removal).

Because of its inherent simplicity, it is possible to scale up this process from laboratory (mg/day) to industrial scales (tons/day).

Flame Assisted Ultrasonic Spray Pyrolysis

In this process, precusrsors are nebulized and then unwanted components are burnt in a flame to get the required material, eg. ZrO_2 has been obtained by this method from a precursor of Zr $(CH_3 CH_2 CH_2O)_4$. Flame hydrolysis that is a variant of this process is used for the manufacture of fused silica. In the process, silicon tetrachloride is heated in an oxy-hydrogen flame to give highly dispersed silica. The resulting white amorphous powder consists of spherical particles with sizes in the range 7-40 nm. The combustion flame synthesis, in which the burning of a gas mixture, e.g. acetylene and oxygen or hydrogen and oxygen, supplies the energy to initiate the pyrolysis of precursor compounds, is widely used for the industrial production of powders in large quantities, such as carbon black, fumed silica and titanium dioxide. However, since the gas pressure during the reaction is high, highly agglomerated powders are produced which is disadvantageous for subsequent processing. The basic idea of low pressure combustion flame synthesis is to extend the pressure range to the pressures used in gas phase synthesis and thus to reduce or avoid the agglomeration. Low pressure flames have been extensively used by aerosol scientists to study particle formation in the flame.

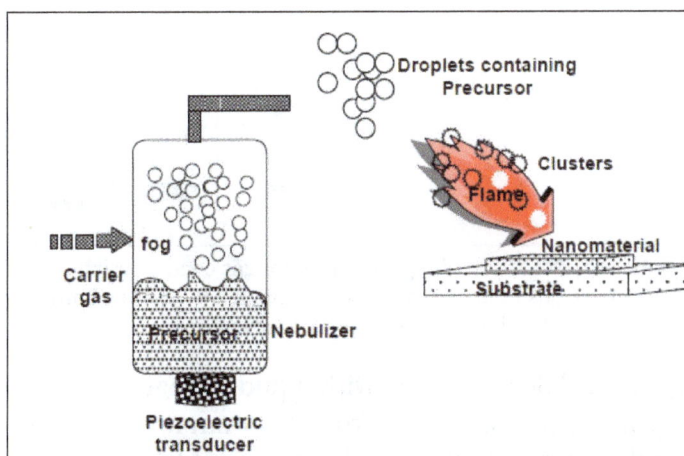

Flame assisted ultrasonic spray pyrolysis.

A key for the formation of nanoparticles with narrow size distributions is the exact control of the flame in order to obtain a flat flame front. Under these conditions the thermal history, i.e. time and temperature, of each particle formed is identical and narrow distributions result. However, due to the oxidative atmosphere in the flame, this synthesis process is limited to the formation of oxides in the reactor zone.

Gas Condensation Processing

In this technique, a metallic or inorganic material, e.g. a suboxide, is vaporised using thermal evaporation sources such as crucibles, electron beam evaporation devices or sputtering sources in an atmosphere of 1-50 mbar He (or another inert gas like Ar, Ne, Kr). Cluster form in the vicinity of the source by homogenous nucleation in the gas phase and grow by coalescence and incorporation of atoms from the gas phase. The cluster or particle size depends critically on the residence time of the particles in the growth system and can be influenced by the gas pressure, the kind of inert gas, i.e. He, Ar or Kr, and on the evaporation rate/vapour pressure of the evaporating material. With increasing gas pressure, vapour pressure and mass of the inert gas used the average particle size of the nanoparticles increases. Lognormal size distributions have been found experimentally and have been explained theoretically by the growth mechanisms of the particles. Even in more complex processes such as the low pressure combustion flame synthesis where a number of chemical reactions are involved the size distributions are determined to be lognormal.

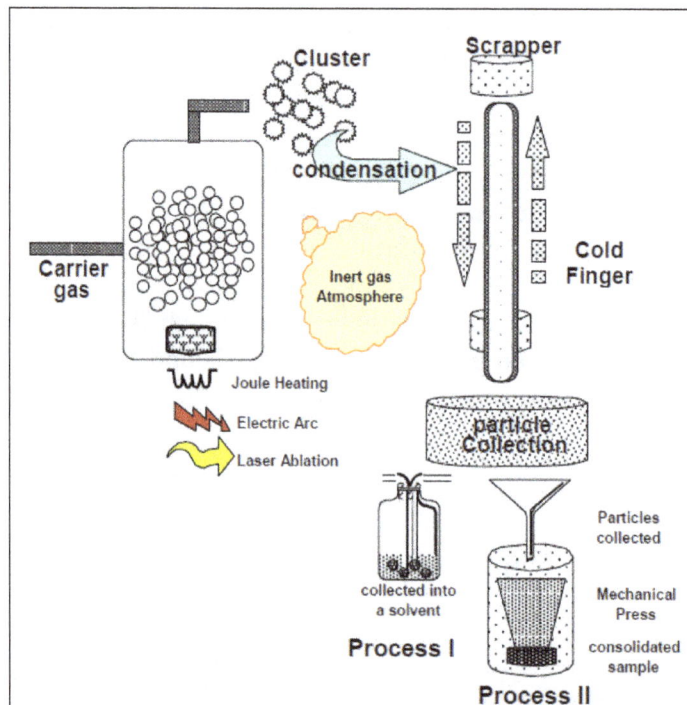

Schematic representation of typical set-up for gas condensation synthesis of nanomaterials followed by consolidation in a mechanical press or collection in an appropriate solvent media.

Originally, a rotating cylindrical device cooled with liquid nitrogen was employed for the particle collection: the nanoparticles in the size range from 2-50 nm are extracted from the gas flow by thermophoretic forces and deposited loosely on the surface of the collection device as a powder of low density and no agglomeration. Subsequently, the nanoparticles are removed from the surface of the cylinder by means of a scraper in the form of a metallic plate. In addition to this cold finger device several techniques known from aerosol science have now been implemented for the use in gas condensation systems such as corona discharge, etc. These methods allow for the continuous operation of the collection device and are better suited for larger scale synthesis of nanopowders. However, these methods can only be used in a system designed for gas flow, i.e. a dynamic vacuum is generated by means of both continuous pumping and gas inlet via mass flow controller. A major advantage over convectional gas flow is the improved control of the particle sizes. It has been found that the particle size distributions in gas flow systems, which are also lognormal, are shifted towards smaller average values with an appreciable reduction of the standard deviation of the distribution. Depending on the flow rate of the He-gas, particle sizes are reduced by 80% and standard deviations by 18%.

The synthesis of nanocrystalline pure metals is relatively straightforward as long as evaporation can be done from refractory metal crucibles (W, Ta or Mo). If metals with high melting points or metals which react with the crucibles, are to be prepared, sputtering, i.e. for W and Zr, or laser or electron beam evaporation has to be used. Synthesis of alloys or intermetallic compounds by thermal evaporation can only be done in the exceptional cases that the vapour pressures of the elements are similar. As an alternative, sputtering from an alloy or mixed target can be employed. Composite materials such as Cu/Bi or W/Ga have been synthesised by simultaneous evaporation from two separate crucibles onto a rotating collection device. It has been found that excellent intermixing on the scale of the particle size can be obtained.

However, control of the composition of the elements has been difficult and reproducibility is poor. Nanocrystalline oxide powders are formed by controlled postoxidation of primary nanoparticles of a pure metal (e.g. Ti to TiO_2) or a suboxide (e.g. ZrO to ZrO_2). Although the gas condensation method including the variations have been widely employed to prepared a variety of metallic and ceramic materials, quantities have so far been limited to a laboratory scale. The quantities of metals are below 1 g/day, while quantities of oxides can be as high as 20 g/day for simple oxides such as CeO_2 or ZrO_2. These quantities are sufficient for materials testing but not for industrial production. However, it should be mentioned that the scale-up of the gas condensation method for industrial production of nanocrystalline oxides by a company called nanophase technologies has been successful.

Chemical Vapour Condensation

As shown schematically in figure, the evaporative source used in GPC is replaced by a hot wall reactor in the Chemical Vapour Condensation or the CVC process. Depending on the processing parameters nucleation of nanoparticles is observed during chemical vapour deposition (CVC) of thin films and poses a major problem in obtaining good film qualities. The original idea of the novel CVC process which is schematically shown below where, it was intended to adjust the parameter field during the synthesis in order to suppress film formation and enhance homogeneous nucleation of particles in the gas flow. It is readily found that the residence time of the precursor in the reactor determines if films or particles are formed. In a certain range of residence time both particle and film formation can be obtained.

Adjusting the residence time of the precursor molecules by changing the gas flow rate, the pressure difference between the precursor delivery system and the main chamber occurs. Then the temperature of the hot wall reactor results in the fertile production of nanosized particles of metals and ceramics instead of thin films as in CVD processing. In the simplest form a metal organic precursor is introduced into the hot zone of the reactor using mass flow controller. Besides the increased quantities in this continuous process compared to GPC has been demonstrated that a wider range of ceramics including nitrides and carbides can be synthesised. Additionally, more complex oxides such as $BaTiO_3$ or composite structures can be formed as well. Appropriate precursor compounds can be readily found in the CVD literature. The extension to production of nanoparticles requires the determination of a modified parameter field in order to promote particle formation instead of film formation. In addition to the formation of single phase nanoparticles by CVC of a single precursor the reactor allows the synthesis of:

- Mixtures of nanoparticles of two phases or doped nanoparticles by supplying two precursors at the front end of the reactor.

- Coated nanoparticles, i.e., n-ZrO_2 coated with n-Al2O_3 or vice versa, by supplying a second precursor at a second stage of the reactor. In this case nanoparticles which have been formed by homogeneous nucleation are coated by heterogeneous nucleation in a second stage of the reactor.

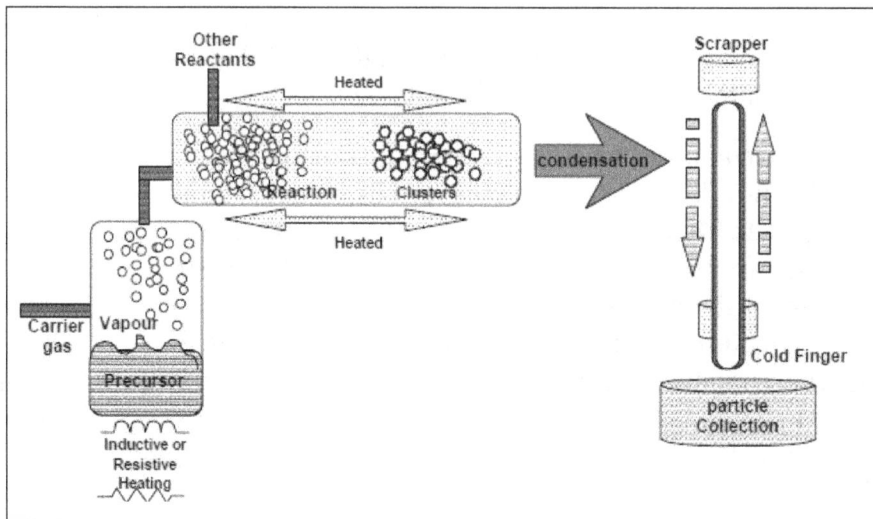

A schematic of a typical CVC reactor.

Because CVC processing is continuous, the production capabilities are much larger than in GPC processing. Quantities in excess of 20 g/hr have been readily produced with a small scale laboratory reactor. A further expansion can be envisaged by simply enlarging the diameter of the hot wall reactor and the mass flow through the reactor.

Sputtered Plasma Processing

In this method is yet again a variation of the gas-condensation method excepting the fact that the source material is a sputtering target and this target is sputtered using rare gases and the constituents are allowed to agglomerate to produce nanomaterial. Both dc (direct current) and rf (radio-frequency) sputtering has been used to synthesize nanoparticles. Again reactive sputtering or multitarget sputtering has been used to make alloys and/or oxides, carbides, nitrides of materials. This method is specifically suitable for the preparation of ultrapure and non-agglomerated nanoparticles of metal.

Microwave Plasma Processing

This technique is similar to CVC method but employs plasma instead of high temperature for decomposition of the metal organic precursors. The method uses microwave plasma in a 50 mm diameter reaction vessel made of quartz placed in a cavity connected to a microwave generator. A precursor such as a chloride compound is introduced into the front end of the reactor. Generally, the microwave cavity is designed as a single mode cavity using the TE10 mode in a WR975 waveguide with a frequency of 0.915 GHz. The major advantage of the plasma assisted pyrolysis in contrast to the thermal activation is the low temperature reaction which reduces the tendency for agglomeration of the primary particles. This is also true in the case of plasma-CVD processes. Additionally, it has been shown that by introducing another precursor into a second reaction zone of the tubular reactor, e.g. by splitting the microwave guide tubes, the primary particles can be coated with a second phase. For example, it has been demonstrated that ZrO_2 nanoparticles can be coated by $Al2O_3$. In this case the inner ZrO_2 core is crystalline, while the $Al2O_3$ coating is amorphous. The reaction sequence can be reversed with the result that an amorphous $Al2O_3$ core is coated with crystalline ZrO_2. While the formation of the primary particles occurs by homogeneous nucleation, it can be easily estimated using gas reaction kinetics that the coating on the primary particles grows heterogeneously and that homogeneous nucleation of nanoparticles originating from the second compound has a very low probability. A schematic representation of the particle growth in plasma's is given below.

Particle Precipitation Aided CVD

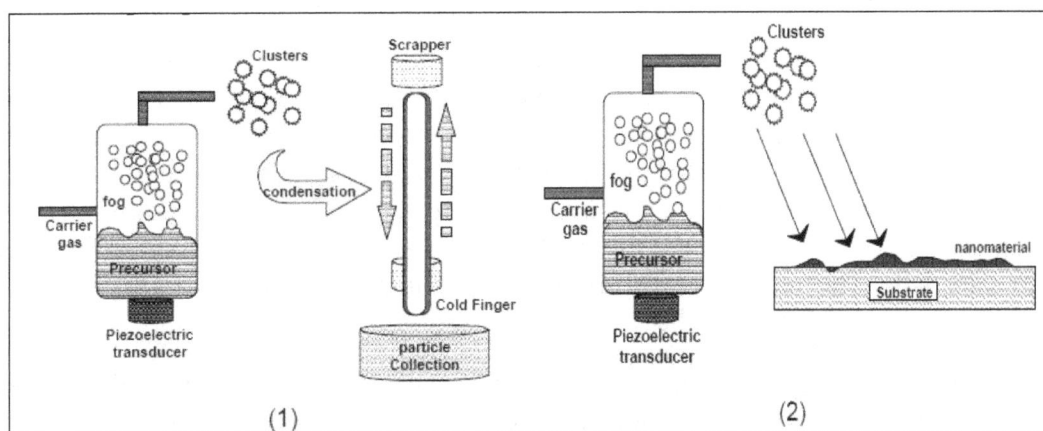

Schematic representation of (1) nanoparticle, and (2) particulate film formation.

In another variation of this process, colloidal clusters of materials are used to prepare nanoparticles. The CVD reaction conditions are so set that particles form by condensation in the gas phase and collect onto a substrate, which is kept under a different condition that allows heterogeneous nucleation. By this method both nanoparticles and particulate films can be prepared. An example of this method has been used to form nanomaterials eg. SnO_2, by a method called pyrosol deposition process, where clusters of tin hydroxide are transformed into small aerosol droplets, following which they are reacted onto a heated glass substrate.

Laser Ablation

Laser ablation has been extensively used for the preparation of nanoparticles and particulate films. In this process a laser beam is used as the primary excitation source of ablation for generating clusters directly from a solid sample in a wide variety of applications. The small dimensions of the particles and the possibility to form thick films make this method quite an efficient tool for the production of ceramic particles and coatings and also an ablation source for analytical applications such as the coupling to induced coupled plasma emission spectrometry, ICP, the formation of the nanoparticles has been explained following a liquefaction process which generates an aerosol, followed by the cooling/solidification of the droplets which results in the formation of fog. The general dynamics of both the aerosol and the fog favors the aggregation process and micrometer-sized fractal-like particles are formed. The laser spark atomizer can be used to produce highly mesoporous thick films and the porosity can be modified by the carrier gas flow rate. ZrO_2 and SnO_2 nanoparticulate thick films were also synthesized successfully using this process with quite identical microstructure. Synthesis of other materials such as lithium manganate, silicon and carbon has also been carried out by this technique.

Properties of Nanomaterials

Nanomaterials have the structural features in between of those of atoms and the bulk materials. While most microstructured materials have similar properties to the corresponding bulk materials, the properties of materials with nanometer dimensions are significantly different from those of atoms and bulks materials. This is mainly due to the nanometer size of the materials which render them: (i) large fraction of surface atoms; (ii) high surface energy; (iii) spatial confinement; (iv) reduced imperfections, which do not exist in the corresponding bulk materials.

Due to their small dimensions, nanomaterials have extremely large surface area to volume ratio, which makes a large to be the surface or interfacial atoms, resulting in more "surface" dependent material properties. Especially when the sizes of nanomaterials are comparable to length, the entire material will be affected by the surface properties of nanomaterials. This in turn may enhance or modify the properties of the bulk materials. For example, metallic nanoparticles can be used as very active catalysts. Chemical sensors from nanoparticles and nanowires enhanced the sensitivity and sensor selectivity. The nanometer feature sizes of nanomaterials also have spatial confinement effect on the materials, which bring the quantum effects.

The energy band structure and charge carrier density in the materials can be modified quite differently from their bulk and in turn will modify the electronic and optical properties of the materials. For example, lasers and light emitting diodes (LED) from both of the quantum dots and quantum wires are very promising in the future optoelections. High density information storage using quantum dot devices is also a fast developing area. Reduced imperfections are also an important factor in determination of the properties of the nanomaterials. Nanosturctures and Nanomaterials favors of a self-purification process in that the impurities and intrinsic material defects will move to near the surface upon thermal annealing. This increased materials perfection affects the properties of nanomaterials. For example, the chemical stability for certain nanomaterials may be enhanced, the mechanical properties of nanomaterials will be better than the bulk materials. The superior mechanical properties of carbon nanotubes are well known. Due to their nanometer size, nanomaterials are already known to have many novel properties. Many novel applications of the nanomaterials rose from these novel properties have also been proposed.

Optical Properties

One of the most fascinating and useful aspects of nanomaterials is their optical properties. Applications based on optical properties of nanomaterials include optical detector, laser, sensor, imaging, phosphor, display, solar cell, photocatalysis, photoelectrochemistry and biomedicine.

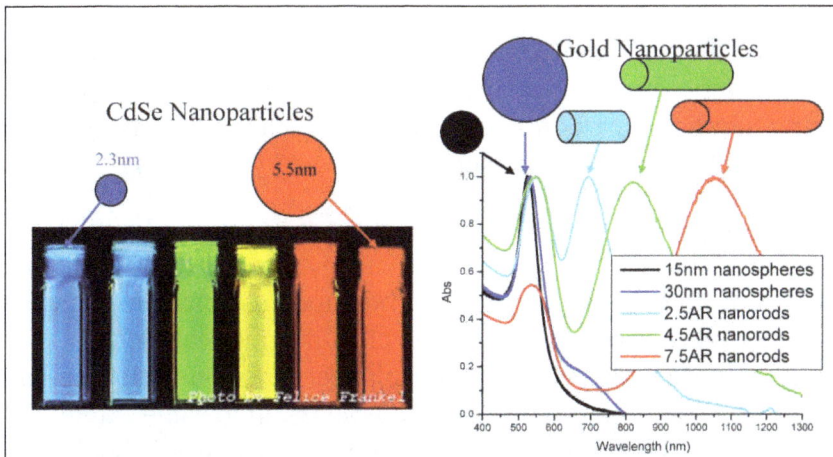

Fluorescence emission of (CdSe) ZnS quantum dots of various sizes
and absorption spectra of various sizes and shapes of gold nanoparticles.

The optical properties of nanomaterials depend on parameters such as feature size, shape, surface characteristics, and other variables including doping and interaction with the surrounding environment or other nanostructures. Likewise, shape can have dramatic influence on optical properties of metal nanostructures. Exemplifies the difference in the optical properties of metal and semiconductor nanoparticles. With the CdSe semiconductor nanoparticles, a simple change in size alters the optical properties of the nanoparticles. When metal nanoparticles are enlarged, their optical properties change only slightly as observed for the different samples of gold nanospheres. However, when an anisotropy is added to the nanoparticle, such as growth of nanorods, the optical properties of the nanoparticles change dramatically.

Electrical Properties

Electrical Properties of Nanoparticles" discuss about fundamentals of electrical conductivity in nanotubes and nanorods, carbon nanotubes, photoconductivity of nanorods, electrical conductivity of nanocomposites. One interesting method which can be used to demonstrate the steps in conductance is the mechanical thinning of a nanowire and measurement of the electrical current at a constant applied voltage. The important point here is that, with decreasing diameter of the wire, the number of electron wave modes contributing to the electrical conductivity is becoming increasingly smaller by well-defined quantized steps.

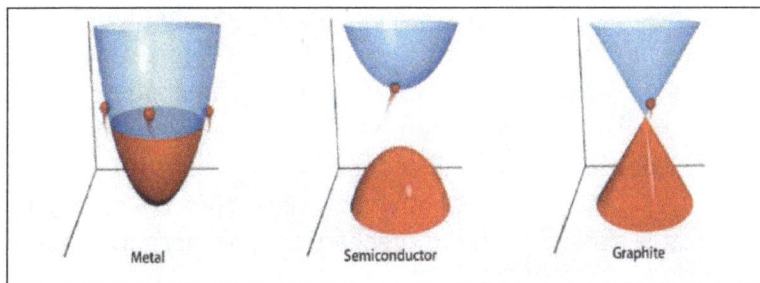

Electrical behavior of naotubes (P. G. Collins and Ph. Avouris, Scientific American.

In electrically conducting carbon nanotubes, only one electron wave mode is observed which transport the electrical current. As the lengths and orientations of the carbon nanotubes are different, they touch the surface of the mercury at different times, which provides two sets of information: (i) the influence of carbon nanotube length on the resistance; and (ii) the resistances of the different nanotubes. As the nanotubes have different lengths, then with increasing protrusion of the fiber bundle an increasing number of carbon nanotubes will touch the surface of the mercury droplet and contribute to the electrical current transport.

Mechanical Properties

Mechanical Properties of Nanoparticles deals with bulk metallic and ceramic materials, influence of porosity, influence of grain size, superplasticity, filled polymer composites, particle filled polymers, polymer-based nanocomposites filled with platelets, carbon nanotube-based composites. The discussion of mechanical properties of nanomaterials is, in to some extent, only of quite basic interest, the reason being that it is problematic to produce macroscopic bodies with a high density and a grain size in the range of less than 100 nm. However, two materials, neither of which is produced by pressing and sintering, have attracted much greater interest as they will undoubtedly achieve industrial importance.

These materials are polymers which contain nanoparticles or nanotubes to improve their mechanical behaviors, and severely plastic-deformed metals, which exhibit astonishing properties. However, because of their larger grain size, the latter are generally not accepted as nanomaterials. Experimental studies on the mechanical properties of bulk nanomaterials are generally impaired by major experimental problems in producing specimens with exactly defined grain sizes and porosities. Therefore, model calculations and molecular dynamic studies are of major importance for an understanding of the mechanical properties of these materials.

Filling polymers with nanoparticles or nanorods and nanotubes, respectively, leads to significant improvements in their mechanical properties. Such improvements depend heavily on the type of the filler and the way in which the filling is conducted. The latter point is of special importance, as any specific advantages of a nanoparticulate filler may be lost if the filler forms aggregates, thereby mimicking the large particles. Particulate-filled polymer-based nanocomposites exhibit a broad range of failure strengths and strains. This depends on the shape of the filler, particles or platelets, and on the degree of agglomeration. In this class of material, polymers filled with silicate platelets exhibit the best mechanical properties and are of the greatest economic relevance. The larger the particles of the filler or agglomerates, the poorer are the properties obtained. Although, potentially, the best composites are those filled with nanofibers or nanotubes, experience teaches that sometimes such composites have the least ductility. On the other hand, by using carbon nanotubes it is possible to produce composite fibers with extremely high strength and strain at rupture. Among the most exciting nanocomposites are the polymerceramic nanocomposites, where the ceramic phase is platelet-shaped. This type of composite is preferred in nature, and is found in the structure of bones, where it consists of crystallized mineral platelets of a few nanometers thickness that are bound together with collagen as the matrix. Composites consisting of a polymer matrix and defoliated phyllosilicates exhibit excellent mechanical and thermal properties.

Magnetic Properties

Bulk gold and Pt are non-magnetic, but at the nano size they are magnetic. Surface atoms are not only different to bulk atoms, but they can also be modified by interaction with other chemical species, that is, by capping the nanoparticles. This phenomenon opens the possibility to modify the physical properties of the nanoparticles by capping them with appropriate molecules. Actually, it should be possible that non-ferromagnetic bulk materials exhibit ferromagnetic-like behavior when prepared in nano range. One can obtain magnetic nanoparticles of Pd, Pt and the surprising case of Au (that is diamagnetic in bulk) from non-magnetic bulk materials. In the case of Pt and Pd, the ferromagnetism arises from the structural changes associated with size effects.

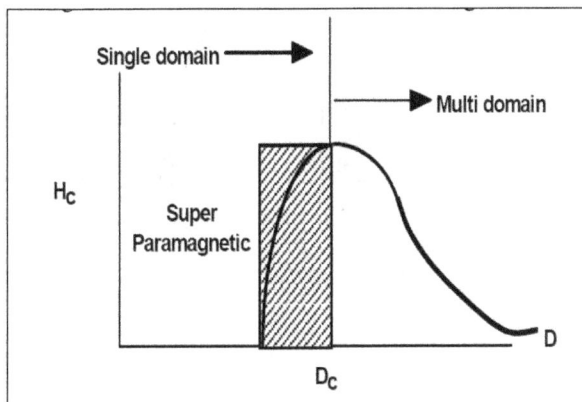

Magnetic properties of nanostrucutred materials.

However, gold nanoparticles become ferromagnetic when they are capped with appropriate molecules: the charge localized at the particle surface gives rise to ferromagnetic-like behavior. Surface and the core of Au nanoparticles with 2 nm in diameter show ferromagnetic and paramagnetic character, respectively. The large spin-orbit coupling of these noble metals can yield to a large anisotropy and therefore exhibit high ordering temperatures. More surprisingly, permanent magnetism was observed up to room temperature for thiol-capped Au nanoparticles. For nanoparticles with sizes below 2 nm the localized carriers are in the 5d band. Bulk Au has an extremely low density of states and becomes diamagnetic, as is also the case for bare Au nanoparticles. This observation suggested that modification of the d band structure by chemical bonding can induce ferromagnetic like character in metallic clusters.

Selected Application of Nanomaterials

Nanomaterials having wide range of applications in the field of electronics, fuel cells, batteries, agriculture, food industry, and medicines, etc. It is evident that nanomaterials split their conventional counterparts because of their superior chemical, physical, and mechanical properties and of their exceptional formability.

Fuel Cells

A fuel cell is an electrochemical energy conversion device that converts the chemical energy from fuel (on the anode side) and oxidant (on the cathode side) directly into electricity. The heart of fuel cell is the electrodes. The performance of a fuel cell electrode can be optimized in two ways; by improving the physical structure and by using more active electro catalyst. A good structure of electrode must provide ample surface area, provide maximum contact of catalyst, reactant gas and electrolyte, facilitate gas transport and provide good electronic conductance. In this fashion the structure should be able to minimize losses.

Carbon Nanotubes - Microbial Fuel Cell

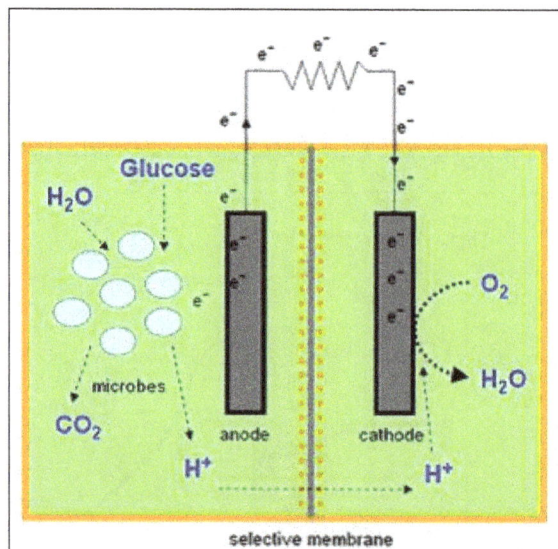

Schematic representation of microbial fuel cell.

Microbial fuel cell is a device in which bacteria consume water-soluble waste such as sugar, starch and alcohols and produces electricity plus clean water. This technology will make it possible to generate electricity while treating domestic or industrial wastewater. Microbial fuel cell can turn different carbohydrates and complex substrates present in wastewaters into a source of electricity. The efficient electron transfer between the microorganism and the anode of the microbial fuel cell plays a major role in the performance of the fuel cell. The organic molecules present in the waste-water posses a certain amount of chemical energy, which is released when converting them to sim-pler molecules like CO_2. The microbial fuel cell is thus a device that converts the chemical energy present in water-soluble waste into electrical energy by the catalytic reaction of microorganisms.

Carbon nanotubes (CNTs) have chemical stability, good mechanical properties and high surface area, making them ideal for the design of sensors and provide very high surface area due to its structural network. Since carbon nanotubes are also suitable supports for cell growth, electrodes of microbial fuel cells can be built using of CNT. Due to three-dimensional architectures and en-larged electrode surface area for the entry of growth medium, bacteria can grow and proliferate and get immobilized. Multi walled CNT scaffolds could offer self-supported structure with large surface area through which hydrogen producing bacteria (e.g., E. coli) can eventually grow and proliferate. Also CNTs and MWCNTs have been reported to be biocompatible for different eu-karyotic cells. The efficient proliferation of hydrogen producing bacteria throughout an electron conducting scaffold of CNT can form the basis for the potential application as electrodes in MFCs leading to efficient performance.

Catalysis

Higher surface area available with the nanomaterial counterparts, nano-catalysts tend to have exceptional surface activity. For example, reaction rate at nano-aluminum can go so high, that it is utilized as a solid-fuel in rocket propulsion, whereas the bulk aluminum is widely used in utensils. Nano-aluminum becomes highly reactive and supplies the required thrust to send off pay loads in space. Similarly, catalysts assisting or retarding the reaction rates are dependent on the surface activity, and can very well be utilized in manipulating the rate-controlling step.

Phosphors for High-definition TV

The resolution of a television, or a monitor, depends greatly on the size of the pixel. These pixels are essentially made of materials called "phosphors," which glow when struck by a stream of electrons inside the cathode ray tube (CRT). The resolution improves with a reduction in the size of the pixel, or the phosphors. Nanocrystalline zinc selenide, zinc sulfide, cadmium sulfide, and lead telluride synthesized by the sol-gel techniques are candidates for improving the resolution of monitors. The use of nanophosphors is envisioned to reduce the cost of these displays so as to render high-definition televisions (HDTVs) and personal computers affordable to be purchase.

Next-generation Computer Chips

The microelectronics industry has been emphasizing miniaturization, whereby the circuits, such as transistors, resistors, and capacitors, are reduced in size. By achieving a significant reduction in their size, the microprocessors, which contain these components, can run much faster, thereby enabling computations at far greater speeds. However, there are several technological impediments to these advancements, including lack of the ultrafine precursors to manufacture these components; poor dissipation of tremendous amount of heat generated by these microprocessors due to faster speeds; short mean time to failures (poor reliability), etc. Nanomaterials help the industry break these barriers down by providing the manufacturers with nanocrystalline starting materials, ultra-high purity materials, materials with better thermal conductivity, and longer-lasting, durable interconnections (connections between various components in the microprocessors).

Nanowires for Junctionless Transistors

Silicon nanowires in junctionless transistors.

Transistors are made so tiny to reduce the size of sub assemblies of electronic systems and make smaller and smaller devices, but it is difficult to create high-quality junctions. In particular, it is very difficult to change the doping concentration of a material over distances shorter than about 10 nm. Researchers have succeeded in making the junctionless transistor having nearly ideal electrical properties. It could potentially operate faster and use less power than any conventional transistor on the market today. The device consists of a silicon nanowire in which current flow is perfectly controlled by a silicon gate that is separated from the nanowire by a thin insulating layer. The entire silicon nanowire is heavily n-doped, making it an excellent conductor. However, the gate is p-doped and its presence has the effect of depleting the number of electrons in the region of the nanowire under the gate. The device also has near-ideal electrical properties and behaves like the most perfect of transistors without suffering from current leakage like conventional devices and operates faster and using less energy.

Elimination of Pollutants

Nanomaterials possess extremely large grain boundaries relative to their grain size. Hence, they are very active in terms of their chemical, physical, and mechanical properties. Due to their enhanced chemical activity, nanomaterials can be used as catalysts to react with such noxious and toxic gases as carbon monoxide and nitrogen oxide in automobile catalytic converters and power generation equipment to prevent environmental pollution arising from burning gasoline and coal.

Sun-screen Lotion

Prolonged UV exposure causes skin-burns and cancer. Sun-screen lotions containing nano-TiO_2 provide enhanced sun protection factor (SPF) while eliminating stickiness. The added advantage of nano skin blocks (ZnO and TiO_2) arises as they protect the skin by sitting onto it rather than penetrating into the skin. Thus they block UV radiation effectively for prolonged duration. Additionally, they are transparent, thus retain natural skin color while working better than conventional skin-lotions.

Sensors

Sensors rely on the highly active surface to initiate a response with minute change in the concentration of the species to be detected. Engineered monolayers (few Angstroms thick) on the sensor surface are exposed to the environment and the peculiar functionality (such as change in potential as the CO/anthrax level is detected) is utilized in sensing.

Disadvantages of Nanomaterials

- Instability of the particles: Retaining the active metal nanoparticles is highly challenging, as the kinetics associated with nanomaterials is rapid. In order to retain nanosize of particles, they are encapsulated in some other matrix. Nanomaterials are thermodynamically metastable and lie in the region of high-energy local-minima. Hence they are prone to attack and undergo transformation. These include poor corrosion resistance, high solubility, and phase change of nanomaterials. This leads to deterioration in properties and retaining the structure becomes challenging.

  Fine metal particles act as strong explosives owing to their high surface area coming in direct contact with oxygen. Their exothermic combustion can easily cause explosion.

- Impurity: Because nanoparticles are highly reactive, they inherently interact with impurities as well. In addition, encapsulation of nanoparticles becomes necessary when they are synthesized in a solution (chemical route). The stabilization of nanoparticles occurs because of a non-reactive species engulfing the reactive nano-entities. Thereby, these secondary impurities become a part of the synthesized nanoparticles, and synthesis of pure nanoparticles becomes highly difficult. Formation of oxides, nitrides, etc can also get aggravated from the impure environment/ surrounding while synthesizing nanoparticles. Hence retaining high purity in nanoparticles can become a challenge hard to overcome.

- Biologically harmful: Nanomaterials are usually considered harmful as they become transparent to the cell-dermis. Toxicity of nanomaterials also appears predominant owing to their high surface area and enhanced surface activity. Nanomaterials have shown to cause irritation, and have indicated to be carcinogenic. If inhaled, their low mass entraps them inside lungs, and in no way they can be expelled out of body. Their interaction with liver/ blood could also prove to be harmful (though this aspect is still being debated on).

Difficulty in synthesis, isolation and application - It is extremely hard to retain the size of nanoparticles once they are synthesized in a solution. Hence, the nanomaterials have to be encapsulated in a bigger and stable molecule/material.

Hence free nanoparticles are hard to be utilized in isolation, and they have to be interacted for intended use via secondary means of exposure. Grain growth is inherently present in nanomaterials during their processing. The finer grains tend to merge and become bigger and stable grains at high temperatures and times of processing.

- Recycling and disposal: There are no hard-and-fast safe disposal policies evolved for nanomaterials. Issues of their toxicity are still under question, and results of exposure experiments are not available. Hence the uncertainty associated with affects of nanomaterials is yet to be assessed in order to develop their disposal policies.

Nanoparticles

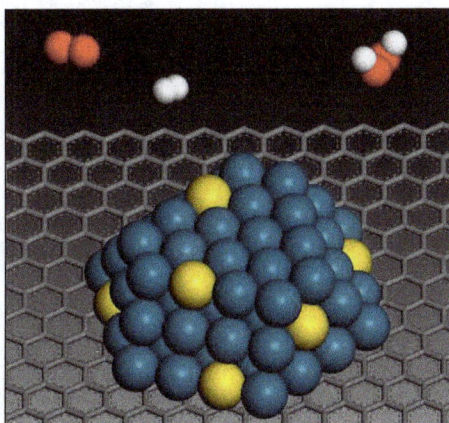

Image showing nanoparticles of an alloy of gold (yellow) and palladium (blue) on an
acid-treated carbon support (gray). These particles were employed as catalysts
for the formation of hydrogen peroxide from hydrogen (white) and oxygen (red).

Nanoparticle is an ultrafine unit with dimensions measured in nanometres (nm; 1 nm = 10^{-9} metre). Nanoparticles exist in the natural world and are also created as a result of human activities.

Because of their submicroscopic size, they have unique material characteristics, and manufactured nanoparticles may find practical applications in a variety of areas, including medicine, engineering, catalysis, and environmental remediation.

Properties of Nanoparticles

In 2008 the International Organization for Standardization (ISO) defined a nanoparticle as a discrete nano-object where all three Cartesian dimensions are less than 100 nm. The ISO standard similarly defined two-dimensional nano-objects (i.e., nanodiscs and nanoplates) and one-dimensional nano-objects (i.e., nanofibres and nanotubes). But in 2011 the Commission of the European Union endorsed a more-technical but wider-ranging definition:

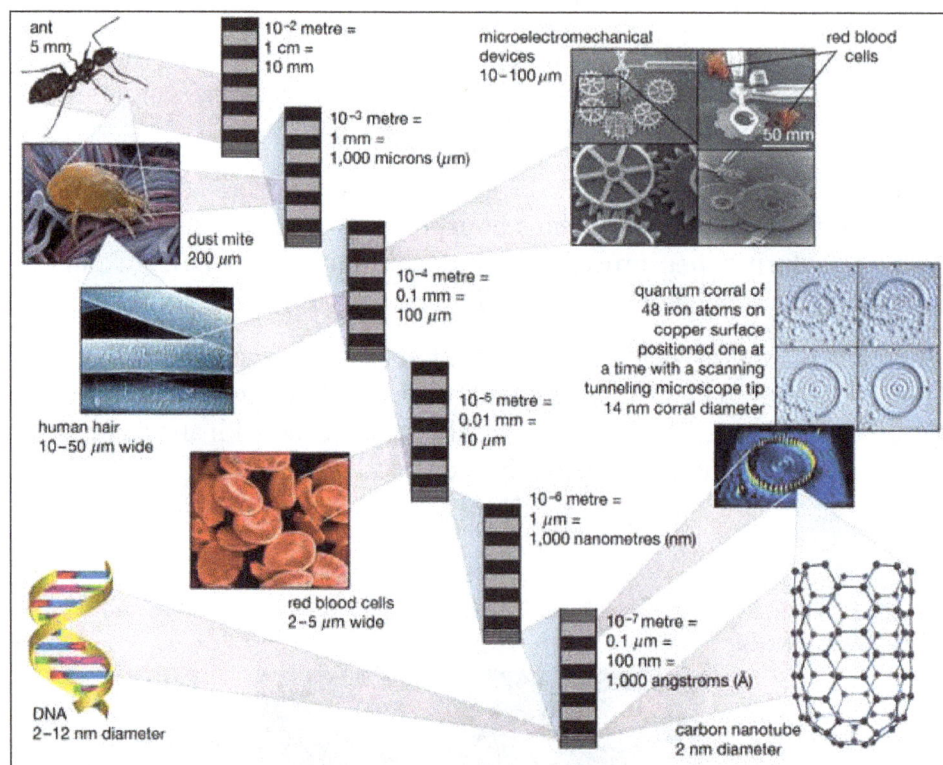

Examples from biological and mechanical realms illustrate various "orders of magnitude" (powers of 10), from 10^{-2} metre down to 10^{-7} metre.

A natural, incidental or manufactured material containing particles, in an unbound state or as an aggregate or as an agglomerate and where, for 50% or more of the particles in the number size distribution, one or more external dimensions is in the size range 1 nm–100 nm.

Under that definition a nano-object needs only one of its characteristic dimensions to be in the range 1–100 nm to be classed as a nanoparticle, even if its other dimensions are outside that range. (The lower limit of 1 nm is used because atomic bond lengths are reached at 0.1 nm).

That size range—from 1 to 100 nm—overlaps considerably with that previously assigned to the field of colloid science—from 1 to 1,000 nm—which is sometimes alternatively called the meso-scale. Thus, it is not uncommon to find literature that refers to nanoparticles and colloidal particles in equal terms. The difference is essentially semantic for particles below 100 nm in size.

Nanoparticles can be classified into any of various types, according to their size, shape, and material properties. Some classifications distinguish between organic and inorganic nanoparticles; the first group includes dendrimers, liposomes, and polymeric nanoparticles, while the latter includes fullerenes, quantum dots, and gold nanoparticles. Other classifications divide nanoparticles according to whether they are carbon-based, ceramic, semiconducting, or polymeric. In addition, nanoparticles can be classified as hard (e.g., titanium dioxide, silica dioxide particles, and fullerenes) or as soft (e.g., liposomes, vesicles, and nanodroplets). The way in which nanoparticles are classified typically depends on their application, such as in diagnosis or therapy versus basic research, or may be related to the way in which they were produced.

There are three major physical properties of nanoparticles, and all are interrelated: (1) they are highly mobile in the free state (e.g., in the absence of some other additional influence, a 10-nm-diameter nanosphere of silica has a sedimentation rate under gravity of 0.01 mm/day in water); (2) they have enormous specific surface areas (e.g., a standard teaspoon, or about 6 ml, of 10-nm-diameter silica nanospheres has more surface area than a dozen doubles-sized tennis courts; 20 percent of all the atoms in each nanosphere will be located at the surface); and (3) they may exhibit what are known as quantum effects. Thus, nanoparticles have a vast range of compositions, depending on the use or the product.

Nanoparticle-based Technologies

In general, nanoparticle-based technologies centre on opportunities for improving the efficiency, sustainability, and speed of already-existing processes. That is possible because, relative to the materials used traditionally for industrial processes (e.g., industrial catalysis), nanoparticle-based technologies use less material, a large proportion of which is already in a more "reactive" state. Other opportunities for nanoparticle-based technologies include the use of nanoscale zero-valent iron (NZVI) particles as a field-deployable means of remediating organochlorine compounds, such as polychlorinated biphenyls (PCBs), in the environment. NZVI particles are able to permeate into rock layers in the ground and thus can neutralize the reactivity of organochlorines in deep aquifers. Other applications of nanoparticles are those that stem from manipulating or arranging matter at the nanoscale to provide better coatings, composites, or additives and those that exploit the particles' quantum effects (e.g., quantum dots for imaging, nanowires for molecular electronics, and technologies for spintronics and molecular magnets).

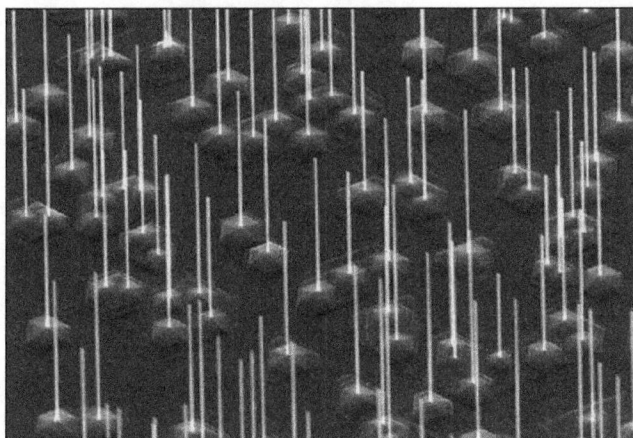

Nanowires as seen by a field-emission microscope.

Nanoparticle Applications in Materials

Many properties unique to nanoparticles are related specifically to the particles' size. It is therefore natural that efforts have been made to capture some of those properties by incorporating nanoparticles into composite materials. An example of how the unique properties of nanoparticles have been put to use in a nanocomposite material is the modern rubber tire, which typically is a composite of a rubber (an elastomer) and an inorganic filler (a reinforcing particle), such as carbon black or silica nanoparticles.

For most nanocomposite materials, the process of incorporating nanoparticles is not straightforward. Nanoparticles are notoriously prone to agglomeration, resulting in the formation of large clumps that are difficult to redisperse. In addition, nanoparticles do not always retain their unique size-related properties when they are incorporated into a composite material.

Despite the difficulties with manufacture, the use of nanomaterials grew markedly in the early 21st century, with especially rapid growth in the use of nanocomposites. Nanocomposites were employed in the development and design of new materials, serving, for example, as the building blocks for new dielectric (insulating) and magnetic materials. The following sections describe some of the many applications of nanoparticles and nanocomposites in materials.

Polymers

Similar to the way in which carbon and silica nanoparticles have been used as fillers in rubber to improve the mechanical properties of tires, such particles and others, including nanoclays, have been incorporated into polymers to improve their strength and impact resistance. In the early 21st century, increasing use of non-petroleum-based polymers that were derived from natural sources drove the development of "all-natural" nanocomposite polymers. Such materials incorporate a biopolymer derived from an alginate (a carbohydrate found in the cell wall of brown algae), cellulose, or starch; the biopolymer is used in conjunction with a natural nanoclay or a filler derived from the shells of crustaceans. The materials are biodegradable and do not leave behind potentially harmful or nonnatural residues.

Food Packaging

Nanoparticles have been increasingly incorporated into food packaging to control the ambient atmosphere around food, keeping it fresh and safe from microbial contamination. Such composites use nanoflakes of clays and claylike particles, which slow down the ingress of moisture and reduce gas transport across the packaging film. It is also possible to incorporate nanoparticles with apparent antimicrobial effects (e.g., nanocopper or nanosilver) into such packaging. Nanoparticles that exhibit antimicrobial activity had also been incorporated into paints and coatings, making those products particularly useful for surfaces in hospitals and other medical facilities and in areas of food preparation.

Flame Retardants

Nanoparticles were explored for their potential to replace additives based on flammable organic halogens and phosphorus in plastics and textiles. Studies had suggested that, in the event of a serious fire, products with nanoclays and hydroxide nanoparticles were associated with fewer emissions of harmful fumes than products containing certain other types of additives.

Batteries and Supercapacitors

The ability to engineer nanocomposite materials to have very high internal surface areas for storing electrical charge in the form of small ions or electrons has made them especially valuable for use in batteries and supercapacitors. Indeed, nanocomposite materials have been synthesized for various applications involving electrodes. Composite materials based on carbon nanotubes and layered-type materials, such as graphene, were also researched extensively, making their first appearances in commercial devices in the early 2000s.

Nanoceramics

A long-term objective in materials science had been to transform ceramics that are brittle and prone to cracking into tougher, more resilient materials. By the early 21st century, researchers had achieved that goal by incorporating an effective blend of nanoparticles into ceramics materials. Other new ceramics materials that were under development included all-ceramic or polymer-ceramic blends, which combined the unique functional (e.g., electrical, magnetic, or mechanical) properties of a nanocomposite material with the properties of ceramics materials.

Light Control

In the 1990s the development of blue light-emitting diodes (LEDs), which had the potential to produce white light at significantly reduced costs, inspired a revolution in lighting. Blue LEDs brought about a need for composite materials that could be used to coat the diodes to convert blue light into other wavelengths (such as red, yellow, or green) in order to achieve white light. One way of obtaining the desired light is by leveraging the size or quantum effect of small semiconducting particles. The application of such particles facilitated the development of nanocomposite polymers for greenhouse enclosures; the polymers optimize plant growth by effectively converting wavelengths of full-spectrum sunlight into the red and blue wavelengths used in photosynthesis. Light conversion in the above cases is achieved with submicron particles of inorganic phosphor materials incorporated into the polymer.

Nanoparticle Applications in Medicine

The small size of nanoparticles is especially advantageous in medicine; nanoparticles can not only circulate widely throughout the body but also enter cells or be designed to bind to specific cells. Those properties have enabled new ways of enhancing images of organs as well as tumours and other diseased tissues in the body. They also have facilitated the development of new methods of delivering therapy, such as by providing local heating (hyperthermia), by blocking vasculature to diseased tissues and tumours, or by carrying payloads of drugs.

Magnetic nanoparticles have been used to replace radioactive technetium for tracking the spread of cancer along lymph nodes. The nanoparticles work by exploiting the change in contrast brought about by tiny particles of superparamagnetic iron oxide in magnetic resonance imaging (MRI). Such particles also can be used to kill tumours via hyperthermia, in which an alternating magnetic field causes them to heat and destroy tissue on a local scale.

Nanoparticles can be designed to enhance fluorescent imaging or to enhance images from positron emission tomography (PET) or ultrasound. Those methods typically require that the nanoparticle be able to recognize a particular cell or disease state. In theory, the same idea of targeting could be used in aiding the precise delivery of a drug to a given disease site. The drug could be carried via a nanocapsule or a liposome, or it could be carried in a porous nanosponge structure and then held by bonds at the targeted site, thereby allowing the slow release of drug. The development of nanoparticles to aid in the delivery of a drug to the brain via inhalation holds considerable promise for the treatment of neurological disorders such as Parkinson disease, Alzheimer disease, and multiple sclerosis.

Nanoparticles and nanofibres play an important part in the design and manufacture of novel scaffold structures for tissue and bone repair. The nanomaterials used in such scaffolds are biocompatible. For example, nanoparticles of calcium hydroxyapatite, a natural component of bone, used in combination with collagen or collagen substitutes could be used in future tissue-repair therapies.

Nanoparticles also have been used in the development of health-related products. For example, a sunscreen known as Optisol, invented at the University of Oxford in the 1990s, was designed with the objective of developing a safe sunscreen that was transparent in visible light but retained ultraviolet-blocking action on the skin. The ingredients traditionally used in sunscreens were based on large particles of either zinc oxide or titanium dioxide or contained an organic sunlight-absorbing compound. However, those materials were not satisfactory: zinc oxide and titanium dioxide are very potent photocatalysts, and in the presence of water and sunlight they generate free radicals, which have the potential to damage skin cells and DNA (deoxyribonucleic acid). Scientists proceeded to develop a nanoparticle form of titanium oxide that contained a small amount of manganese. Studies indicated that the nanoparticle-based sunscreen was safer than sunscreen products manufactured by using traditional materials. The improvement in safety was attributed to the introduction of manganese, which changed the semiconducting properties of the compound from n-type to p-type, thus shifting its Fermi level, or oxidation-reduction properties, and making the generation of free radicals less likely.

Treatments and diagnostic approaches based on the use of nanoparticles are expected to have important benefits for medicine in the future, but the use of nanoparticles also presents significant challenges, particularly regarding impacts on human health. For example, little is known about the fate of nanoparticles that are introduced into the body or whether they have undesirable effects on the body (see below Health effects of nanoparticles). Extensive clinical trials are needed in order to fully address concerns about the safety and effectiveness of nanoparticles used in medicine. There also are manufacturing problems to be overcome, such as the ability to produce nanoparticles under sterile conditions, which is required for medical applications.

Manufacture of Nanoparticles

Nanoparticles are made by one of three routes: by comminution (the pulverization of materials), such as through industrial milling or natural weathering; by pyrolysis (incineration); or by sol-gel synthesis (the generation of inorganic materials from a colloidal suspension). Comminution is known as a top-down approach, whereas the sol-gel process is a bottom-up approach. Examples of those three processes (comminution, pyrolysis, and sol-gel synthesis) include the production of titania nanoparticles for sunscreens from the minerals anatase and rutile, the production of fullerenes or fumed silica (not to be confused with silica fume, which is a different product), and the production of synthetic (or Stöber) silica, of other "engineered" oxide nanoparticles, and of quantum dots. For the generation of small nanoparticles, comminution is a very inefficient process.

Detection, Characterization and Isolation

The detection and characterization of nanoparticles present scientists with particular challenges. Being of a size that is at least four to seven times smaller than the wavelength of light means that individual nanoparticles cannot be detected by the human eye, and they are observable under optical microscopes only in liquid samples under certain conditions. Thus, in general, specialized techniques are required to see them, and none of those approaches is currently field-deployable.

Techniques to detect and characterize nanoparticles fall into two categories: direct, or "real space," and indirect, or "reciprocal space". Direct techniques include transmission electron microscopy (TEM), scanning electron microscopy (SEM), and atomic force microscopy (AFM). Those techniques can image nanoparticles, directly measure sizes, and infer shape information, but they are limited to studying only a few particles at a time. There are also significant issues surrounding sample preparation for electron microscopy. In general, however, those techniques can be quite effective for obtaining basic information about a nanoparticle.

Indirect techniques use X-rays or neutron beams and obtain their information by mathematically analyzing the radiation scattered or diffracted by the nanoparticles. The techniques of greatest relevance to nanoscience are small-angle X-ray scattering (SAXS) and small-angle neutron scattering (SANS), along with their surface-specific analogues GISAXS and GISANS, where GI is "grazing incidence," and X-ray or neutron reflectometry (XR/NR). The advantage of those techniques is that they are able to simultaneously sample and average very large numbers of nanoparticles and often do not require any particular sample preparation. Indirect techniques have many applications. For example, in studies of nanoparticles in raw sewage, scientists used SANS measurements, in which neutrons readily penetrated the turbid sewage and scattered strongly from the nanoparticles, to follow the aggregation behaviour of the particles over time.

The isolation of nanoparticles from colloidal and larger matter involves specialized techniques, such as ultra centrifugation and field-flow fractionation. Such laboratory-based techniques are normally coupled to standard spectroscopic instrumentation to enable particular types of chemical characterization.

Nanoparticles in the Environment

Nanoparticles occur naturally in the environment in large volumes. For example, the sea emits an aerosol of salt that ends up floating around in the atmosphere in a range of sizes, from a few nanometres upward, and smoke from volcanoes and fires contains a huge variety of nanoparticles, many of which could be classified as dangerous to human health. Dust from deserts, fields, and so on also has a range of sizes and types of particles, and even trees emit nanoparticles of hydrocarbon compounds such as terpenes (which produce the familiar blue haze seen in forests, from which the Great Smoky Mountains in the United States get their name).

Human-made (anthropogenic) nanoparticles are emitted by large industrial processes, and in modern life it is particles from power stations and from jet aircraft and other vehicles (namely, those powered by internal-combustion engines; car tires are also a factor) that constitute the major fraction of nanoparticle emissions. Types of nanoparticles that are emitted include partially burned hydrocarbons (in soot), ceria (cerium oxide; from vehicle exhaust catalysts), metallic dust (from brake linings), calcium carbonate (in engine lubricating oils), and silica (from car tires). Other sources of nanoparticles to the environment include the semiconductor industry, domestic and industrial wastewater discharges, the health care industry, and the photographic industry. However, all those emission levels are still considered to be lower than the levels of nanoparticles produced through natural processes. Indeed, recent human-made particles contribute only a small amount to air and water pollution.

Understanding the relationship between nanoparticles and the environment forms an important area of research. There are several mechanisms by which nanoparticles are believed to affect the environment negatively. Two scenarios that are under investigation are the possibilities (1) that the mobility and sorptive capacity of nanoparticles (natural or human-made) make them potent vectors (carriers) in the transport of chemical pollutants (e.g., phosphorus from sewage and agriculture), particularly in rivers and lakes, and (2) that some nanoparticles are able to reduce the functioning of (and may even disrupt or kill) naturally occurring microbial communities, as well as microbial communities that are employed in industrial processes (e.g., those that are used in sanitation processes, including sewage treatment).

Nanoparticles also can have beneficial impacts on the environment and appear to contribute to natural processes. Thus, in addition to the potential use of nanoparticles to remove chemical contaminants from the environment, scientists are investigating how nanoparticles interact with all life-forms—from fungi to microbes, algae, plants, and higher-order animals. That type of study is essential not only to improving scientists' knowledge of nanoparticles but also to gaining a more complete understanding of life on Earth, since the soil is naturally full of nanoparticles, in a richly diverse environment.

Health Effects of Nanoparticles

Humans have evolved to cope with most naturally occurring nanoparticles. However, some nanoparticles, generated as a result of certain human activities such as tobacco smoking and fires, account for many premature deaths as a result of lung damage. For example, fires from the types of cooking stoves used in developing countries are known to emit fine particles and lead to early mortality, especially among women who routinely work near the stoves.

Laboratory and clinical investigations of the effects of nanoparticles on health have been somewhat controversial and remain largely inconclusive. Most studies in animals have involved nanoparticle inhalation, and the dosages have been very large. The results of those studies have indicated that large quantities of nanoparticles can cause cellular damage in the lungs, with lung cells absorbing the particles and becoming damaged or undergoing genetic mutation. Animal studies involving the ingestion of nanoparticles in food or water suggest that nanoparticles can also affect health in other ways. For example, consumption of the food additive E171, which consists of titanium dioxide nanoparticles, is associated with changes in gut microbiota (bacteria occurring in the gut), potentially contributing to the development of conditions such as inflammatory bowel disease.

In humans, the health effects of typical exposure levels—those that are encountered by most persons during daily activities—remain unknown. Nonetheless, there is a general awareness of the problems that might occur upon excess exposure to nanoparticles, and, thus, most manufacturers of such particles take serious precautions to avoid exposure of their workers. Efforts have been made to educate the public in the use of nanoparticle-containing products. The existence of pressure groups has also helped to ensure nanoparticle safety compliance among manufacturers. However, nanoparticles offer tremendous potential for new or improved forms of health care treatment. That has spawned a new field of science called nanomedicine.

Types of Nanoparticles

Nanoparticles can be classified into different types according to the size, morphology, physical and chemical properties. Some of them are carbon-based nanoparticles, ceramic nanoparticles, metal nanoparticles, semiconductor nanoparticles, polymeric nanoparticles and lipid-based nanoparticles.

Carbon-based Nanoparticles

Carbon-based nanoparticles include two main materials: carbon nanotubes (CNTs) and fullerenes. CNTs are nothing but graphene sheets rolled into a tube. These materials are mainly used for the structural reinforcement as they are 100 times stronger than steel.

CNTs can be classified into single-walled carbon nanotubes (SWCNTs) and multi-walled carbon nanotubes (MWCNTs). CNTs are unique in a way as they are thermally conductive along the length and non-conductive across the tube.

Fullerenes are the allotropes of carbon having a structure of hollow cage of sixty or more carbon atoms. The structure of C-60 is called Buckminsterfullerene, and looks like a hollow football. The carbon units in these structures have a pentagonal and hexagonal arrangement. These have commercial applications due to their electrical conductivity, structure, high strength, and electron affinity.

Ceramic Nanoparticles

Ceramic nanoparticles are inorganic solids made up of oxides, carbides, carbonates and phosphates. These nanoparticles have high heat resistance and chemical inertness. They have applications in photocatalysis, photodegradation of dyes, drug delivery, and imaging.

By controlling some of the characteristics of ceramic nanoparticles like size, surface area, porosity, surface to volume ratio, etc, they perform as a good drug delivery agent. These nanoparticles have been used effectively as a drug delivery system for a number of diseases like bacterial infections, glaucoma, cancer, etc.

Metal Nanoparticles

Metal nanoparticles are prepared from metal precursors. These nanoparticles can be synthesized by chemical, electrochemical, or photochemical methods. In chemical methods, the metal nanoparticles are obtained by reducing the metal-ion precursors in solution by chemical reducing agents. These have the ability to adsorb small molecules and have high surface energy.

These nanoparticles have applications in research areas, detection and imaging of biomolecules and in environmental and bioanalytical applications. For example gold nanoparticles are used to coat the sample before analyzing in SEM. This is usually done to enhance the electronic stream, which helps us to get high quality SEM images.

Semiconductor Nanoparticles

Semiconductor nanoparticles have properties like those of metals and non-metals. They are found in the periodic table in groups II-VI, III-V or IV-VI. These particles have wide bandgaps, which on tuning shows different properties. They are used in photocatalysis, electronics devices, photo-optics and water splitting applications.

Some examples of semiconductor nanoparticles are GaN, GaP, InP, InAs from group III-V, ZnO, ZnS, CdS, CdSe, CdTe are II-VI semiconductors and silicon and germanium are from group IV.

Polymeric Nanoparticles

Polymeric nanoparticles are organic based nanoparticles. Depending upon the method of preparation, these have structures shaped like nanocapsular or nanospheres. A nanosphere particle has a matrix-like structure whereas the nanocapsular particle has core-shell morphology. In the former, the active compounds and the polymer are uniformly dispersed whereas in the latter the active compounds are confined and surrounded by a polymer shell.

Some of the merits of polymeric nanoparticles are controlled release, protection of drug molecules, ability to combine therapy and imaging, specific targeting and many more. They have applications in drug delivery and diagnostics. The drug deliveries with polymeric nanoparticles are highly biodegradable and biocompatible.

Lipid-based Nanoparticles

Lipid nanoparticles are generally spherical in shape with a diameter ranging from 10 to 100nm. It consists of a solid core made of lipid and a matrix containing soluble lipophilic molecules. The external core of these nanoparticles is stabilized by surfactants and emulsifiers. These nanoparticles have application in the biomedical field as a drug carrier and delivery and RNA release in cancer therapy.

Thus, the field of nanotechnology is far from being saturated and it is, as the statistic says, sitting on the staircase of an exponential growth pattern. It is basically at the same stage as the information technology was in the 1960s and biotechnology in the year of 1980s. Thus it can easily be predicted that this field would witness a same exponential growth as the other two technological field witnessed earlier.

Nanostructured Materials

Nanostructured Materials (NsM) are materials with a microstructure the characteristic length scale of which is on the order of a few (typically 1–10) nanometers. NsM may be in or far away from thermodynamic equilibrium. NsM synthesized by supramolecular chemistry are examples of NsM in thermodynamic equilibrium. NsM consisting of nanometer-sized crystallites (e.g. of Au or NaCl) with different crystallographic orientations and/or chemical compositions are far away from thermodynamic equilibrium. The properties of NsM deviate from those of single crystals (or coarse-grained polycrystals) and/or glasses with the same average chemical composition. This deviation results from the reduced size and/or dimensionality of the nanometer-sized crystallites as well as from the numerous interfaces between adjacent crystallites. An attempt is made to summarize the basic physical concepts and the microstructural features of equilibrium and non-equilibrium NsM.

Categories of Nanostructured Materials

One of the very basic results of the physics and chemistry of solids is the insight that most properties of solids depend on the microstructure, i.e. the chemical composition, the arrangement of the atoms (the atomic structure) and the size of a solid in one, two or three dimensions. In other words, if one changes one or several of these parameters, the properties of a solid vary. The most well-known example of the correlation between the atomic structure and the properties of a bulk material is probably the spectacular variation in the hardness of carbon when it transforms from diamond to graphite. Comparable variations have been noted if the atomic structure of a solid deviates far from equilibrium or if its size is reduced to a few interatomic spacings in one, two or three dimensions. An example of the latter case is the change in color of CdS crystals if their size is reduced to a few nanometers.

The synthesis of materials and/or devices with new properties by means of the controlled manipulation of their microstructure on the atomic level has become an emerging interdisciplinary field based on solid state physics, chemistry, biology and materials science. The materials and/or devices involved may be divided into the following three categories.

The first category comprises materials and/or devices with reduced dimensions and/or dimensionality in the form of (isolated, substrate-supported or embedded) nanometer-sized particles, thin wires or thin films. CVD, PVD, inert gas condensation, various aerosol techniques, precipitation from the vapor, from supersaturated liquids or solids (both crystalline and amorphous) appear to be the techniques most frequently used to generate this type of microstructure. Well-known examples of technological applications of materials the properties of which depend on this type of microstructure are catalysts and semiconductor devices utilizing single or multilayer quantum well structures.

The second category comprises materials and/or devices in which the nanometer-sized microstructure is limited to a thin (nanometer-sized) surface region of a bulk material. PVD, CVD, ion implantation and laser beam treatments are the most widely applied procedures to modify the chemical composition and/or atomic structure of solid surfaces on a nanometer scale. Surfaces with enhanced corrosion resistance, hardness, wear resistance or protective coatings (e.g. by diamond) are examples taken from today's technology in which the properties of a thin surface layer are improved by means of creating a nanometer-sized microstructure in a thin surface region. An important subgroup of this category are materials, the surface region of which are structured laterally on a nanometer scale by "writing" a nanometer-sized structural pattern on the free surface. For example, patterns in the form of an array of nanometer-sized islands (e.g., quantum dots) connected by thin (nanometer scale) wires. Patterns of this type may be synthesized by lithography, by means of local probes (e.g. the tip of a tunneling microscope, near-field methods, focussed electron or ion beams) and/or surface precipitation processes. Processes and devices of this sort are expected to play a key role in the production of the next generation of electronic devices such as highly integrated circuits, terrabit memories, single electron transistors, quantum computers, etc.

Two classes of such solids may be distinguished. In the first class, the atomic structure and/or the chemical composition varies in space continuously throughout the solid on an atomic scale. Glasses, gels, supersaturated solid solutions or implanted materials are examples of this type. In many cases these types of solids are produced by quenching a high-temperature (equilibrium) structure, e.g., a melt or a solid solution to low temperatures at which the structure is far away from equilibrium.

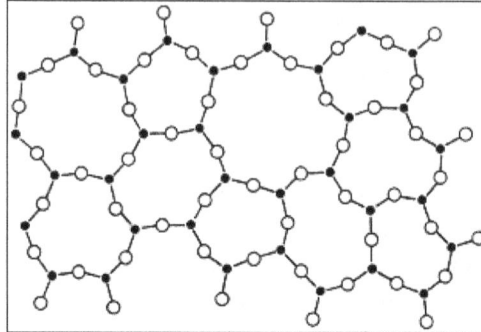

Two-dimensional model of an Al_2O_3 glass.

In the last two decades a second class of materials with a nanometer-sized microstructure has been synthesized and studied. These materials are assembled of nanometer-sized building blocks—mostly crystallites—as displayed in figure. These building blocks may differ in their atomic structure, their crystallographic orientation and/or their chemical composition. If the building blocks are crystallites, incoherent or coherent interfaces may be formed between them, depending on the atomic structure, the crystallographic orientation and/or the chemical composition of adjacent crystallites. In other words, materials assembled of nanometer-sized building blocks are microstructurally heterogeneous consisting of the building blocks (e.g. crystallites) and the regions between adjacent building blocks (e.g. grain boundaries). It is this inherently heterogeneous structure on a nanometer scale that is crucial for many of their properties and distinguishes them from glasses, gels, etc. that are microstructurally homogeneous. Materials with a nanometer-sized microstructure are called "Nanostructured Materials" (NsM) or—synonymously—nanophase materials, nanocrystalline materials or supramolecular solids.

Two-dimensional model of a nanostructured material. The atoms in the centers of the crystals are indicated in black. The ones in the boundary core regions are represented as open circles.

The synthesis, characterization and processing of such NsM are part of an emerging and rapidly growing field referred to as nanotechnology. R&D in this field emphasizes scientific discoveries in generation of materials with controlled microstructural characteristics, research on their processing into bulk materials with engineered properties and technological functions, and introduction of new device concepts and manufacturing methods.

Effects Controlling the Properties of Nanostructured Materials

As the properties of solids depend on size, atomic structure and chemical composition, NsM exhibit new properties due to one or several of the following effects.

Size Effects

Size effects result if the characteristic size of the building blocks of the microstructure is reduced to the point where critical length scales of physical phenomena (e.g. the mean free paths of electrons or phonons, a coherency length, a screening length, etc.) become comparable with the characteristic size of the building blocks of the microstructure. An example is shown in figure. If the thickness of the layers of a superlattice is comparable with the wavelength of the electrons at the Fermi edge, discrete energy levels for electrons and holes are formed in the quantum wells. Such size effects modifying the mechanical and optical properties are displayed in figure.

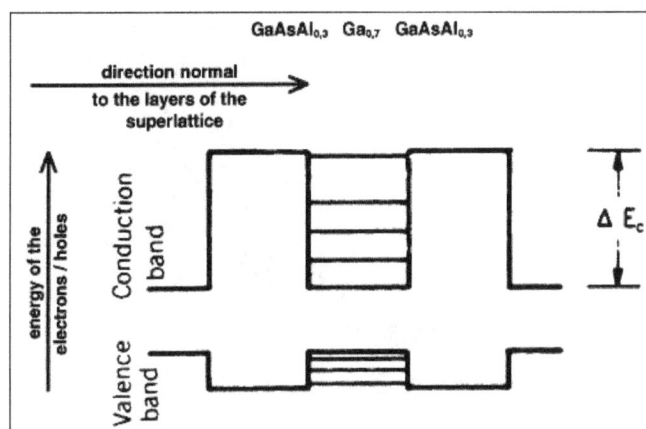

Energy-band diagrams for undoped GaAs–$Al_xGa_{1-x}As$ superlattices showing conduction and valence-band edges with heterostructure potential wells at x=0.3, $\Delta E_c \approx 300 meV$. The horizontal lines represent quantum-well discrete energy levels for electrons and holes confined in the GaAs layers.

Change of the Dimensionality of the System

If a NsM consists of thin needle-shaped or flat, two-dimensional crystallites, only two or one dimension of the building blocks becomes comparable with the length scale of a physical phenomenon. In other words, in these cases the NsM becomes a two- or one-dimensional system with respect to this phenomenon.

Changes of the Atomic Structure

Changes in the atomic structure result if a high density of incoherent interfaces —or other lattice defects such as dislocations, vacancies, etc.—is incorporated. The cores of lattice defects represent a constrained state of solid matter differing structurally from (unconstrained) crystals and/or glasses. As a consequence, a solid containing a high density of defect cores differs structurally from a defect-free solid with the same (average) chemical composition. The boundaries in figure represent an example of this effect: the misfit between adjacent crystallites changes the atomic structure (e.g. the average atomic density, the nearest-neighbor coordination, etc.) in the boundary regions relative to the perfect crystal. At high defect densities the volume fraction of defect cores becomes comparable with the volume fraction of the crystalline regions. In fact, this is the case if the crystal diameter becomes comparable with the thickness of the interfaces, i.e. for crystal sizes on the order of one or a few nanometers as is the case in NsM.

Alloying of Components that are Immiscible in the Solid and the Molten State

The following cases of this type of immiscible components in NsM may be distinguished: Solute atoms with little solubility in the lattice of the crystallites frequently segregate to the boundary cores (e.g., the free energy of the system in several alloys is reduced if large solute atoms segregate to the boundary core). The second case of nanostructured alloys results if the crystallites of a NsM have different chemical compositions. Even if the constituents are immiscible in the crystalline and/or molten state (e.g., Fe and Ag), the formation of solid solutions in the boundary regions of the NsM has been noticed.

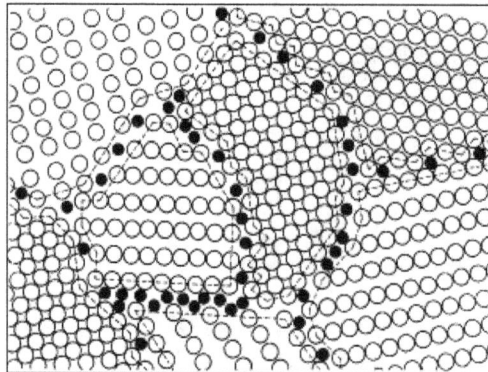

Schematic model of the structure of nanostructured Cu–Bi and W–Ga alloys. The open circles represent the Cu or W atoms, respectively, forming the nanometer-sized crystals. The black circles are the Bi or Ga atoms, respectively, incorporated in the boundaries at sites of enhanced local free volume. The atomic structure shown was deduced from EXAFS and X-ray diffraction measurements.

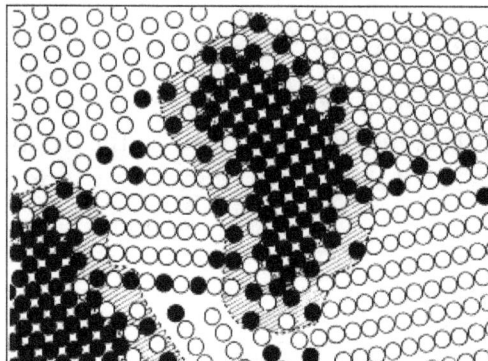

Schematic model of nanocrystalline Ag–Fe alloys according to the data of Mössbauer spectroscopy. The alloys consist of a mixture of nanometer-sized Ag and Fe crystals (represented by open and full circles, respectively). In the (strained) interfacial regions between Ag and Fe crystals, solid solutions of Fe atoms in Ag crystallites, and Ag atoms in the Fe crystallites are formed although both components are immiscible in the liquid as well as in the solid state. Similar effects may occur in the grain boundaries between adjacent Fe and Ag crystals.

Finally, it may be pointed out that NsM are by no means limited to polycrystalline materials consisting of the type displayed in figure. In semicrystalline polymers, nanometer-sized microstructures are formed that consist of crystalline and non-crystalline regions differing in molecular structure and/or chemical composition. Polymeric NsM will be discussed in Sections. NsM synthesized by supramolecular chemistry result if different types of molecular building blocks are self-assembled into a large variety of one-, two- or three-dimensional arrays.

The remarkable potential the field of NsM offers in the form of bulk materials, composites or coating materials to optoelectronic engineering, magnetic recording technologies, micro-manufacturing, bioengineering, etc. is recognized by industry. Large-scale programs, institutes and research networks have been initiated recently on these and other topics in the United States, Japan, EC, China and other countries.

Microstructures

Materials with nanometer-sized microstructures may be classified according to their free energy into equilibrium NsM and NsM far away from thermodynamic equilibrium which will be called "non-equilibrium NsM".

Non-equilibrium Nanostructured Materials

Non-equilibrium NsM are materials composed of structural elements—mostly crystallites—with a characteristic size (at least in one direction) of a few nanometers. In other words, non-equilibrium NsM are inherently heterogeneous on a nanometer scale consisting of nanometer-sized building blocks separated by boundary regions. The various types of non-equilibrium NsM differ by the characteristic features of their building blocks (e.g. crystallites with different or identical chemical composition, different or identical atomic structure, different or identical shape, size, etc). However, the size, structure, etc. of the building blocks are not the only microstructural features distinguishing different NsM. In fact, the boundary regions between them play a similar role. The chemical composition, atomic structure, thickness, etc. of the boundary regions are equally crucial for the properties of NsM. In other words, even if the building blocks, e.g. the crystallites of two NsM, have comparable size, chemical composition, etc., the properties of both NsM may deviate significantly if their interfacial structures differ. Different interfacial structures may result if the two NsM have been synthesized by different procedures. For example, nanocrystalline Ni (crystal size about 10 nm, density about 94%) prepared by consolidation of Ni powder exhibited little (<3%) ductility whereas nanocrystalline Ni (similar grain size and chemical composition) obtained by means of an electro-deposition process could be deformed extensively (>100%). The major difference noticed between both materials was the energy stored in the interfacial regions suggesting different interfacial structures.

One of the technologically attractive features of non-equilibrium NsM is the fact that their microstructure (and properties) can be manipulated—as in all non-equilibrium systems—by the mode of preparation. This allows a wide variety of microstructures (and hence properties) to be generated. Naturally, the other side of the coin is that any technological application of NsM is only possible if one is able to fully characterize and control their microstructure, and if the correlation between their properties and their microstructure is well understood so that NsM with controlled properties can be produced reproducibly.

Classification of Nanostructured Materials

Non-polymeric NsM consisting of nanometer-sized crystallites and interfaces may be classified according to their chemical composition and the shape (dimensionality) of their microstructural constituents. According to the shape of the crystallites, three categories of NsM may be distinguished: layer-shaped crystallites, rod-shaped crystallites (with layer thickness or rod diameters in the order of a few nanometers), and NsM composed of equiaxed nanometer-sized crystallites. Depending on the chemical composition of the crystallites, the three categories of NsM may be grouped into four families. In the simplest case, all crystallites and interfacial regions have the same chemical composition. Examples of this family of NsM are semicrystalline polymers (consisting of stacked crystalline lamellae separated by non-crystalline regions; first category in figure.) or NsM made up of equiaxed nanometer-sized crystals, e.g., of Cu (third category). NsM belonging to the second family consist of crystallites with different chemical compositions (indicated in figure by different thickness of the lines used for hatching). Quantum well (multilayer) structures are probably the most well-known examples of this type (first category). If the compositional variation occurs primarily between crystallites and the interfacial regions, the third family of NsM is obtained. In this case one type of atoms (molecules) segregates preferentially to the interfacial regions so that the structural modulation (crystals/interfaces) is coupled to the local chemical modulation. NsM consisting of nanometer-sized W crystals with Ga atoms segregated to the grain boundaries are an example of this type (third category). An interesting new type of such materials was recently produced by co-milling Al_2O_3 and Ga. It turned out that this procedure resulted in nanometer-sized Al_2O_3 crystals separated by a network of non-crystalline layers of Ga. Depending on the Ga content, the thickness of the Ga boundaries between the Al_2O_3 crystals varies between less than a monolayer and up to about seven layers of Ga. The fourth family of NsM is formed by nanometer-sized crystallites (layers, rods or equiaxed crystallites) dispersed in a matrix of different chemical composition. Precipitation-hardened alloys belong in this group of NsM. Nanometer-sized Ni_3Al precipitates dispersed in a Ni matrix—generated by annealing a supersaturated Ni–Al solid solution—are an example of such alloys. Most high-temperature materials used in jet engines of modern aircraft are based on precipitation-hardened Ni_3Al/Ni alloys.

Classification schema for NsM according to their chemical composition and the dimensionality (shape) of the crystallites (structural elements) forming the NsM. The boundary regions of the first and second family of NsM are indicated in black to emphasize the different atomic arrangements in the crystallites and in the boundaries. The chemical composition of the (black) boundary regions and the crystallites is identical in the first family. In the second family, the (black) boundaries are the regions where two crystals of different chemical composition are joined together causing a steep concentration gradient.

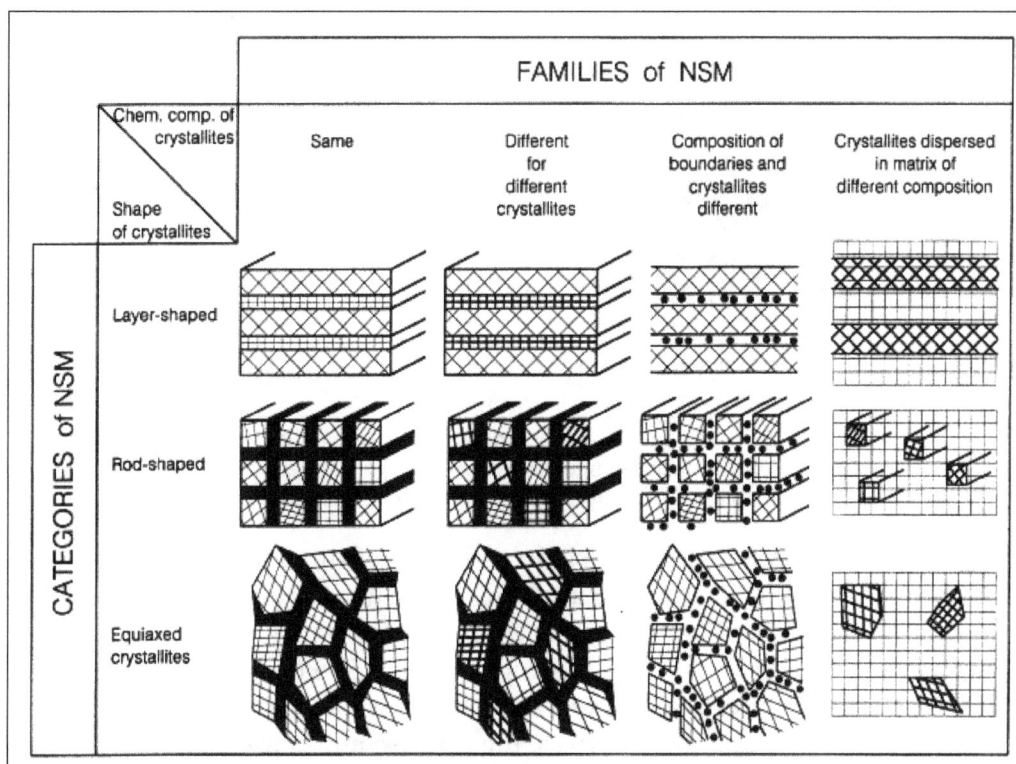

The NsM considered so far consisted mostly of crystalline components. However, in addition, NsM are known in which one or all constituents are non-crystalline. For example, semicrystalline polymers consist of alternating (nanometer thick) crystalline and non-crystalline layers. The various types of microstructures that may be formed in polymeric NsM. Other NsM consisting of a crystalline and a non-crystalline structural component are partially crystallized glasses and nanocrystalline metal nitrides, carbides of the type M_nN, M_nC (metal=Ti, Zr, Nb, W, V) embedded in an amorphous matrix, e.g. a Si_3N_4 matrix. Metal nitrides embedded in amorphous Si_3N_4 have been prepared by high-frequency discharge, direct current discharge or plasma-induced chemical vapor deposition. The remarkable feature of these materials is their hardness which seems to be comparable with or higher than that of diamond. Elastic image forces are argued to require a very high stress to force dislocations to cut through the nanometer-sized nitride crystallites. This high stress may, however, not lead to fracture because any crack formed in one of the crystallites is suggested to be stopped by the ductile amorphous Si_3N_4 matrix surrounding the cracked crystallite. Another family of technologically interesting NsM consisting of nanometer-sized crystallites embedded in an amorphous matrix is nanocrystalline magnetic materials. They are derived from crystallizing amorphous ribbons of (Fe, B)-based metallic glasses. Their microstructure is characterized by 10–25 nm-sized grains of a b.c.c.-α-FeX phase consuming about 70–80% of the total volume. This phase is homogeneously dispersed in an amorphous matrix. The two families of alloys showing the best performance characteristics are Fe–Cu–Nb–B–Si (FINEMET) and Fe–Zr–Cu–B–Si (NANOPERM). "Finemet" alloys have a saturation induction of about 1.2 T and their properties at high frequencies are comparable with the best Co-based amorphous metals. The outstanding features of "Nanoperm" alloys are the very low losses exhibited at low frequencies (<100Hz) offering potential for applications in electrical power distribution transformers.

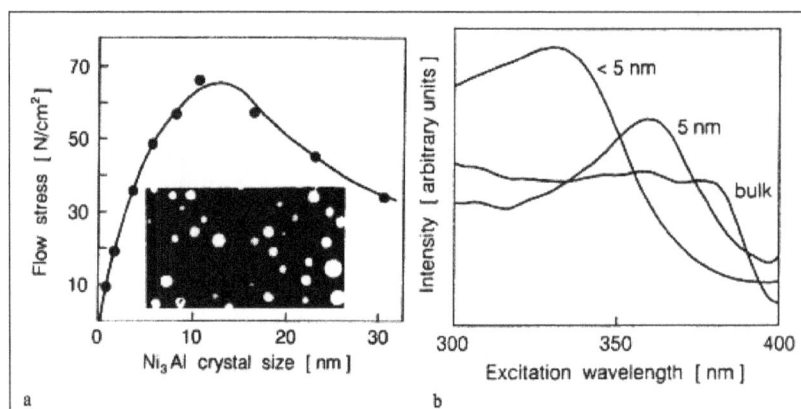

(a) Flow stress of Ni–13 at.% Ni alloys as a function of the size of the Ni_3Al precipitates.
(b) Photoluminescence spectra of nanocrystalline ZnO with different crystal sizes in comparison with the bulk material. The detection wavelength was 550 nm.

Spinodally decomposed glasses represent NsM in which all constituents are non-crystalline. Finally, crystalline or non-crystalline materials containing a high density of nanometer-sized voids (e.g. due to the α-particle irradiation) are NsM, one component of which is a gas or vacuum. A well-known example of a NsM with a void-type structure is porous Si. Porous Si has attracted considerable attention because of its strong photoluminescence in visible light. Two fundamental features of bulk Si limit its use in optoelectronic devices: the centrosymmetric crystal structure prevents a linear electro-optical effect. Hence, Si cannot be used for light modulation. Secondly, the band gap of Si is indirect and lies in the infrared region ($Eg\sim1.170eV$). As a consequence, Si was considered unsuitable for light-emitting technologies. In 1990 it was reported that porous Si could luminesce. Two lines of thought were put forward to explain the effect: quantum confinement and luminescence of chemical complexes attached to the free surface of the silicon crystallites. The quantum confinement model proposes the carriers in porous Si to be confined to microcrystallites with a size of 1–4 nm formed due to the porosity of the Si. The chemical complexes capable of luminescing in the observed spectral range were proposed to be siloxene compounds, a complex of Si, H and O. Recently, "hybrid models" where both, the interior and the surface of the porous Si are involved in the photoluminescence.

Obviously, the model of a non-equilibrium NsM considered so far is highly simplified in the sense that it is based on a hard-sphere approach. Nonetheless, two characteristic features of a NsM are already borne out by this approach: The nanometer-sized crystallites are expected to exhibit size and/or dimensionality effects. Moreover, several properties (e.g. diffusion, internal friction, etc.) of a NsM should be controlled by the presence of a high density of grain and/or interphase boundaries.

Size Effects

The dependence of the flow stress of a nanocomposite consisting of nanometer-sized Ni_3Al crystallites dispersed in a matrix made up of a NiAl solid solution. The total volume fraction of Ni_3Al crystallites is the same for all Ni_3Al crystal sizes shown in figure; the only parameter varied is the size of the Ni_3Al crystallites. Figure displays the blue shift in the luminescence spectra as a function of the crystal size for nanocrystalline ZnO (consisting of consolidated ZnO crystals separated by grain boundaries). The blue shift is a quantum size effect. If the crystallite size becomes comparable or smaller than the de Broglie wavelength of the charge carriers generated by the absorbed light, the confinement increases the energy required for absorption. This energy increase shifts the absorption/luminescence spectra towards shorter wavelengths (blue).

Another effect related to the reduced size of the crystallites in NsM concerns the atomic structure of the interfaces. More precisely, the question: is the atomic structure of the interfaces between nanometer-sized crystallites different from the structure of the interfaces between crystals of infinite size (same chemical composition, orientation, relationship, etc.)? So far, only a few specific cases have been studied experimentally and theoretically by means of molecular dynamics computations. The results of these studies may be summarized as follows: in metallic NsM, the low-temperature atomic structure of the boundaries of a NsM differs from the structure of the boundaries in a coarse polycrystal primarily by the rigid body translation. The deviating rigid body relaxation of both types of boundaries results from the different constraints in both materials: in coarse-grained polycrystals adjacent crystallites are free to minimize the boundary energy by a translational motion relative to one another (called rigid body relaxation). In a NsM the constraints exerted by the neighboring nanometer-sized crystallites limit the rigid body relaxation more the smaller the crystallites are. Another crystallite size effect concerns the structural stability of NsM 28, 29. The vibrational densities of state of a NsM and of the related glass, determined from lattice-dynamics simulations, exhibit low- and high-frequency modes not seen in the perfect crystal. The low-frequency modes give rise to a low-temperature peak in the excess specific heat in both types of metastable microstructures. Free-energy simulations of NsM and the related glass suggest that a phase transition from the nanocrystalline state to the glass should occur below a critical grain size. Figure displays the dependence of the free energy of various NsM with different grain sizes. Obviously, below a crystal size of about 1.4 nm, NsM are unstable relative to the glass as they exhibit a higher free energy. A structural transformation consistent with these results was, in fact, reported for nanocrystalline Si prepared by glow discharge decomposition of silane: nanostructured Si was noticed by Raman spectroscopy to transform into amorphous Si if the crystal size was reduced below a critical value of a few nanometers.

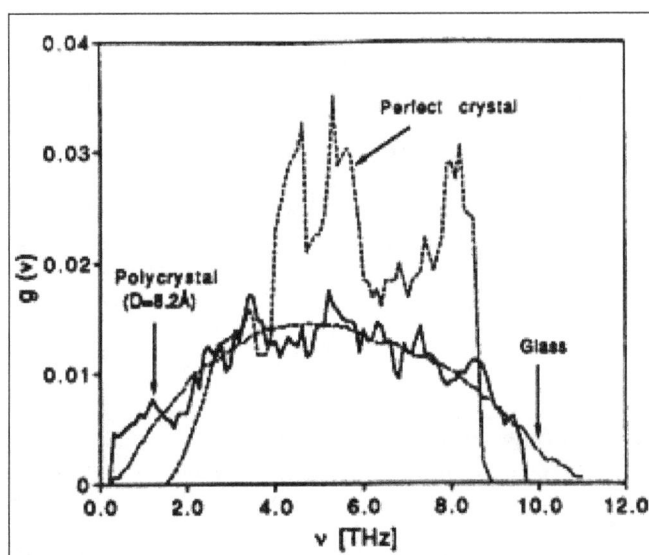

Comparison of the vibrational density of states g(ν) for nanocrystalline Cu (crystal size 8.2 Å, solid line) with those for a Lennard–Jones glass (molecular-dynamics simulation, dash–dotted line) and for the perfect f.c.c. crystal (dashed line); ν is the phonon frequency.

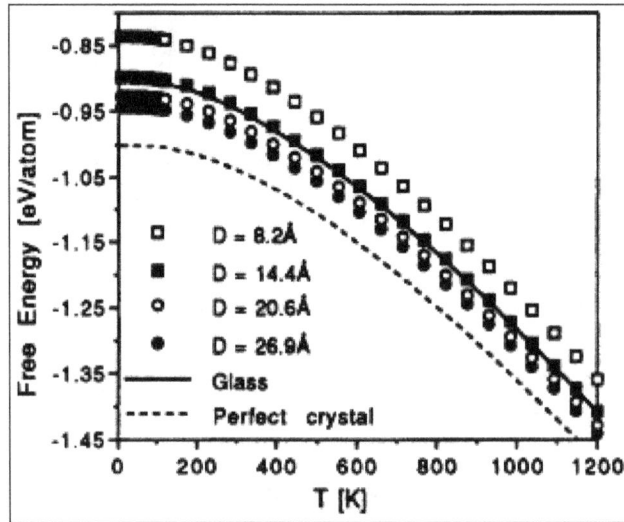

Comparison of the temperature dependence of the total free energy of
nanocrystalline Cu (for different grain sizes as indicated in the figure)
with a Lennard–Jones glass and the perfect crystal.

Reduced Density and Coordination in the Boundaries

In the core of incoherent interfaces, the misfit between crystallites joined together locally modifies
the atomic structure by reducing the atomic density and by altering the coordination between nearest-neighbor atoms relative to the perfect crystal. The reduced density (or enhanced free volume)
in the boundaries is directly visible in high-resolution electron micrographs and has also been evidenced by Mössbauer spectroscopy. The Mössbauer spectra of the interfacial component of nanocrystalline Fe exhibit a pressure-induced reversible change in the isomer shift that is about one order
of magnitude larger than that of the α-Fe crystals and of glassy iron alloys. The enhanced isomer shift
indicates an enhanced compressibility of the boundary regions and thus a reduced interfacial density.

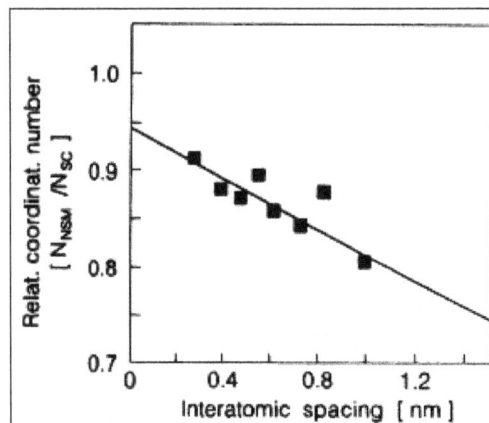

Coordination number (measured by X-ray scattering) for nanocrystalline Pd (12 nm crystal size)
relative to a Pd single crystal as a function of the interatomic spacings. N_{NSM} and N_{SC} are the
measured coordination numbers of the nanocrystalline Pd and of a Pd single crystal.

The modified nearest-neighbor coordination in the boundary regions relative to a perfect crystal (with
the same chemical composition) has been revealed by measuring (X-ray diffraction) the pair correlation
functions of nanostructured Pd and of a Pd single crystal. The same result was obtained by Mössbauer

studies of FeF_2, α-Fe_2O_3 and γ-Fe_2O_3. The grain boundary structure of these materials was found to consist of structural units the coordinations of which differ from the ones in the crystalline state.

Chemical Binding Effects

The atomic structures of the boundary regions in NsM are expected to depend on the type of chemical binding forces. The following picture seems to emerge from the presently available experimental and theoretical studies on the correlation between chemical binding and the nature of the boundaries in NsM.

In materials with directional bonds (e.g., Si, C), the boundary structure depends significantly on the competition between local structural disorder in the boundary and localized variation in the hybridization of the bonds in the region of the interfaces. Silicon and carbon provide relatively simple cases for the physical understanding of the coupling between structural disorder and bonding modifications. Silicon is a purely sp^3 bonded material. Diamond exhibits greater bond stiffness combined with the ability to change hybridization in a disordered environment from sp^3 to sp^2. The interplay between these two factors may be elucidated by comparing the different ways in which the two materials respond to structural disorder. Figures compare the atomic structures of two grain boundaries in diamond and Si. In both materials, the (111) boundary is clearly more ordered than the (100) boundary shown in figure. The different degrees of disorder in both types of boundaries are evidenced by the much lower energy of the (111) grain boundary relative to the (100) interface ($\approx$30% in diamond and $\approx$47% in Si).

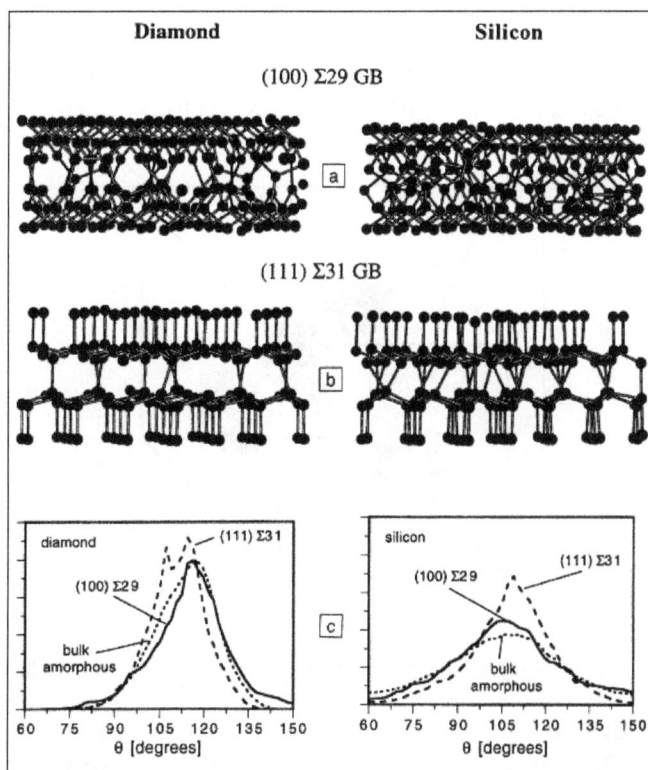

Projected structures of the high-temperature relaxed (a) (100) Σ29 and (b) (111) Σ30 twist boundaries in diamond and Si. All the nearest-neighbor bonds between grain boundary atoms are shown. (c)

Distribution of bond angles (in arbitrary units) for the atoms in the two center planes of the above grain boundaries. For comparison the distributions for bulk amorphous carbon and silicon are also shown.

The average nearest-neighbor coordination, $\langle C \rangle$, of the atoms in the two center planes of the diamond (100) and (111) grain boundaries are 3.16 and 3.50, respectively. These low values are indicative of a significant fraction of grain boundary atoms being only threefold-coordinated (i.e. by sp^2 bonded). Eighty percent of all (100) grain boundary atoms are threefold-coordinated compared to "only" 50% in the (111) grain boundary, whereas practically all other atoms are tetrahedrally coordinated. By contrast, in Si these two grain boundaries have $\langle C \rangle \approx 4.02$ and 4.06, respectively—that is, close to the perfect tetrahedral coordination with only a few three- and fivefold-coordinated Si atoms in the grain boundary unit cell.

These differences are strikingly apparent in the related bond-angle distribution functions shown in figure. For example, the presence of equal fractions of three- and fourfold-coordinated C atoms and the high degree of structural ordering in the diamond (111) grain boundary give rise to two distinct peaks, one near the sp^3 bond angle of $109.47°$ and the other near the sp^2 bond angle of $120°$. By comparison, in Si the sp^2 peak is completely absent. In contrast to the (111) grain boundary, the bond-angle distribution function of the high-energy (100) grain boundary in both diamond and Si is similar to that of the corresponding bulk amorphous material. However, whereas—in diamond—the peak is centered near the sp^2 bond angle, indicating the presence of mostly threefold-coordinated C atoms and significant structural disordering, in Si the peak is centered at the tetrahedral bond angle.

This comparison reveals that because in Si sp^2-type bonding is not allowed, a large driving force exists for the initially threefold-coordinated atoms in the unrelaxed grain boundary to recover as much as possible their full fourfold coordination—even at the cost of severe grain boundary disordering. In contrast, diamond has only a small driving force for structural disordering, at the cost of significant bond disordering.

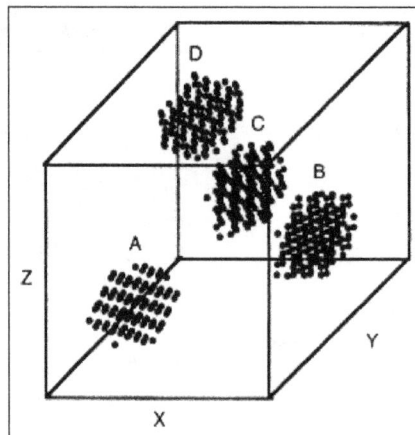

Cubic, three-dimensional periodic simulation cell containing four randomly oriented seed grains arranged on a f.c.c. lattice and embedded in the melt (schematic).

Based on these results, the number of threefold-coordinated C atoms was estimated and was found to agree with recent Raman scattering experiments on nanocrystalline diamond grown from fullerene precursors. The physical significance of the similarity of the bond-angle distribution of the amorphous Si and the (100) $\Sigma29$ boundary was investigated further comparing the atomic arrangement in nanocrystalline Si averaged over many boundaries between nanometer-sized Si grains with different orientation relationships. The fully dense nanostructured Si was synthesized (molecular dynamics) by inserting small crystalline seeds with randomly preselected crystallographic orientations into a Si melt. Subsequent cooling below the melting point of Si resulted in the growth of the inserted seed crystals to form equilibrated grain boundaries in a fully dense polycrystalline Si. The boundary structure may be compared with the structure of amorphous Si by comparing the radial and angular distribution functions of the various interfacial structural components (boundaries, triple lines, etc.) of the nanometer-sized Si with the radial and angular distribution functions of amorphous Si. The results obtained indicate that the atomic arrangement in the interfaces of Si is similar to the atomic arrangement of amorphous Si. In fact, these results suggest that nanometer-sized Si may be treated as a two-phase system consisting of an ordered crystalline phase (in the crystal interiors) connected by an amorphous-like intergranular phase.

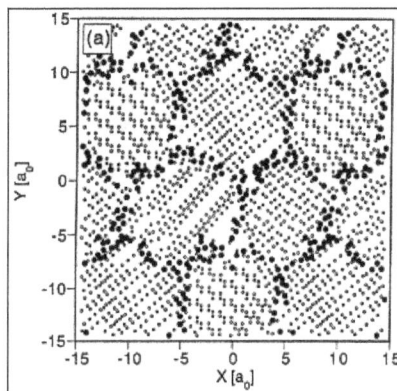

Positions of the atoms within a slice of thickness $0.5a_0$ cut out (parallel to the X–Yplane, of a nanocrystalline material. The nanocrystalline material was generated by the procedure described. The solid circles represent atoms with excess energies larger than 0.1 eV.

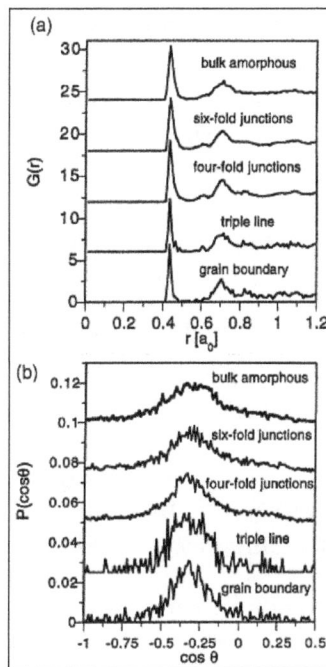

(a) Typical local radial distribution functions, G(r), for the nanocrystalline Si with a grain size of 5.4 nm. Shown is a comparison of these local radial distribution functions for the atoms in the grain boundary regions, the triple lines, the fourfold and sixfold point grain junctions, with the overall radial distribution function of bulk amorphous silicon. (b) Angular distribution functions, P(cos θ), for the same defected regions as in (a).

Temperature Effects

Elevated temperatures seem to affect the microstructure of NsM by one or both of the following two types of processes:

- Grain growth.

- Temperature-induced variations of the atomic structure.

Grain Growth

Grain growth in NsM is primarily driven by the excess energy stored in the grain or interphase boundaries. Analogous to the growth of cells in soap froths, the boundaries move toward their centers of curvature and the rate of movement varies with the amount of curvature. The earliest theoretical considerations of the kinetics of normal grain growth assume a linear relationship between the rate of grain growth and the inverse grain size, which in turn is proportional to the radius of curvature of the grain boundaries 30, 31. This assumption yields, under ideal conditions, the following equation for grain growth:

$$D^2 - D_0^2 = kt$$
,

where D_0 and D are the grain sizes at the beginning of the experiment and at time, t, respectively. K is a constant that depends on temperature. A number of more recent theoretical treatments came to the same conclusion that normal grain growth should ideally occur in a parabolic manner. However, this is rarely observed except for high-purity metals at high homologous temperatures. For practical purposes, the most widely used relationship incorporates the empirical time exponent n$\leq$0.5 which allows the description of isothermal grain growth that often does not fit the ideal relationship modeled by $D^2 - D_0^2 = kt$:

$$D^{1/n} - D_0^{1/n} = k`t \ or \ D = \left( k`t + D_0^{1/n} \right)^n$$

The rate constant k (or k′) can be expressed in an Arrhenius-type equation:

$$k = k_0 \exp\{-Q/RT\}$$
,

where Q is the activation enthalpy for isothermal grain growth, R the molar gas constant and k_0 a constant that is independent of the absolute temperature T. The activation enthalpy, Q, is often used to determine the microscopic mechanism which dominates the grain growth.

Grain growth studies have been carried out for various NsM using TEM, DSC, X-ray diffraction and Raman spectroscopy. The materials studied were prepared by crystallizing glasses, sliding wear, inert gas condensation, electrodeposition, electron gun evaporation, mechanical milling and CVD. Studies of the grain growth process in NsM produced by the crystallization of glasses have the attractive feature that pore-free nanocrystalline materials are obtained. Obviously, the synthesis of

NsM by crystallization of glasses is limited to the specific chemical compositions that permit the preparation of the glassy state, e.g. by rapidly cooling, by a sol–gel process, etc. Moreover, only those glasses are suitable for grain growth studies in NsM that convert the glassy phase directly into a crystalline phase of the same chemical composition. For example, a stable tetragonal (Fe, Co)Zr$_2$ phase forms directly from the ternary Fe–Co–Zr amorphous phase, while in the binary Fe–Zr alloys, the amorphous phase first results in a metastable f.c.c. FeZr$_2$ phase which later transforms to the equilibrium tetragonal FeZr$_2$ phase. Grain growth studies were performed for both the stable and metastable phases and it was found that the grain size increases with annealing time. It has also been noted that grain growth starts at a lower temperature in the nanocrystalline sample with smaller grains and that grain growth is rapid above a certain temperature and becomes negligible for longer annealing times.

As grain growth involves the transport of atoms across and presumably also along the boundaries, the activation energy of the process is frequently compared with that of grain boundary diffusion. The two activation energies agree reasonably well in most systems studied so far.

Ganapathi et al. tried to fit their grain growth data on nanocrystalline Cu produced by sliding wear and observed an excellent fit for values of n of 1/2, 1/3 or 1/4. Thus, they concluded that it is difficult to identify the grain growth mechanism on the basis of the exponent n alone, and that grain growth in nanocrystalline materials probably occurs in a manner similar to that in conventional polycrystalline materials.

In most of the studies involving nanocrystalline materials, the value of n is different from the value of 0.5, deduced from the parabolic relationship for grain growth. Thus, in addition to Zener drag (where a particle interacts with the grain boundary to reduce the energy of the boundary–particle system and restrains the boundary movement), other mechanisms such as pinning of grain boundaries by pores, solute atoms or inclusions may also be operative. The fact that pores 33, 54 and impurity doping have considerable effect on the grain growth characteristics was demonstrated in TiO$_2$. For an initial grain size of 14 nm, when the porosity was about 25%, the grain size (after annealing for 20 h at 700°C) was 30 nm. When the porosity was reduced to about 10%, the grain size for a similar annealing treatment was dramatically increased to 500 nm. The same authors have also demonstrated that sintering the same nanocrystalline material under pressure (1 GPa), or with appropriate dopants such as Y, can suppress the grain growth.

In general, n seems to change during grain growth and tends toward the ideal value of 0.5 as is found in high-purity materials or at high annealing temperatures. The values obtained for n from grain growth measurements seem to depend—at least in some systems—on the evaluation of the experimental data. For example, Krill and co-workers re-evaluated Marlow and Koch's results. The data fit used by Marlow and Koch yielded n=0.32. Krill and co-workers showed that the same measurements can be equally well matched by an impurity drag model with a growth exponent of n=0.5. Another problem associated with grain growth in nanocrystalline samples containing impurities has recently been re-emphasized although it is known to exist in principle in coarse-grained polycrystals as well. During grain growth, the area available to the segregant is reduced. Thus, if all impurity atoms remain in (or close to) the boundaries, their concentration must increase which should manifest itself in an enhanced drag force (rather than being independent of grain size as is commonly assumed). Recent measurements using Pd–Zr solute solutions seem to confirm the expected impurity drag enhancement. Naturally, in NsM this effect will be enhanced relative to a coarse-grained polycrystal due to the large reduction of the boundary area during grain growth.

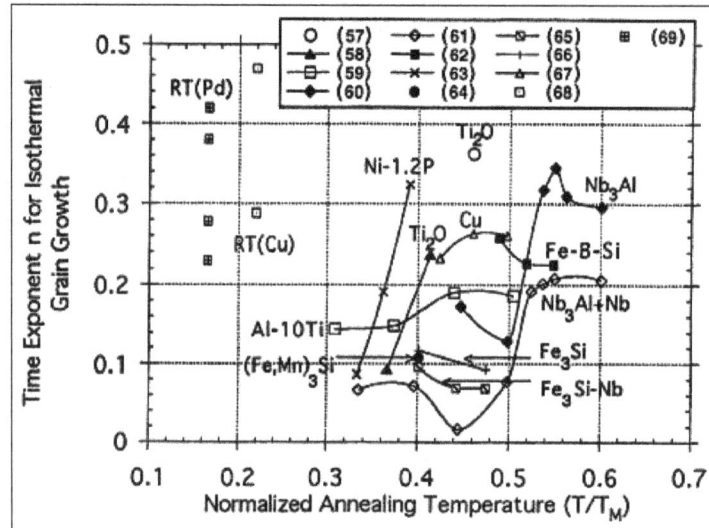

Time exponent for isothermal grain growth of various nanocrystalline materials as a function of the reduced annealing temperature.

Abnormal grain growth in NsM has been observed at room temperature or slightly above in some instances, e.g. in Cu, Ag, Pd and crystallized metallic glasses of the FINEMET type. Similar to the observation of Hahn et al., Gertsman and Birringer also noted that grain growth occurs preferentially in the denser materials. Anomalous grain growth has been suggested to be due to: (a) a certain non-uniformity of the grain size distribution in the as-prepared samples (so that the larger grains act as nuclei); and (b) impurity segregation. If the impurity distribution is spatially non-uniform, enhanced grain growth may occur in regions of lowest impurity content. The reason why such abnormal grain growth does not occur in many coarse-grained polycrystalline materials has been attributed to the enhanced grain boundary enthalpy (leading to high driving forces) and/or non-equilibrium grain boundary structures (leading to increased mobility of grain boundaries) in the nanocrystalline materials.

Stability against Grain Growth

Several approaches for preventing grain growth have been proposed. On the one hand are those that aim to slow down the growth kinetics by reducing the driving force (the grain boundary free energy) or the grain boundary mobility. In these cases the material remains in an unstable state where small local rearrangements of the grain boundary planes can reduce the material's free energy, but the time interval and temperature required for significant grain growth to take place are increased. The second type of stabilization aims at achieving a truly metastable state where each small variation of the total grain boundary area increases the free energy of the material. In this case a large energy barrier has to be overcome, e.g. by thermal activation, in order to start the evolution towards the equilibrium state, the single crystal.

Inclusions of a second phase act as pinning sites for grain boundaries in essentially the same way as do pores during sintering: the total free energy of a segment of boundary intersecting an inclusion is reduced by the product of the cross-section of the inclusion and the specific boundary free energy. Zener derived a relation between the stable grain radius R and the radius r and volume fraction f of the inclusions: $R/r \approx 3/4f$. This relationship implies that when a fine dispersion of small inclusions can be generated, then small volume fractions of inclusions can stabilize a microstructure with a very fine grain size. In the stable microstructure the location of each boundary corresponds to a local energy minimum, and the material is therefore in a metastable state. When the temperature is increased, grain growth will remain suppressed until the inclusions dissolve in the matrix or until they become mobile. Retarded grain growth will also result from solute drag effects. In many solid solutions, solute atoms are known to segregate to the boundaries forming a solute cloud in the vicinity of the boundary.

If the boundary migrates, three modes of motion may occur depending on the relative rates of boundary and solute-cloud mobility:

- If the boundary migrates slowly[†], it drags its solute cloud along with it, thus reducing the boundary mobility and, hence, grain growth.

- If the boundary migrates very fast, it breaks away from the solute atoms and moves freely.

- At intermediate migration rates, the boundary breaks loose locally from its cloud and this impurity-free segment bulges out. The resulting increase of the boundary area reduces the rate of motion of the impurity-free boundary segment and permits the impurity cloud to be formed again ("jerky motion").

All three modes of boundary motion have been observed experimentally in coarse-grained polycrystals. The first two cases are likely to occur in NsM as well. In fact, pinning of the grain boundaries in nanocrystalline Ni solid by the Ni_3P precipitates in a crystallized Ni–P amorphous alloy and segregation of Si to grain boundaries in a Ni–Si solid solution have been found to be responsible for preventing grain growth in nanocrystalline phases. In addition to the kinetic factors discussed so far, energetic effects may also affect the growth rate of the crystallites in NsM. For example, Lu studied the thermal stability of 7–48 nm grains in a Ni–P alloy and concluded that samples with smaller grain sizes have enhanced thermal stabilities, suggesting that the grain growth temperatures and the activation energy for growth in a nanocrystalline solid are higher in comparison with coarser grains. This is attributed to the configuration and the energetic state of the interfaces in the nanocrystalline materials.

In general, the solute solubility in the core of grain boundaries differs considerably from the solubility in the interior of the crystals. Therefore, in thermodynamic equilibrium, the grain boundaries are enriched or depleted in solute. This can have two beneficial effects on the stability of the microstructure. The first effect is solute drag. The second effect is a reduction of the driving force for grain growth. According to the Gibbs adsorption equation, the grain boundary free energy decreases when solute segregates to the boundary. Experimental evidence shows that the decrease can be substantial, and the theory indicates that in alloy systems with a large atomic size mismatch, the grain boundary free energy may even be reduced to zero.

As a consequence of the solute enrichment at the grain boundaries and of the large specific grain boundary area, theory and experiment show that nanocrystalline materials also have an enhanced overall solubility for solute with a large heat of segregation.

The potential existence of alloy systems with a vanishing grain boundary free energy has led to speculations on the existence of a metastable nanocrystalline state in which grain growth requires that the nucleation barrier for the formation of a second phase be overcome by thermal activation. While there is as yet no definite experimental proof of the existence of a metastable nanocrystalline state, there are a number of experimental observations that favor its existence. Y–Fe is an alloy system with a large atomic size difference, suggesting a large enthalpy of grain boundary segregation. In Y–Fe alloys (prepared by inert gas condensation) Fe segregates to the grain boundaries. In agreement with the theoretical predictions, the grain size of the alloy samples decreases as the alloy concentration is enhanced, and reaches values as small as 2 nm. Although alloys with a low Fe molar fraction, x_{Fe}, undergo grain growth upon annealing, alloys with higher x_{Fe} show little grain growth before the equilibrium phase YFe_2 nucleates, indicating that energetic rather than kinetic factors are responsible for the suppression of growth. A similar correlation between the onset of grain growth and the nucleation of the stable intermetallic phase has also been observed in Nb–Cu alloy prepared by high-energy ball milling. The grain size of mechanically alloyed Pd–Zr solid solutions has also been found to decrease with increasing solute concentration, and the heat release upon annealing indicates that solute (Zr) segregates to the grain boundaries, thereby reducing the specific grain boundary energy and impeding grain growth. It has also been demonstrated that alloying solute to ceramic nanometer-sized particles results in a drastic change in the grain-size density trajectory, with a substantially lower grain size in the densely sintered body. Finally, the existence of stable liquid microstructures with a nanometer-scale structure and a large number of internal interfaces, the microemulsions, lends support to the expectation that solid microstructures can also be stabilized against growth at very fine grain sizes and elevated temperatures.

In NsM consisting of nanometer-sized crystallites (TiN) embedded in an amorphous matrix (of amorphous Si_3N_4), the rate of crystal growth was observed to decrease with crystal size. In fact, if the crystal size was about 1 nm no measurable crystal growth occurred at temperatures below 1200°C (which is about 80% of the decomposition temperature). If the crystal size was about 10 nm, the grain growth started at 800°C. The physical reasons for this "inverse" grain growth kinetics are not yet fully understood. An attempt to rationalize the surprising stability in terms of the high cohesive energy of the amorphous/crystalline interface has been proposed.

Temperature Induced Variation in the Atomic Structure of the Boundaries

A variation in the atomic structure of the boundaries of NsM as a function of temperature was recently reported for nanocrystalline Si., the boundaries in nanocrystalline Si exhibit an amorphous-like structure. This structure was found to represent an equilibrium structure by contrast with bulk amorphous Si. If the nanometer-sized Si is heated to elevated temperatures, the amorphous structure seems to undergo (above a glass transition temperature T_g) a reversible and dynamical structural transformation from the structure of amorphous Si to liquid Si. By contrast with the bulk glass transition, however, this transition is continuous, fully reversible and thermally activated, starting at T_g and being complete at the equilibrium melting point T_m of Si, at which the entire nanometer-sized Si sample is liquid. Figure shows the reversibility of the structural transition. A reversible temperature variation from 1600 to 900 K and back to 1600 K results in a reversible variation of the free volume (δV) of the boundary. The temperature-dependent variation of δV is summarized in figure. Between T_g and the melting point of Si (T_m), δV varies continuously and reversibly between the amorphous and the molten state of Si. In other words, a continuous, reversible phase transition exists between the amorphous Si and the liquid Si (continuous melting and solidification). Figure compares the angular correlation functions, P(θ), of the boundaries in Si with the ones of bulk amorphous and liquid Si: at and below T_g the angular distribution function of the boundaries is similar to that of bulk amorphous Si. The same applies at 1600 K for liquid Si and the structure of the boundaries.

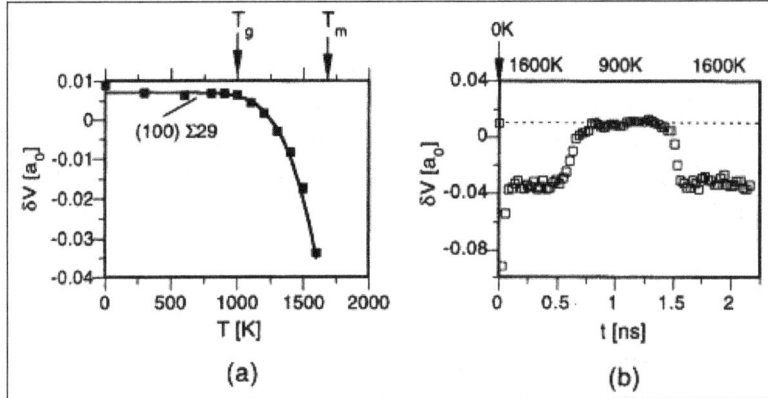

(a)   (b)

Temperature dependence of the volume expansion, δV (in units of the zero-temperature lattice parameter) per unit grain boundary area for the high-energy (100), Σ=29 grain boundary in silicon. The bulk glass transition temperature T_g and melting point T_m are indicated on the top axis. (b) Response to thermal cycling of the volume expansion δV for the (100), Σ=29 twist grain boundary, illustrating the reversibility of the transition between the confined amorphous and liquid grain boundary phases; t is the simulation time.

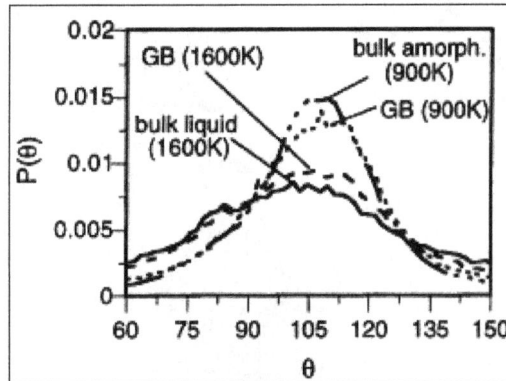

Comparison of the bond-angle distribution functions, P(θ), for the confined amorphous and confined grain boundary phases with those for bulk amorphous and supercooled liquid silicon, respectively. In perfect-crystal silicon at T = 0k , P(θ) exhibits a single δ-function peak at the tetrahedral angle θ_t=109.47°.

Formation of Non-equilibrium Alloys

In several nanostructured alloys, the solute solubility in the boundary regions was noticed to deviate from the solute solubility in the crystal lattice. The different solubilities (and presumably other effects as well) lead to the formation of alloys in nanocrystalline materials which do not exist in coarse-grained polycrystals.

Time–temperature History and Preparation Effects

The NsM discussed so far are a non-equilibrium state of condensed matter. Hence, their structure and properties depend not only on the chemical composition and the size/shape of the crystallites but also on the mode of preparation and the previous time–temperature history of the material. For example, the enthalpy stored in nanocrystalline Pt may be reduced during annealing up to 50% without grain growth (i.e. at constant crystal size and chemical composition). The reduction is presumably caused by atomic rearrangements in the boundary regions. Measurements of other properties of NsM (e.g. thermal expansion, specific heat, compressibility) and spectroscopic studies (e.g. by Mössbauer or positron lifetime spectroscopy) indicate structural differences between chemically identical NsM with comparable crystallite sizes if these materials were prepared by different methods and/or if their previous time–temperature history was different. In fact, similar effects have been reported for other non-equilibrium states of condensed matter (e.g. glasses). The non-equilibrium character of NsM implies that any comparison of experimental observations is meaningful only if the specimens used have comparable crystal size, chemical composition, preparation mode and time–temperature history. Moreover, the non-equilibrium character of NsM renders them susceptible to structural modifications by the methods applied to study their structure. An example is shown in figure.

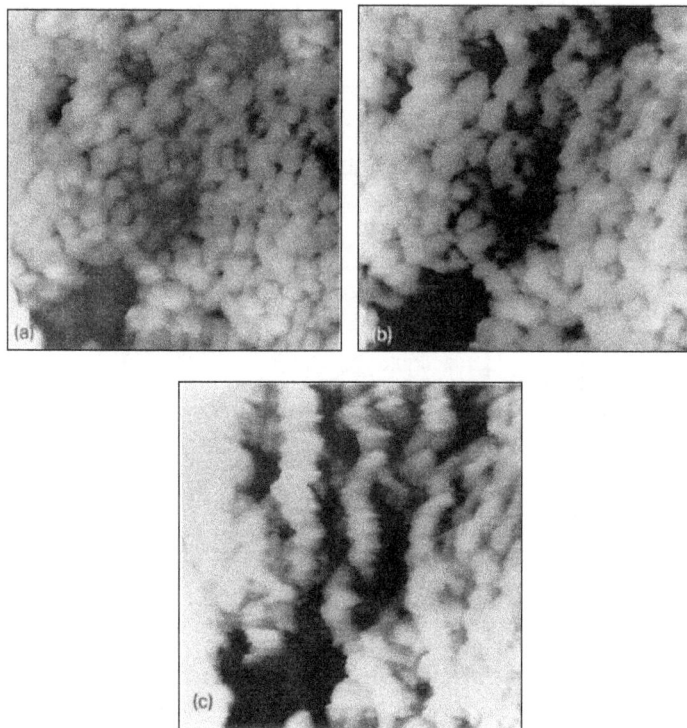

Sequential STM images of a nanostructured Pd surface, imaged with a tunneling voltage of -40 mV (tip negative) and a tunneling current of 6 nA. The area of $400 \times 400 \, nm^2$ is scanned at 2.5 min/image. Typical roughness data for the as-prepared samples as shown here are: peak to valley $= 400 \, nm$ r.m.s roughness $= 80 \, nm$, average roughness $= 65 \, nm$. (a) Image obtained from the first scan. (b) Image from the fifth scan (taken 10 min after the first scan), indicating the initial movements of some randomly distributed grains around the voids. (c) Image from the seventh scan, taken 15 min after the first scan. Grains were pictured to be moving dynamically in a worm-like fashion to yield channel-like grain boundaries.

Polymeric Nanostructured Materials

So far, the considerations have been limited to elemental or low molecular weight NsM, i.e. NsM formed by atoms/molecules that are more or less spherical in shape. A different situation arises if NsM are synthesized from high molecular weight polymers, i.e. long, flexible molecular chains.

It is one of the remarkable features of semicrystalline polymers that a nanostructured morphology is always formed if these polymers are crystallized from the melt or from solution, unless crystallization occurs under high pressure or if high pressure annealing is applied subsequent to crystallization. However, if a polymer is crystallized from solution or from the melt under ambient pressure, multilayer structures consisting of stacks of polymer single crystals result. Inside the crystals, the atoms forming the polymer chains arrange in a periodic three-dimensional (crystalline) fashion. The disordered interfacial regions between neighboring crystals consist of macromolecules folding back into the same crystal and of tie molecules that meander between neighboring crystals. The typical thicknesses of the crystal lamellae are of the order of 10–20 nm. These relatively small crystal thicknesses have been interpreted in terms of a higher nucleation rate of chain-folded crystals relative to extended chain crystals or in terms of a frozen-in equilibrium structure: at the crystallization temperature, the excess entropy associated with the chain folds may reduce the Gibbs free energy of the chain-folded crystal below that of the extended-chain crystal. Hence, at the crystallization temperature, crystallization will result in chain-folded crystals rather than in extended-chain crystals. Estimates of the excess entropy associated with the chain folds lead to a thickness of the nucleating crystals of about 10–20 nm. It may be pointed out that the nucleation of imperfect crystals during crystallization is not limited to polymeric materials. The excess entropy associated with vacancies, e.g. in elemental crystals, results in an equilibrium vacancy concentration at the melting temperature, i.e. in the nucleation of imperfect crystals. In metals, this equilibrium vacancy concentration at the melting temperature is typically about 10^{-4}.

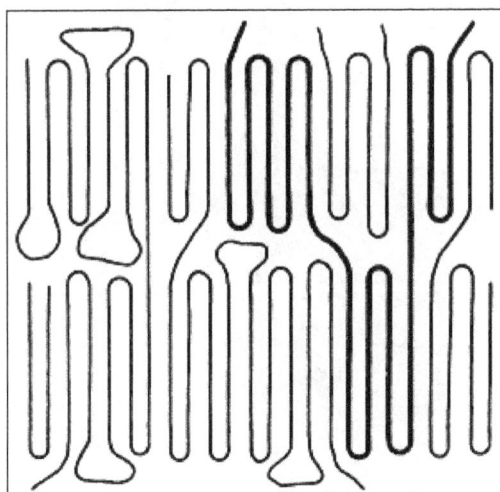

Molecular folding in semicrystalline polymers resulting in stacks of lamellar crystals with a thickness of about 10–20 nm separated by "amorphous" regions.

Chain folding may lead to rather complex nanometer-sized microstructures, depending on the crystallization conditions. Spherulites consisting of radially arranged twisted lamellae are preferred in unstrained melts. However, if the melt is strained during solidification, different morphologies may result, depending on the strain rate and the crystallization temperature (i.e. the undercooling).

High crystallization temperatures and small strain rates favor a stacked lamellar morphology, high temperatures combined with high strain rates result in needle-like arrangements. Low temperatures and high strain rates lead to oriented micellar structures. The transition between these morphologies is continuous and mixtures of them may also be obtained under suitable conditions. The way to an additional variety of nanostructured morphologies was opened when multicomponent polymer systems, so-called polymer blends, were prepared. Polymer blends usually do not form spacially homogeneous solid solutions but separate on length scales ranging from a few nanometers to many micrometers. The following types of nanostructured morphologies of polymer blends are formed in blends made up of one crystallizable and one amorphous (non-crystallizable) component: (I) The spherulites of the crystallizable component grow in a matrix consisting mainly of the non-crystallizable polymer. (II) The non-crystallizable component may be incorporated into the interlamellar regions of the spherulites of the crystallizable polymer. The spherulites are space-filling. (III) The non-crystallizable component may be included within the spherulites of the crystallizable polymer forming domains having dimensions larger than the interlamellar spacing. For blends of two crystallizable components, the four most common morphologies are: (I) Crystals of the two components are dispersed in an amorphous matrix. (II) One component crystallizes in a spherulitic morphology while the other crystallizes in a simpler mode, e.g. in the form of stacked crystals. (III) Both components exhibit a separate spherulitic structure. (IV) The two components crystallize simultaneously resulting in so-called mixed spherulites, which contain lamellae of both polymers.

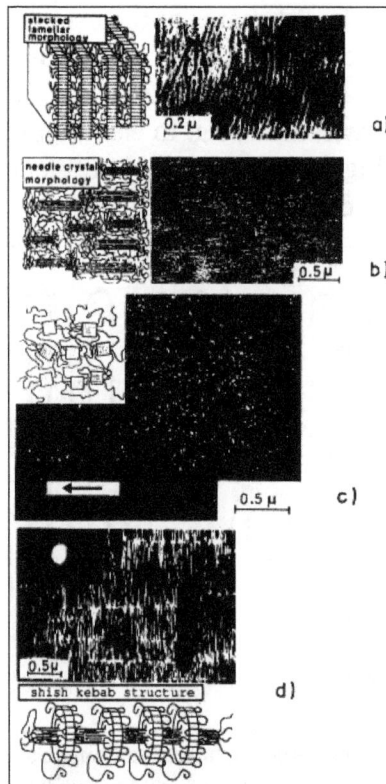

(a) Stacked lamellar morphology in polyethylene (TEM bright field). (b) Needle-like morphology in polybutene-1 (TEM bright field). (c) Oriented micellar morphology in polyethylene terephthalate (TEM dark field micrograph). (d) Shish-kebab morphology in isotactic polystyrene (TEM dark field micrograph).

Self-organized Nanostructured Arrays

Non-polymeric NsM

A modified Stranski–Krastanov growth mechanism has been noticed to result in self-organized (periodic) arrays of nanometer-sized crystallites. If a thin InGaAs layer is grown on a AlGaAs substrate, the InGaAs layer disintegrates into small islands once it is thicker than a critical value. These islands are spontaneously overgrown by a AlGaAs layer so that nanometer-sized InGaAs crystals buried in AlGaAs result. The observations reported indicate that the size, morphology and the periodic arrangement of the buried islands are driven by a reduction in the total free energy of the system. The driving force for the periodic arrangement of the crystallites seems to be the reduction in the strain energy of the system.

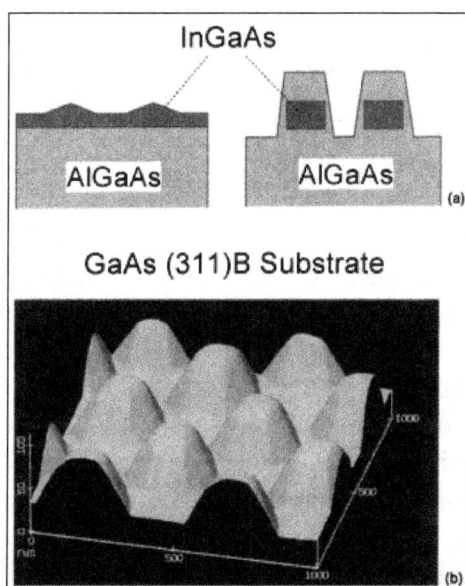

(a) Growth model of buried quantum dots of InGaAs in AlGaAs. (b) STM of the surface of a AlGaAs crystal. Underneath the surface small quantum dot crystals of $In_{0.2}Ga_{0.4}$ As are buried (a). The crystallites are periodically arranged.

Polymeric NsM

Block copolymers constitute a class of self-organized nanostructured materials. The macromolecules of a block copolymer consist of two or more chemically different sections that may be periodically or randomly arranged along the central backbone of the macromolecules and/or in the form of side branches. An example of a block copolymer is atactic polystyrene blocks alternating with blocks of polybutadiene or polyisoprene. The blocks are usually non-compatible and aggregate in separate phases on a nanometer scale if the copolymer is crystallized.

As an example of the various self-organized nanostructured morphologies possible in such systems, figure displays the morphologies formed in the system polystyrene/polybutadiene as a function of the relative polystyrene fraction. The large variety of nanostructured morphologies that may be obtained in polymers depending on the crystallization conditions and the chemical structure of the macromolecules causes the properties of polymers to vary dramatically depending on the processing conditions.

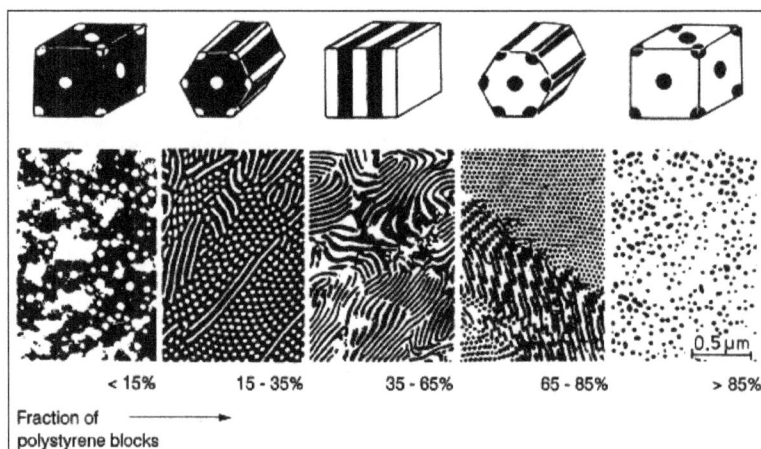

Electron micrographs of the morphologies of a copolymer consisting of polystyrene and polybutadiene blocks, as a function of the fraction of polystyrene blocks. The spacial arrangements of the polystyrene and polybutadiene in the solidified polymer are indicated in the drawings above the micrographs.

NsM formed by block copolymers seem to represent (metastable) equilibrium structures despite the high excess energy stored in the interfaces between the structural constituents, e.g., the polystyrene and the polybutadiene regions. The formation of these interfaces results from the local accumulation of the compatible segments of the macromolecules. Hence, the only way to remove these interfaces would be to generate a solid solution of the different segments forming the block copolymer, e.g. a solid solution of polystyrene and polybutadiene. However, the solid solution has a higher free energy than the nanometer-scaled microstructure. Hence, due to the block structure of the macromolecule, the microstructure of lowest free energy that the system can form during crystallization, is a nanometer-sized arrangement of regions formed by chemically identical block segments. These regions are separated by interfaces. In other words, the nanometer-sized microstructure is already "implanted" into the system by way of the block copolymer synthesis of the macromolecules. The only way to avoid the high density of interfaces between the constituents would be to break the (covalent) bonds of the backbone of the polymer at the points where the polystyrene and polybutadiene blocks are joined together and by joining the segments of the same chemical structure into new macromolecules of pure polystyrene or polybutadiene.

Equilibrium Nanostructured Materials

Supramolecular Self-assembled Structures

Supramolecules are oligomolecular species that result from the intermolecular association of a few components (receptors and substrates) following an inherent assembling pattern based on the principles of molecular recognition. Supramolecular self-assembly† concerns the spontaneous association of either a few or a large number of components resulting in the generation of either discrete oligomolecular supermolecules or of extended polymolecular assemblies such as molecular layers, films, membranes, etc. In other words, specific phases having well-defined microscopic molecular arrangements and related macroscopic characteristics.

Self-assembly seems to open the way to nanostructures, organized and functional species of nanometer-sized dimensions that bridge the gap between molecular events and macroscopic features of bulk materials.

Self-assembled Inorganic Architectures

Multiple Helical Metal Complexes

Self-assembled supramolecular structures may be generated if linear oligobipyridine ligands formed by two or up to five 2,2′-bipyridine units are brought together with Cu(I) ions. In the presence of Cu(I) ions, the ligands spontaneously assemble into double-stranded di- to pentahelicates consisting of two ligand strands wrapped around one another, Cu(I) holding them together. An important feature of this nanometer-sized structure is that it allows the attachment of substituents to the bipy units arranged in a helical fashion. If the Cu(I) ions are replaced by Ni(II) ions, a triple helix results consisting of three strands held together by three Ni(II) ions.

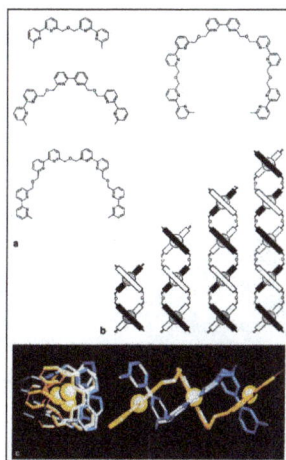

(a) Oligopyridine ligands with the ability to form helical structures. The ligands shown consist of two, three, four or five 2,2-bipyridine units. (b) Formation of enantiomeric double-stranded helicates from two to five tetrahedrally coordinated metal ions [Cu(I), Ag(I), dotted circles]. (c) Structural model deduced from X-ray diffraction studies.

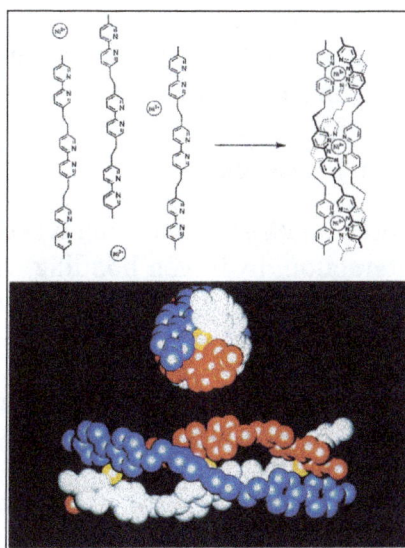

Self-organized triple-helical structure. The structure comprises three ligand molecules each of which contains three 2,2′-bipyridine units and three octahedrally coordinated Ni(II) ions 98, 100, 101. Bottom: Structure of a trihelicate deduced by X-ray diffraction.

Multicomponent Self-assembly of Nanometer-sized Structures: Racks, Ladders and Grids

Multicomponent self-assembly allows the spontaneous generation of well-defined three-dimensional molecular architectures in the form of racks, ladders or grids. They are formed by the complexation of linear ligands or extended units with metal ions in tetrahedral or octahedral sites. Figure displays (as an example) a 3×3nm-sized grid made up of two pyridine groups and one bipyridazine unit connected by Ag(I) ions.

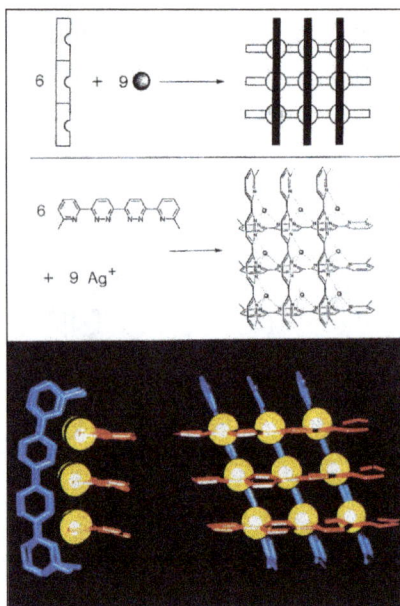

Top: Schematic diagram of the self-assembly of an inorganic lattice. The lattice consists of six linear molecules each of which contains three bonding sites. The molecules are held together by nine metal atoms attached to the bonding sites. Middle: Spontaneous formation of a 3×3 lattice comprising six molecules each of which consists of two pyridine and two pyridazine groups. The bonding sites contain two nitrogen atoms. The molecules are held together by nine tetrahedrally binding Ag(I) ions. Bottom: Structure of a lattice of this type deduced from X-ray diffraction data.

Self-assembled Organic Architectures

Self-assembly of organic architectures utilizes the following types of interaction between the components involved: electrostatic interaction, hydrogen bonding, van de Waals or donor–acceptor effects. If the self-assembling molecules incorporate specific optical, electrical, magnetic, etc. properties, their ordering on a nanometer scale induces a range of novel features.

Self-assembly by hydrogen bonding leads to two- or three-dimensional molecular architectures which often have a typical length scale of a few nanometers. The self-assembly of structures of this type requires the presence of hydrogen-bonding subunits the disposition of which determines the topology of the architecture. Ribbon, tape, rosette, cage-like and tubular morphologies have been synthesized. For example, figure displays a supramolecular ribbon structure; with increasing control being achieved over the molecular design of the building subunits, a large variety of new two- and three-dimensional architectures will be realized.

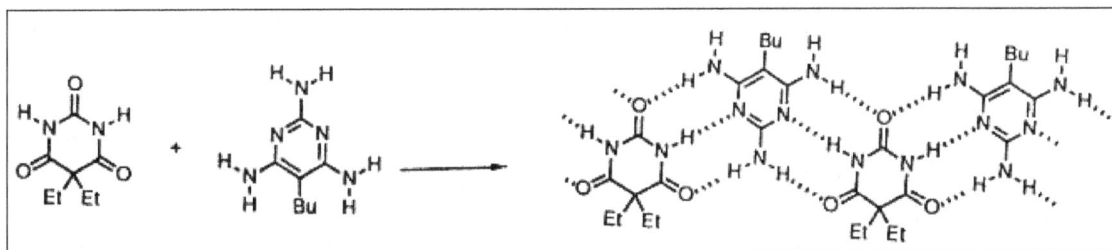

Self-assembly of a supramolecular ribbon from barbituric acid and 2,4,6-triaminopyrimidine units.

Supramolecular interactions play a crucial role in the formation of liquid crystals and in supramolecular polymer chemistry. The latter involves the designed manipulation of molecular interactions (e.g. hydrogen bonding, etc.) and recognition processes (receptor–substrate interaction) to generate main-chain or side-chain supramolecular polymers by self-assembly of complementary monomeric components.

Figure displays some of the different types of polymeric superstructures that represent supramolecular versions of various species and procedures of supramolecular polymer chemistry leading to materials with nanometer-sized microstructures. Recognition effects are expected to play a major role in the assembly and self-organization processes. In the case of macromolecules, the supramolecular association may be either intermolecular occurring between the large molecules, or intramolecular involving recognition sites located either in the main chain or in side-chain appendages. The controlled manipulation of the intermolecular interaction opens the way to the supramolecular engineering of NsM.

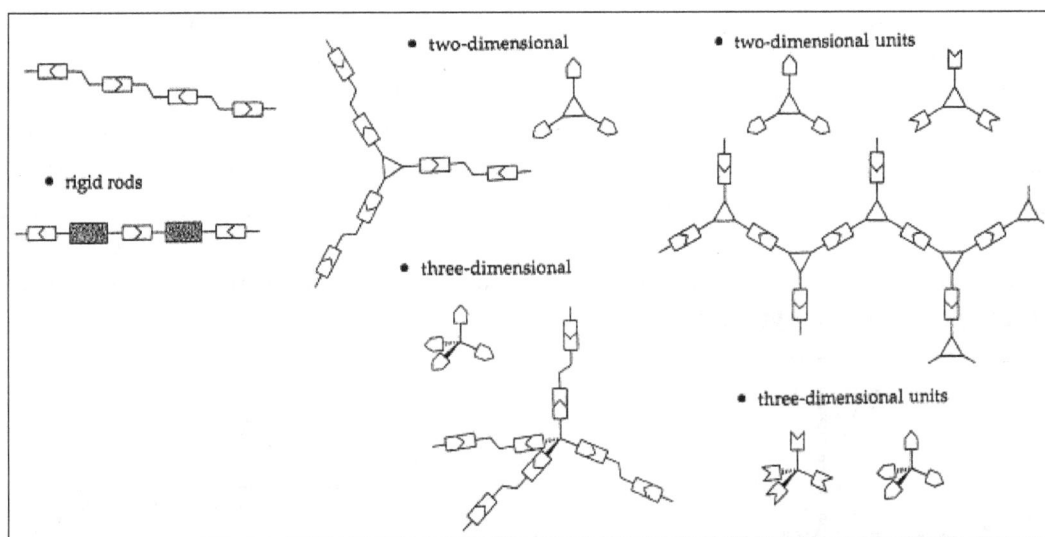

Schematic diagram indicating some of the (many) possible nanometer-sized
molecular structures to be synthesized by supramolecular polymer chemistry.

Supramolecular Materials and Nanochemistry

The ability to control the way in which molecules associate allows the design of nanometer-sized molecular architectures. Some implications for nanotechnology appear to be obvious. For example, surfaces with molecular recognition units will display selective surface binding leading to recognition-controlled adhesion. Components derived from biological structures are likely to yield biomaterials such as biomesogens, biominerals obtained by using supramolecular assemblies as support for inorganic particles in protein cages. Solid-state inorganic self-assembled structures present tunnels, cages and micropores where size, shape and spacing may be tailored to serve as selective hosts for nanometer-sized crystals, nano-wires or related entities. Self-assembly of inorganic architectures based on organometallic building blocks yield various types of frameworks such as Sb or Te chains, chains of metal complexes, honeycomb or diamond arrays, frameworks of metal chalcogenides with helical structures, networks of interlocked rings of inorganic and organic nature.

By increasing the size of the entities, nanochemistry approaches the length scale of lithography and may thus turn out to be an important tool in producing the next generation of devices.

DNA Self-assembled Nanostructures

It is one of the basic results of organic chemistry that intermolecular interaction is based on fixation, molecular recognition and coordination. In other words, molecular binding is highly selective implying a complementary geometry that lays the basis for molecular recognition. The control of newly synthesized molecular structures relies on this specificity and geometric constraints between the partners held together by intermolecular interactions. With this criterion in mind, DNA is an extremely favorable "construction material" for nanoscale structures. It permits the informational character of macromolecules of biological systems to be utilized. In fact, the construction of sticky figures using branched DNA molecules as building blocks has been demonstrated to open the way to the synthesis of a large variety of DNA arrangements. The edges of these arrangements consist of double-helical DNA and the vertices correspond to the branch points of stable DNA branched junctions. This strategy is illustrated in figure. On the left-hand side the stable branched DNA molecule is displayed. The figure in the middle indicates the sticky ends. Four of these sticky-ended molecules are assembled into a quadrilateral. The same technique has been applied to synthesize two- and three-dimensional periodic nanometer-sized DNA structures with predefined topologies, e.g., cubes, truncated octahedrons, etc. In order to synthesize macroscopic periodic arrays made up of cubes, truncated octahedrons, etc. DNA motifs that are more rigid than branched junctions are required. Suitable structures of this type seem to be double crossover molecules. By attaching such molecules to the sides of DNA triangles and deltahedra, two- and three-dimensional nanometer-sized structures may be synthesized.

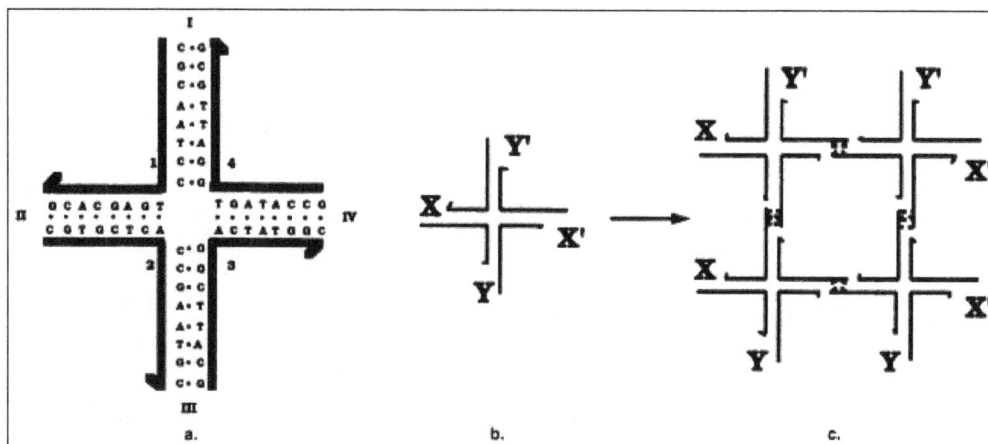

(a) Stable branched DNA molecule. (b) Sticky ends of the DNA molecules.
(c) Assembly of four sticky-ended DNA molecules into a square-shaped pattern.

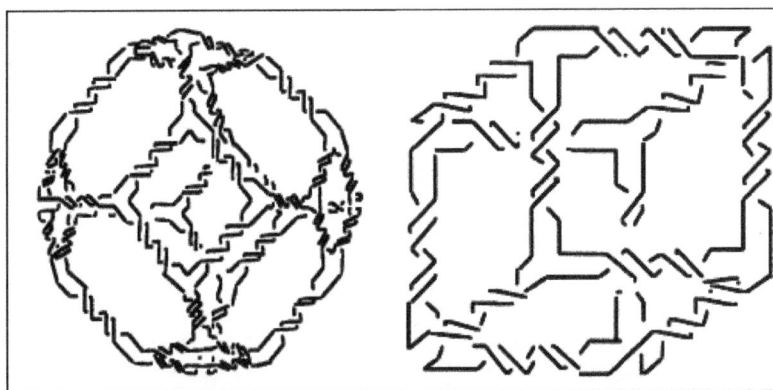

Cube and truncated octahedron assembled of DNA molecules.

Template-assisted Nanostructured Materials and Self-replication

The basic idea of templating is to position the components into predetermined configurations so that subsequent reactions, deliberately performed on the pre-assembled species or occurring spontaneously within them, will lead to the generation of the desired nanoscale structure. The templating process may become self-replication if spontaneous reproduction of one of the initial species takes place by binding, positioning and condensation.

Inorganic and organic templating has been used for the generation of nanometer-sized polymer arrangements displaying molecular recognition through imprinting, i.e. a specific shape and size-selective mark on the surface or in the bulk of the polymer. Imprinting into polymeric materials has been achieved by either a covalent or a noncovalent approach. The former uses the reversible covalent binding of the substrate to the monomer. In the latter, suitable functionalized monomers are left to prearrange around the substrate. Removal of the imprint molecule from the polymer leaves recognition sites that are complementary in geometry and functionality.

Mesophase templating represents a special case that appears to be of considerable significance for the development of this area. Silica precursors when mixed with surfactants result in polymerized silica "casts" or "templates" of commonly observed surfactant–water liquid crystals. Three different mesoporous geometries have been reported, each mirroring an underlying surfactant–water mesophase. These mesoporous materials are constructed of walls of amorphous silica, only about 1 nm thick, organized about a repetitive arrangement of pores up to 10 nm in diameter. The resulting materials are locally amorphous (on atomic length scales) and periodic on larger length scales.

The availability of highly controlled pores on the 1–10 nm scale offers opportunities for creating unusual composites, with structures and properties unlike any that have been made to date. However, the effective use of mesoporous silicates requires two critical achievements: (i) controlling the mesophase pore structure; and (ii) synthesizing large monolithic and mesoporous "building blocks" for the construction of larger, viable composite materials. Although important information exists on some aspects of controlling the mesoporous structure, large-scale structures have not yet been constructed.

The synthesis scheme of silica-based mesostructured materials using assemblies of surfactant molecules to template the condensation of inorganic species has been extended to include a wide variety of transition metal oxides and, recently, cadmium sulfide and selenide semiconductors. Although the exact mechanism for this type of mineralization is still controversial, this technique holds great promise as a synthetic scheme to produce nanostructured materials with novel thermal, electronic, optical, mechanical and selective molecular transport properties. Continuous mesoporous silicate films can be grown on a variety of substrates, e.g. mica, graphite or block copolymers. In fact, nano-structured $BaTiO_3$ films have been grown on a polybutadiene–polystyrene triblock copolymer.

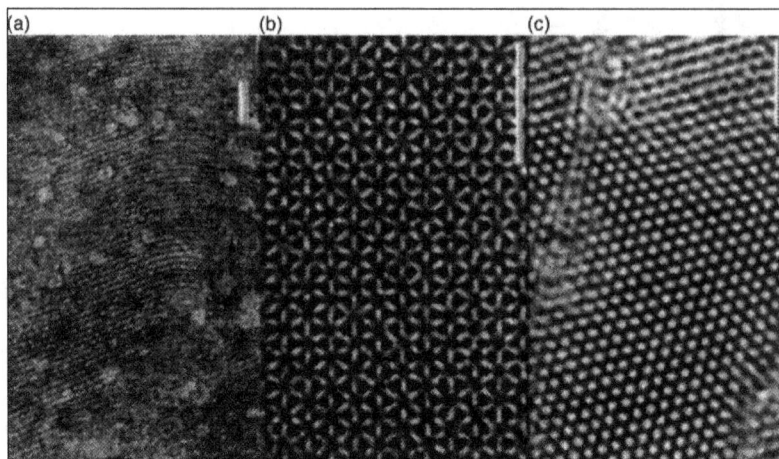

Transmission electron micrograph images of (a) the lamellar morphology, (b) the cubic phase with Ia3d symmetry viewed along its (111) zone axis, and (c) the hexagonal phase viewed along its (001) zone axis of the silica/surfactant nanostructured composites by co-assembly (bars = 30 nm).

A special case is the reproduction of the template itself by self-replication. Reactions occurring in organized media (molecular layers, mesophases, vesicles) offer an entry into the field. Molecular imprinting processes represent a way of copying the information required for recognition of the template. Self-replication takes place when a molecule catalyses its own formation by acting as a template for the constituents, which react to generate a copy of the template. Such systems display autocatalysis and may be termed informational or non-informational depending on whether or not replication involves the conservation of a sequence of information. Several self-replicating systems have been developed in which the template is generated from two components. The first one consists of the replication of a self-complementary or palindromic hexanucleotide CCGCGG from two trinucleotides CCG and CGG in the presence of a condensing agent. The more recent ones involve: (i) the formation of an amide bond between two building blocks undergoing selective hydrogen bonding with the template; and (ii) an amine and aldehyde to imine condensation between components interacting with the template via ion-pairing between an amidinium cation and a carboxylate anion. Self-replication of oligonucleotides in reverse micelles has also been reported.

Supramolecular templating processes seem to provide an efficient route for the synthesis of nanoporous materials used as molecular sieves, catalysts, sensors, etc. In fact, mesoporous bulk and thin-filmsilicates with pore sizes of 2–10 nm have been synthesized by using micellar aggregates of long-chain organic surfactant molecules as templates to direct the structure of the silicate network. Potential applications of these molecular-sieve materials are catalysts, separation membranes and components of sensors. Mesoporous oxides have been synthesized by similar means. In these mesoporous oxides, transition metals partially and/or fully substitute silicon. Templating with organic molecules has also been long used for the synthesis of microporous materials—synthetic zeolites—with pore sizes as small as 0.4–1.5 nm. In this case, the organic molecules are shorter-chain amphiphiles which act as discrete entities around which the framework crystallizes.

It was recently shown that such short-chain molecules can aggregate into supramolecular templates when they form bonds with transition-metal (niobium) alkoxides, and that in this way they can direct the formation of transition-metal oxides with pore sizes of less than 2 nm. These pore sizes, which result from the smaller diameter of micellar structures of the short-chain amines relative to the longer-chain surfactants used for the synthesis of mesoporous materials, qualify the resulting molecular sieves as microporous, even though the supramolecular templating mechanism is similar to that used to make the mesoporous materials. This approach extends the supramolecular templating method to afford microporous transition-metal oxides.

Figure illustrates schematically the synthesis of hexagonally packed transition-metal oxide mesoporous molecular sieves for Nb. It involves the following five steps: (1) partial hydrolysis of niobium ethoxide at low temperatures; (2) introduction of hexylamine; (3) self-assembly of hexylamines as supramolecular templates; (4) condensation and crystallization of the inorganic framework at high temperatures; and (5) amine removal by acidic washes. Figure illustrates the surfactant removal process and the final mesoporous structure for a Ta metal oxide mesoporous material. The N–Ta bonds are cleaved by protolysis at −78°C in the first step. The protonated surfactant is then removed by washing the material in dry 2-propanol (IPA) at ambient temperature for 24 h. Washing with water gives the final hydrated product.

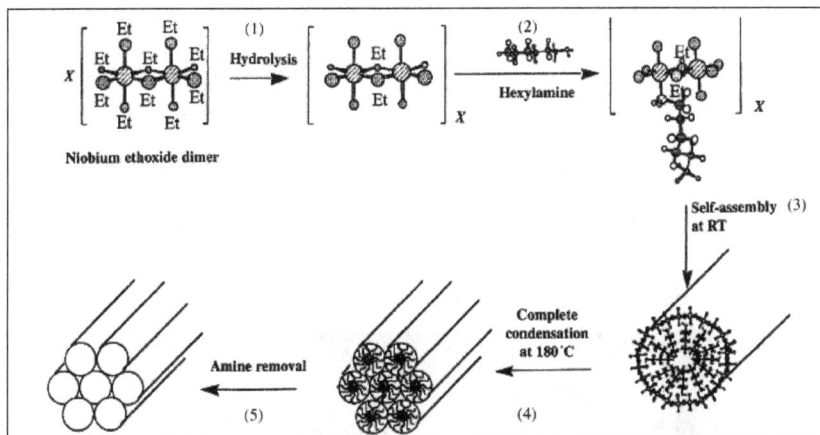

Schematic illustration of the synthesis of microporous transition-metal (Nb) oxide molecular sieves by supramolecular templating: (1) partial hydrolysis of niobium ethoxide at low temperatures; (2) introduction of hexylamine; (3) self-assembly of hexylamines as supramolecular templates at ambient temperature (RT); (4) condensation and crystallization of inorganic framework at 180°C; and (5) amine removal by acidic washes.

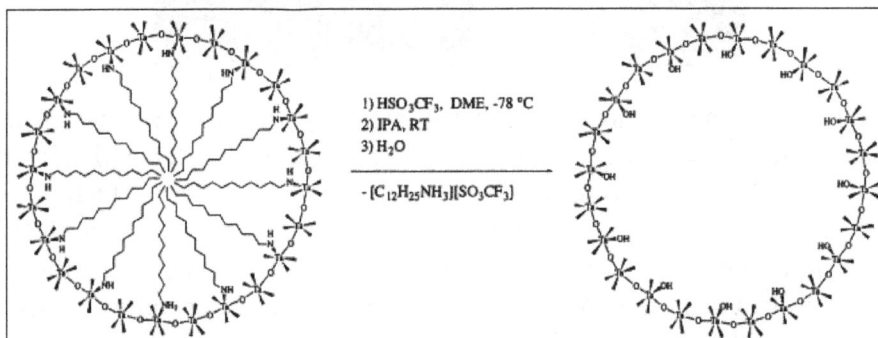

Illustration of the surfactant removal process from hexagonally packed mesoporous Ta oxide molecular sieves by a treatment with triflic acid. The N–Ta bond is cleaved by protolysis at −78°C in the first step. The protonated surfactant is then removed by washing the material in dry 2-propanol (IPA) at ambient temperature for 24 h. Washing with water gives the final hydrated product.

Nanochemistry

Nanochemistry has uses in chemical, physical and materials science, engineering and biological and medical applications. Using single atoms as building blocks offers new ways to create innovative materials, the opportunity to create the smallest features possible in integrated circuits and the chance to explore quantum computing for example.

It might seem relatively new, but nanochemistry has been employed for many years, for example in sunscreens that absorb UV light, in clear coatings for cars which protect the bright paint colors underneath, or in carbon nanotubes for lightweight car parts or sporting equipment. It has been used to study the health and safety effects of airborne and waterborne nanosized particulates, and nanoparticles have been used to clear up or neutralizes pollutants.

Larger molecular assemblies such as dendrimers – highly branched three-dimensional nanoscale molecular objects of the same size and weight as traditional polymers – are under investigation for use in molecular recognition, nanosensing, light harvesting and optoelectrochemical devices. They can be synthesized in a step-by-step fashion, allowing for incredibly precise control of their size and geometry. Since they are created layer by layer, the properties of any single layer can be controlled through selecting specific monomers, making them ideal building blocks in nanochemistry for the creation of three-dimensional structures.

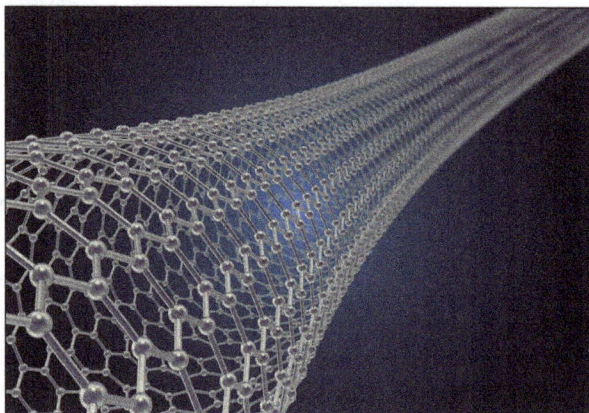

Nanotubes are a relatively new form of carbon and can range from a few microns to a few nanometres with many being only a single atom thick, i.e., single-walled. They display beneficial behavior relative to properties such as electrical and thermal conductivity, strength, stiffness, and toughness. They can be functionalized – that is given a specific purpose – with the addition of molecular recognition agents which bind specifically to molecular tags, making then ideal for used as high-resolution probes in atomic force microscopy, as channels for material separation and as selective gates for molecular sensing.

Nanocomposites constitute the wide variety of systems composed of dissimilar components mixed at the nanometer level. Their behavior is dependent on the properties of the components, their morphology and the interactions between the individual components, which are often novel properties not seen in the parent material.

Nanocrystals and clusters is another promising area of research. Crystals of nano-sized proportions can be combined into clusters and show potential in high-density data storage and optoelectrical applications. They might also find uses as biochemical tags, as chemical catalysts or as laser and optical components.

Small rods of atoms or nanowires are also of interest. These solid, dense structures are built atom by atom in a controlled manner, allowing for impurity doing at for control of the wire's electrical conduction properties.

Nanochemistry also has a use in lab-on-a-chip technologies which are designed to carry out complex chemical processes on an ultra-small scale. Applications might include synthesizing chemical efficiently, combinatorial chemistry, and biological, chemical and clinical analyses. It might also find applications in medicine in drug delivery and wound and tissue engineering.

Applications of Nanomaterials

The term "Nanotechnology" was first defined by Norio Taniguchi of the Tokyo Science University in 1974. Nanotechnology[1-3], shortened to "Nanotech", is the study of manipulating matter on an atomic and molecular scale. Generally nanotechnology deals with structures sized between 1 to 100 nm and involves developing materials or devices within that size. For comparison, 10 nanometers is 1000 times smaller than the diameter of a human hair. Nanotechnology has the capacity to improve our ability to prevent, detect, and remove environmental contaminants in air, water, and soil in a cost effective and environmentally friendly manner. Nanoscience and nanotechnologies are revolutionizing our understanding of matter and are likely to have profound implications for all sectors[4-6] of the economy, including agriculture and food, energy production and efficiency, the automotive industry, cosmetics, medical appliances and drugs, household appliances, computers, and weapons.

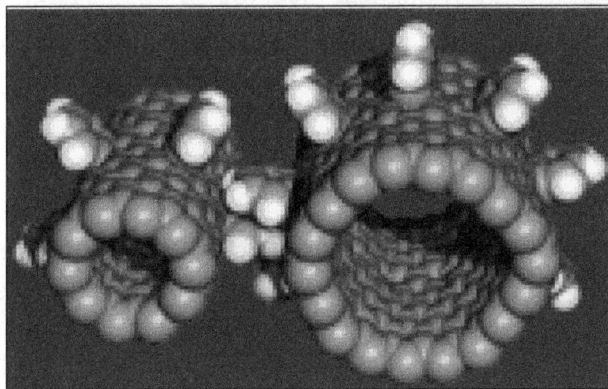

The branch of nanoscience and technology is truly multi-disciplinary and is an emerging technology with full of promises to have an impact on virtually every spectrum of civilization including communications, computing, textiles, cosmetics, sports, therapy, automotives, environmental monitoring, fuel cells and energy devices, water purification, food and beverage industry, etc. The ability to construct tiny objects atom- by-atom or molecule- by- molecule forms one of the exciting prospects of the research field in nano science. It shows great promise for providing us in the near future with many breakthroughs that will change the direction of nanotechnology advances in a wide range of applications.

The application of nanomaterials can be historically traced back to even before the generation of modern science and technology. In 1857, Michael Faraday published a paper which explained how metal nano particles affect the colour of church windows. In 1959, Richard Feynman (awarded Nobel prize in Physics in 1965) gave a lecture titled "There's Plenty of Room at the Bottom", suggesting the possibility of manipulating things at atomic level. He speculated on the possibility and potential of nanosized materials. This is generally considered to be the foreseeing of nanotechnology. Many of his speculations[12-16] have become reality now. However, the real burst of nanotechnology didn't come until the early 1990s. In the past decades, sophisticated instruments for characterization and manipulation[17-20] such as scanning electron microscopy (SEM), transmission electron microscopy (TEM) and scanning probe microscopy (SPM) became more available for researchers to approach the nanoworld.

Synthesis of Nanomaterials

Nanomaterials can be synthesised[27-31] by any one of the following methods:

- Pyrolysis: It involves pyrolysis of hydrocarbons such as acetylene at 7000C in the presence of Fe-silica or Fegraphite catalyst under inert conditions.

- Carbon arc method: It is carried out by applying direct current (60- 100 A and 20-25 V) arc between graphite electrodes of 10-20μm diameter.

- Laser evaporation method: It involves vapourisation of graphite containing small amount of Co and Ni, by exposing it to laser beam at 12000C in a quartz tube reactor. An inert gas like argon is allowed to pass into the reactor to sweep the evaporated carbon atoms from the furnace to the copper collector, on which the nanomaterials condense.

- Chemical vapour deposition: It involves decomposition of vapour of hydrocarbons such as methane, ethylene, acetylene, etc., at 11000C in presence of catalysts like Ni, Co, Fe suported on MgO.

Properties of Nanomaterials

Two principal factors cause the properties[32-34] of nanomaterials to differ significantly from other materials: increased relative surface area, and quantum effects. To understand the effect of particle size on surface area, consider a U.S. silver dollar. The silver dollar contains 26.96 grams of coin silver, has a diameter of about 40 mm, and has a total surface area of approximately 27.70 square centimeters. If the same amount of coin silver were divided into tiny particles – say 1 nanometer in diameter – the total surface area of those particles would be 11,400 square meters. When the amount of coin silver contained in a silver dollar is rendered into 1 nm particles, the surface area of those particles is 4.115 million times greater than the surface area of the silver dollar.

Electrical Properties

The electrical properties of nanomaterials vary between metallic to semiconducting materials. It depends on the diameter of the nanomaterials. The very high electrical conductivity of nanomaterial is due to minimum defects in the structure.

Thermal Conductivity

The thermal conductivity of nanomaterials are very high, is due to the vibration of covalent bonds. Its thermal conductivity is 10 times greater than the metal. The very high thermal conductivity of nanomaterial is also due to minimum defects in the structure.

Mechanical Properties

Nanomaterials are very strong and withstand extreme strain. Most of the materials fracture on bending because of the presence of more defects, but nanomaterials possesss only few defects in the structure.

Applications of Nanomaterials

Below we list some key applications of nanomaterials. Most current applications represent evolutionary developments of existing technologies: for example, the reduction in size of electronics devices.

Sunscreens and Cosmetics

The traditional chemical UV protection approach suffers from its poor long-term stability. A sunscreen based on mineral nanoparticles such as titanium dioxide offer several advantages. Titanium oxide nanoparticles have a comparable UV protection property. Nanosized titanium dioxide and zinc oxide are currently used in some sunscreens, as they absorb and reflect ultraviolet (UV) rays and yet are transparent to visible light and so are more appealing to the consumer. Nanosized iron oxide is present in some lipsticks as a pigment. The use of nanoparticles in cosmetics has raised a number of concerns about consumer safety.

Paints

Incorporating nanoparticles in paints could improve their performance, for example by making them lighter and giving them different properties. Thinner paint coatings ('lightweighting'), used for example on aircraft, would reduce their weight, which could be beneficial to the environment.

Displays

The huge market for large area, high brightness, flat-panel displays, as used in television screens and computer monitors, is driving the development of some nanomaterials. Nanocrystalline zinc selenide, zinc sulphide, cadmium sulphide and lead telluride synthesized by sol gel techniques are candidates for the next generation of light-emitting phosphors.

Batteries

With the growth in portable electronic equipment (mobile phones, laptop computers, remote sensors), there is great demand for lightweight, high-energy density batteries. Nanocrystalline materials synthesized by sol–gel techniques are candidates for separator plates in batteries because of their foam-like (aerogel) structure, which can hold considerably more energy than conventional ones. Nickel–metal hydride batteries made of nanocrystalline nickel and metal hydrides are envisioned to require less frequent recharging and to last longer because of their large surface area.

Catalysis

In general, nanoparticles have a high surface area, and hence provide higher catalytic activity. Catalysis is important for the production of chemicals. Nanoparticles serve as an efficient catalyst for some chemical reaction, due to the extremely large surface to volume ratio. Platinum nanoparticles are now being considered in the next generation of automotive catalytic converters because the very high surface area of nanoparticles could reduce the amount of platinum required. Some chemical reactions are also carried out using nanomaterials. For example, reduction of nickel oxide to the base metal Ni.

Medicine

Nanotechnology has been a boon in medical field by delivering drugs to specific cells using nanoparticles. The overall drug consumption and side effects can be lowered significantly by depositing the active agent in the morbid region only and in no higher dose than needed. This highly selective approach reduces costs and human suffering. Nanotechnology can also help to reproduce or to repair damaged tissue. "Tissue engineering" might replace today's conventional treatments like organ transplants or artificial implants. For example bones can be regrown on carbon nanotube scaffolds.

The use of gold in medicinal preparations is not new. In the Indian medical system called Ayurveda, gold is used in several preparations. One popular preparation is called Saraswatharishtam, prescribed for the memory enhancement. Gold is also added in certain medical preparations for babies in order to enhance their mental capability. Over 5000 years ago, Egyptians used gold in dentistry. In Alexandria, alchemists developed a powerful colloidal elixir known as liquid gold, a preparation that was meant to restore youth. In china, people cook their rice with a gold coin in order to help replenish gold in their bodies.

Sensors of Gases

The gases like NO_2 and NH_3 can be detected on the basis of increase in electrical conductivity of nanomaterials. This is attributed to increase in hole concentration in nanomaterials due to charge transfer from nanomaterials to NO_2 as the gas molecules bind the nanomaterials.

Food

Nanotechnology can be applied in the production, processing, safety and packaging of food. A nanocomposite coating process could improve food packaging by placing anti-microbial agents directly on the surface of the coated film. New foods are among the nanotechnology created consumer products coming onto the market at the rate of 3 to 4 per week. According to company information posted on PEN's Web site, the canola oil, by Shemen Industries of Israel, contains an additive called "nanodrops" designed to carry vitamins, minerals and phytochemicals through the digestive system and urea.

Construction

Nanotechnology has the potential to make construction faster, cheaper and safer. Automation of nanotechnology construction can allow for the creation of structures from advanced homes to massive skyscrapers much more quickly and at much lower cost. The Silica ($SiO2$) is present in conventional concrete as part of the normal mix. When nano silica is added to concrete the particle packing can be improved mechanical properties. The addition of nano silica to cement based materials can also control the degradation of the fundamental C-S-H (calcium silicate hydrate) reaction of concrete caused by calcium leaching in water as well as block water penetration and therefore lead to improvements in durability. The strength of concrete can also be increase by adding haematite (Fe_2O_3) nanoparticles.

Steel has been widely available material and has a major role in the construction industry. The use of nanotechnology in steel helps to improve the properties of steel. The nano size steel produce stronger steel cables which can be used in bridge construction.

The glass is also an important material in construction. There is a lot of research being carried out on the application of nanotechnology to glass. Titanium dioxide ($TiO2$) nanoparticles are used to coat glazing since it has sterilizing and anti-fouling properties. The particles catalyze powerful reactions which breakdown organic pollutants, volatile organic compounds and bacterial membranes. Most of glass in construction is on the exterior surface of buildings. So the light and heat entering the building through glass has to be prevented. The nanotechnology can provide a better solution to block light and heat coming through windows.

Coating is an important area in construction. Coatings are extensively used to paint the walls, doors and windows. Coatings should provide a protective layer which is bound to the base material to produce a surface of the desired protective or functional properties. Nanotechnology is being applied to paints to obtain the coatings having self-healing capabilities and corrosion protection under insulation. Since these coatings are hydrophobic and repel water from the metal pipe and can also protect metal from salt water attack.

Agriculture

Applications of nanotechnology have the potential to change the entire agriculture sector and food industry from production to conservation, processing, packaging, transportation, and even waste treatment.

Energy

The most advanced nanotechnology projects related to energy are: storage, conversion, manufacturing improvements by reducing materials and process rates, energy saving and enhanced renewable energy sources. Today's best solar cells have layers of several different semiconductors stacked together to absorb light at different energies but they still only manage to use 40 percent of the Sun's energy. Commercially available solar cells have much lower efficiencies (15-20%). Nanotechnology could help increase the efficiency of light conversion by using nanostructures.

Other Applications

Some commercial products on the market today utilizing nanomaterials include stain resistant textiles and reinforced tennis rackets. Companies like Kraft foods are heavily funding nanomaterials based plastic packaging. Food will stay fresh longer if the packaging is less permeable to atmosphere. Coors Brewing Company has developed new plastic beer bottles that stay cold for longer periods of time.

Nanoscale Building Blocks to Hybrid Nanomaterials

Controlled organization of materials with micrometer scale precision has led to the miniaturization of devices, which has been the key to major technological innovations in the last decade. Design of materials in the nanometer scale is not merely another step towards miniaturization; it is the exploration of a new size regime wherein materials exhibit properties which are different from that of the bulk. In this regime, size and shape play a dominant role in tuning their properties rather than the constituting elements and their chemical composition. Examples of nanostructured materials include metal and semiconductor nanoparticles and nanostructured carbon based systems such as fullerenes, single and multi-walled carbon nanotubes. The noble metal nanoparticle based systems possess an attractive property in the nanoscale: strong optical absorption arising from the localized surface plasmon resonance (yellow, red and blue color for Ag, Au and Cu, respectively). The larger surface to volume ratio is yet another unique feature of nanostructured materials, which make them a suitable candidate in catalytic processes.

Nanoscale materials exhibit size and shape dependent physical, chemical, electronic and magnetic properties, which are different from the bulk and their isolated atoms/molecules. Classical laws of physics fail to explain the origin of the novel properties of materials in this size regime. Electrons experience a confinement of motion in space, when one or more dimensions of a crystal are in the nanoscale. This situation can be described in terms of particles in a box and under this condition, quantum mechanics is more suitable to explain the size and shape dependent properties (vide infra). Dimensional confinement of electrons in materials leads to dramatic modifications in their density of states (DOS) giving rise to shape dependent properties. Based on dimensional confinement of electrons, nanomaterials are generally classified as 2D, 1D or 0D. The confinement of electrons is imposed only along one axis in the case of 2D nanomaterials (e.g., quantum wells). In the case of one-dimensional nanomaterials, electrons experience confinement along two dimensions and are free to move only in one dimension; the best examples are semiconductor nanowires and carbon nanotubes. When the material experiences confinement of electrons along all the three dimensions, it is often termed as quantum dots (QDs). The dramatic blue shift observed in the absorption and emission spectra of semiconductor quantum dots, with decrease in size, is one of the direct observations of quantum size effect (vide infra).

The synthesis, characterization of various nanomaterials and investigation of their properties are well documented in recent reviews. Newer methods of characterization based on electron and probe microscopic methods have enabled better understanding on the crystallographic properties of nanomaterials. Various spectroscopic methods have been used for understanding the optical and electronic properties of these materials. Among various nanomaterials, gold nanoparticles and nanorods, semiconductor quantum dots and single walled carbon nanotubes are widely used as components for the design of hybrid nanostructures.

Nanomaterials and their Functional Properties

Gold Nanoparticles

Metal particles in its colloidal state have attracted mankind even centuries ago due to their medicinalvalue and fascinating colors. Colloidal gold has a long therapeutic history, which is well rooted in Eastern traditions particularly in the Indian subcontinent. The medicinal value of colloidal gold is well documented in the books of ancient Indian ayurveda like 'Charaka Samhita' and the 'Vedas.' The fascinating colors of metal nanoparticles have been utilized for decorating glass windows in many cathedrals in Europe (stained glass windows). Another interesting example is the famous Lycurgus cup of 4th century, which is now displayed in the British Museum. This glass cup appears green when viewed in reflected light and transmits red color, when illuminated from inside. Analysis of the glass reveals that it contains small quantity of an alloy of gold and silver having a diameter of ~70 nm in the molar ratio of (3:7). The surface plasmon resonance (vide infra) of the alloy is responsible for the special color display in the Lycurgus cup.

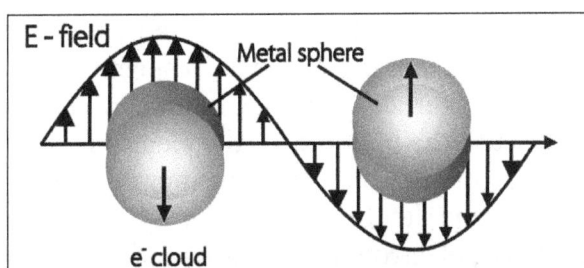

Schematic representation of the oscillation of the electron cloud in presence of an electromagnetic radiation.

The light absorption in metal nanoparticles originates from an interesting phenomenon called localized surface plasmon resonance. According to the Drude–Lorentz model, the atoms in metals exist in a plasma state, having a core of positively charged nuclei surrounded by a pool of negatively charged electrons and hence named as 'plasma electrons'. In the presence of an electromagnetic radiation, the electric vector displaces the free electrons and the columbic electrostatic attraction of the nuclei will restore the electrons to the original position. As a result of the oscillating nature of the electric field of light, the electron cloud coherently oscillates over the surface with a resonance frequency 'ω_p'. When the frequency of this oscillation matches with that of incident radiation, a resonance condition is established which results in the intense absorption, often termed as 'surface plasmon (SP) absorption'. For example, spherical gold nanoparticles exhibit a single surface plasmon band at around 520 nm, attributed to the collective dipolar oscillation of the electron cloud.

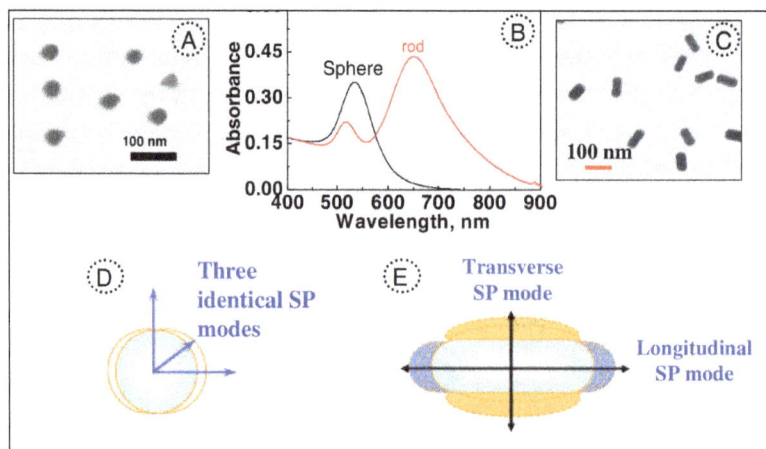

(A) TEM images of gold nanoparticles (obtained by reducing HAuCl4 using gallic acid at room temperature); (B) absorption spectrum of Au nanoparticle and nanorod; (C) TEM images of gold nanorod (obtained by photochemical reduction); (D) surface plasmon oscillation of spherical gold nanoparticles and (E) the surface plasmon oscillation of gold nanorods in the transverse and longitudinal axes.

For a bulk metal with infinite dimension in all the three directions, the resonance frequency of plasma electrons (ω_p) can be expressed as:

$$\omega_p = \left(Ne^2 / \varepsilon_0 m_e \right)^{1/2},$$

where N is the number density of the electrons, ε_0 is the permittivity of vacuum, and e and me are the charge and effective mass of the electron, respectively. Depending on the dimensionality of the nanostructured materials, different boundary conditions can be imposed on the electron plasma. In contrast to the bulk metal, the electron cloud in nanoparticles is confined to a finite volume, which is smaller than the wavelength of light. Hence, the frequency of oscillation of metal nanoparticles is determined mainly by four parameters: number density of electrons, effective mass of the electron and the size and shape distribution of the charge. This allows the tuning of the optical properties of noble metal nanoparticles by varying the size, shape and dielectric environment. In 1908, Gustay Mie provided a quantitative description for the resonance in spherical particles, by solving Maxwell's equations, with appropriate boundary condition. According to Mie theory, the total cross-section consists of scattering and absorption (often termed extinction) and given as summation over all electric and magnetic oscillations. The contribution of absorption and scattering mainly depends on the size and shape of the particles.

Nanostructured metallic systems of noble metal such as silver, gold and copper are of great interest since their localized surface plasmons resonate at the visible range of the electromagnetic radiation. Typically, the surface plasmon (SP) band for spherical Au nanoparticles having diameter of ~20 nm is observed at around 530 nm. The surface plasmon absorption of Au nanoparticle obtained by reducing $HAuCl_4$ using gallic acid at room temperature and gold nanorod prepared by photochemical reduction and their corresponding TEM images are presented in figure. A bathochromic shift in the λ_{max} was observed from 530 to 550 nm on increasing the diameter of the nanosphere from 20 to 80 nm, which is attributed to the electromagnetic retardation in larger nanoparticles. The dependence of the nanoparticle diameter on the maximum of plasmon resonance band was theoretically calculated by adopting Mie theory and discrete dipole approximation (DDA) method and these aspects were reviewed extensively. As the shape of the nanoparticle changes, the surface electron density and hence the electric field on the surface varies. This causes dramatic variations in the oscillation frequency of the electrons, generating different optical cross-sections (including absorption and scattering). For example, nanorods of gold and silver split the dipolar resonance into two surface plasmon bands wherein the induced dipole oscillates along the transverse and longitudinal axes. The longitudinal surface plasmon band shifts to longer wavelengths with increase in aspect ratio, while the position of transverse surface plasmon band remains more or less unaffected. These results indicate that the dimensionality plays a crucial role in determining the plasmon resonance of metal nanoparticles. Au nanoparticles in the size range of 2–5 nm show interesting quantum size effects and can behave as conductor, semiconductor or insulator depending on its dimension. The surface plasmon band of Au nanoparticle, which is characteristic of its metallic nature, undergoes broadening and dampening on decreasing the size below 5.0 nm. A sharp decrease in the intensity of surface plasmon band is observed for nanoparticles having diameter below 3.2 nm due to the transition from metallic to semiconductor/insulator behavior. This effect is attributed to the onset of quantum size effect and was further established by theoretical investigations. Surface plasmon band is absent for Au nanoparticles having core diameter less than 2 nm. The dampening of the SP mode is attributed to the surface scattering of conduction electrons that follows inverse radius (1/radius) dependence. It was found that Au nanoparticles <5 nm in diameter are catalytically active for several reactions, for example, catalysts for the oxidation of CO (an automobile exhaust) at room temperature. Another interesting property of Au nanoparticles/nanorods is the high optical cross-section of the surface plasmon absorption, which is typically 4–5 orders of magnitude higher than conventional dyes. These unique features of Au nanoparticles/nanorods provide excellent opportunities for their use in biomedical field for diagnosis, and imaging. The absorbed light in Au nanoparticles/nanorods is efficiently converted into localized heat and this strategy has been successfully used for the laser photothermal destruction of cancer cells. Au nanoparticles/nanorods can be targeted to tumor site by conjugating them with bioactive molecules.

One of the most widely investigated nanostructured systems is spherical gold nanoparticles owing to their (i) stablity, (ii) ease of synthesis, (iii) excellent optical and electronic properties and (iv) their ability to bind with thiols, which allows functionalization with various molecular systems. In 1857, Michael Faraday provided the first systematic investigation and documentation on the synthesis of colloidal Au nanoparticles. An aqueous solution of sodium tetrachloroaurate, ($Na[AuCl_4]$), was reduced with phosphorus in carbon disulfide. The solution turned to deep ruby color and Faraday concluded that the gold was dispersed in the liquid in a very finely divided form. Several synthetic procedures have been developed over the years for the design of Au nanomaterials having varying size and shape. The notable ones include (i) Turkevich method by boiling an aqueous solution of HAuCl4 and sodium citrate, and (ii) two-phase reduction method developed by Schiffrin, Brust and coworkers. In the last decade we have witnessed a plethora of scientific activities related to the development of newer synthetic methods, which led to the size and shape controlled synthesis of noble metal nanoparticle having tunable optoelectonic properties. Control on the shape and size of metal nanoparticles have been achieved by varying the reaction conditions such as reduction technique, reaction time and concentration of capping agent. Recent developments in the classical wet chemistry methods have enabled the synthesis of anisotropic nanostructures possessing well defined shapes (for example, ellipsoids, rods, cubes, disks, tetrahedra, cylinders, pyramids, triangular prisms, and multipods) and these aspects are discussed extensively. The functional properties of various nanomaterials were correlated with their size and shape and these aspects are well documented.

Semiconductor Quantum Dots

Photoexcitation of a bulk semiconductor results in the transfer of an electron from valence band to conduction band creating an electron-hole pair, called 'exciton', bound by a weak columbic interaction. The minimum energy required to generate such an exciton is called the band gap energy . Excitons can be treated as hydrogen-like system and the spatial separation between the charge carrier pair is termed as exciton Bohr radius (a_B) which can be deduced from the Bohr approximation. When the physical dimensions of matter become comparable or lower to the exciton Bohr diameter (2aB), the functional properties of a semiconductor becomes sensitive to the size and shape due to confinement of excitons. This phenomenon is often expressed as quantum size effect or quantum confinement effect. For example, CdSe has a Bohr exciton radius of ~56 °A; the electron and hole cannot achieve their desired distance when the diameter of nanocrystal is smaller than 112 °A.

Many theoretical models have been reported which correlate the size of quantum dots to their band gap energy and these models were further compared with the experimentally observed data. An earlier theoretical calculation for semiconductor nanoparticles (using CdS and CdSe QDs as examples) was reported by Brus, based on 'effective mass approximation' (EMA). In this approximation, an exciton is considered to be confined to a spherical volume of the crystallite and the mass of electron and hole is replaced with effective masses (m_e and m_h) to define the wave function. Kayanuma accounted the electron-hole spatial correlation effect and modified the Brus equation . Based on the modified equation, the size dependence on the band gap energy of QDs can be quantified as follows:

$$E_g\left(QD\right) = E_g\left(\text{bulk}\right) + \left(h^2\pi^2 / 2\mu R^2\right)$$
$$- \left(1.786e^2 / \varepsilon R\right) - 0.248E_{Ry}^*$$

where 'R' is the radius of QD, μ is the reduced mass, 'ε' is the permittivity of the vacuum and E* Ry is the effective Rydberg energy. The first term on the right hand side represents the band gap of bulk materials, which is characteristic of the material: For example, 2.53 eV for CdS, 1.74 eV for CdSe and 1.50 eV for CdTe. The second additive term of the equation represents the additional energy due to quantum confinement having a $1/R^2$ dependence on band gap energy, Eg(QD). The third subtractive term stands for the columbic interaction energy of exciton having 1/R dependence; often neglected due to high dielectric constant of the material. The last subtractive term, stands for spatial correlation effect (independent of radius), and significant only in case of semiconductor materials with low dielectric constant.

Variation in the band gap energy of semiconductor with size is directly reflected on their optical responses. The confinement of electrons in semiconductor QDs influences their electronic structure in two ways. The energy gap of semiconductor QDs increases dramatically with decrease in its radius (Eg $\propto$ 1/R^2). Also the continuum observed in the conduction band and valence band in the case of bulk materials is replaced with discrete atomic like energy levels as the particle size decreases.

Earlier attempts to produce extremely small particles included template-assisted synthesis using zeolites, micelles, lipid bilayers, molecular sieves and polymers. Most of the initial efforts were concentrated on the synthesis of cadmium chalcogenide such as CdS, CdSe, CdTe (II–VI semiconductors). A breakthrough in producing high quality monodispersed QDs of II–VI semiconductors was reported by Bawendi and coworkers in 1993 by adopting a high temperature organometallic reaction in presence of a coordinating solvent. The reaction was carried out at elevated temperatures (~360°C) under vacuum, using selenium precursor (TOPSe; selenium coordinated to trioctylphosphine) which was injected rapidly into a solution of dimethylcadmium in TOPO. An excellent control over crystal growth was achieved by controlling the reaction parameters such as precursor concentration, temperature and duration of reaction. A marked improvement in the above synthesis strategy was achieved by Peng and coworkers, where the authors have used a non-pyrophoric and stable cadmium precursor, cadmium oxide instead of dimethylcadmium. Later several modifications have been reported with the aim of improving the monodispersity and increasing the photoluminescence quantum yield and these methods were further extended for the synthesis of III–V semiconductor QDs (GaAs, InP, etc).

A hypsochromic shift in the absorption as well as emission band was observed on reducing the size of semiconductor nanocrystal, which is accompanied by the appearance of sharp absorption peaks originating from the first and second excitonic transitions. For example, bulk CdSe has an absorption onset at 720 nm (Eg = 1.74 eV) which is blue shifted to 450 nm (Eg = 2.8 eV) by reducing the size down to about 2 nm. Many other properties of semiconductor nanomaterials also depend on the physical dimension of the material. It is reported that the redox potential of the valence and conduction bands are also sensitive to size of the QDs, which shifts to more positive values for valence band and negative values for conduction band as the size decreases.

Bare QDs possess low emission yield due to the presence of the dangling bonds and deep trap states on their surface, which enhances the nonradiative channels. One of the successful strategies adopted for imparting photostability and improving the photoluminescence efficiency of QD is to overcoat with an inorganic shell material having similar lattice parameters and higher band gap energy (Eg). Such core-shell systems wherein the core is overcoated with large band gap shell material are called as type I heterostucture QDs. In such systems, the conduction band of the shell possesses higher energy than that of the core and the valance band of the shell possesses lower energy than that of the core. As a result, both electrons and holes (exciton) are confined in the core. Shell materials used for overcoating CdSe QDs include zinc sulphide (ZnS), zinc selenide (ZnSe) and cadmium sulphide (CdS). Among these, ZnS is widely used as overcoating material owing to its higher band gap energy, allowing efficient confinement of excitons in the core. Overcoating prevents photoinduced charge transfer in bare QDs retaining PL efficiency. It also prevents the leakage of core materials, avoiding the toxicity inside of the cell when used in biological applications. However, the core-shell QDs with large shell thickness are less effective for biological applications due to the decrease in luminescence quantum yield. Studies from our group have shown that two monolayers (corresponding to ~0.65 nm) of ZnS shell is the optimum shell thickness for a 4.2 nm diameter CdSe quantum dot, which inhibits charge transfer processes and provides maximum PL quantum yield.

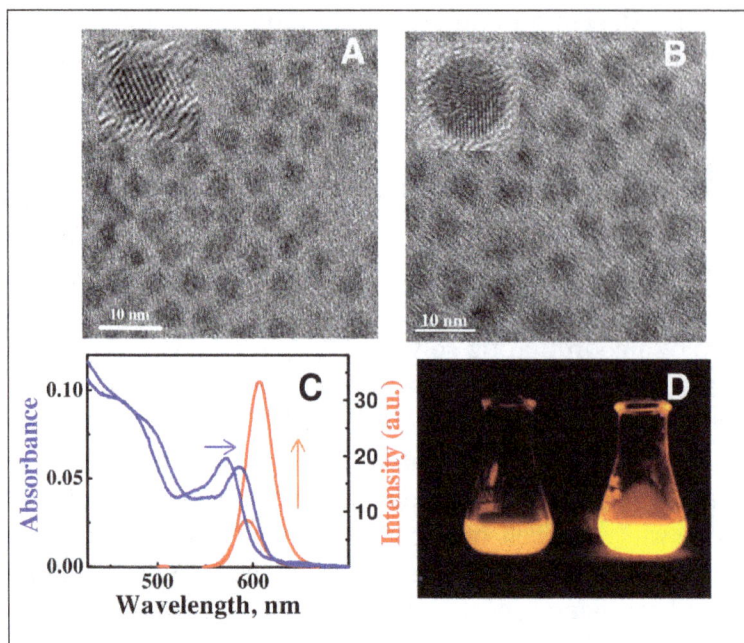

HRTEM images of (A) CdSe QDs and (B) CdSe QDs overcoated with 3.9 monolayers of ZnS; (C) Absorption and PL spectra of bare CdSe QDs and overcoated with 3.9 monolayers of ZnS (bathochromic shift in absorption band and enhancement in PL) along with (D) photographs of bare (left flask) and overcoated (right flask) CdSe QDs.

Carbon Nanotubes

Another class of nanoscale building blocks which received much attention in recent years is the carbon nanotube (CNT), an allotrope of carbon. They are molecular scale tubes of graphene sheets; an ideal CNT is a hexagonal network of carbon atoms rolled as a seamless cylinder, with diameter in the order of few nanometers (0.8– 2.0 nm) and a tube length varying in thousands of nanometer. This class of quasi one-dimensional nanostructures possess unique electrical properties which make them an attractive candidate for the fabrication of nanodevices and these aspects are reviewed in detail. The chemical bonding of nanotubes is similar to that of graphite, composed entirely of sp2 hybridized carbon atoms. Carbon nanotubes are mainly classified as two types: single-walled carbon nanotubes (SWCNT) and multi-walled carbon nanotubes (MWCNT). Based on the theoretical studies, it was proposed that the electronic properties of 'ideal' carbon nanotubes depend on their diameter and chirality. However, experimental studies have showed that various defects in carbon nanotubes such as pentagons, heptagons, vacancies or dopant can drastically modify the electronic properties of CNTs.

The fascinating electronic properties of carbon nanotubes originates from the quantum confinement of electrons normal to the nanotube axis. In the radial direction, electrons are confined by the monolayer thickness of the graphene sheet and periodic boundary conditions exist around the circumference of the nanotube. Electrons can only propagate along the nanotube axis due to quantum confinement and their wave vectors point in this direction. The resulting number of onedimensional conduction and valence bands depends on the standing waves around the circumference of the nanotube.

The single walled carbon nanotubes are usually described using the chiral vector, $C_h = n\hat{a}_1 + m\hat{a}_2$, which connects two crystallographically equivalent sites, on a graphene sheet where $\hat{a}_1$ and $\hat{a}_2$ are unit vectors of the hexagonal honeycomb lattice and n and m are integers. When the graphene sheet is rolled up into a nanotube, the ends of the chiral vector meet each other to form the circumference of the nanotube. The chiral vector Ch also defines chiral angle (θ), the angle between the chiral axis and the zigzag axis of the graphene sheet. The various values of n, m and θ, defines the different nanotube structures such as chiral (n, m), armchair (n, n) and zigzag (n, 0 or 0, m). Armchair nanotubes are formed when n = m having a chiral angle of 30° whereas zigzag nanotubes are formed when either n or m are zero and the chiral angle is 0°. All other nanotubes with chiral angle between 0° and 30° are known as chiral nanotubes. The electronic properties of a nanotube vary in a periodic way between metallic and semiconductor by following a general rule of n − m = 3i, where i is an integer or noninteger. If i is an integer, i.e., (n − m) is a multiple of 3, then the tube exhibits a metallic behavior and possesses a finite value of carriers in the density of states at the Fermi energy level. In contrast, if i is a non-integer, then the tube exhibits a semiconducting behavior and has no charge carriers in the density of states (DOS) at the Fermi energy level. The DOS of various types of CNTs were experimentally investigated and compared with theoretical models. Although there have been many interesting and successful attempts to grow CNTs by various methods, most widely used techniques are (i) arc discharge, (ii) laser furnace and (iii) chemical vapor deposition (CVD). Commercially available nanotubes contain a mixture of metallic and semiconductor tubes; one third of them are metallic and the rest are semiconducting. Even though CNTs are considered as versatile building blocks for the design of optoelectronic devices, the availability of carbon nanotubes in a scalable quantity having (i) uniform diameter and bandgap, and (ii) metallic/semiconducting character remains a major hurdle.

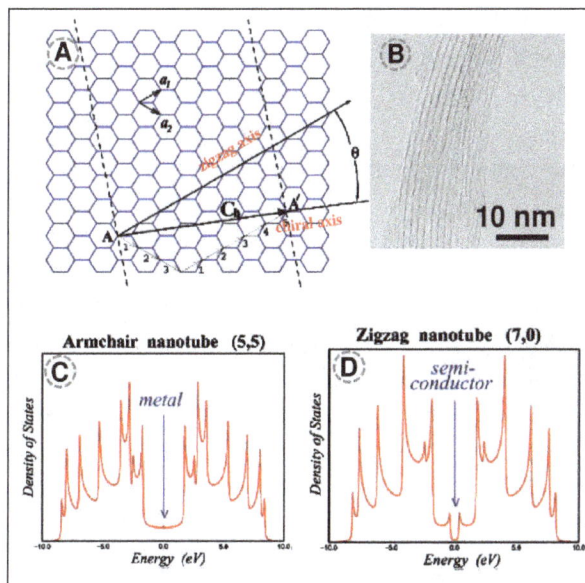

(A) Schematic diagram of two-dimensional graphene sheet illustrating chiral axis, zigzag axis and chiral angle; (B) HRTEM images of bundled SWCNT; (C) electronic properties of an armchair nanotube exhibiting metallic behavior and (D) zigzag nanotube exhibiting semiconductor behaviour.

Functional Properties of Hybrid Nanomaterials

The functional properties of nanomaterials can be further tuned by the stepwise integration of nanoscale building blocks (nanoparticles, nanorods, nanotubes, etc.) into hybrid nanomaterials.

When integrated as hybrid nanomaterials, the functional properties of nanoscale building blocks may couple each other to yield newer properties. Thus, hybrid nanomaterials may possess novel properties different from that of isolated components or possess complementary properties useful for performing specific functions (vide infra). Several strategies have been reported for designing hybrid nanostructures and representative systems include organic-inorganic and inorganic-inorganic hybrid systems. Design of organic-inorganic hybrid systems can be achieved by functionalizing photoor electroactive molecules on to metal or semiconductor nanomaterials wherein electron/energy transfer processes may occur. Inorganic-inorganic hybrid nanomaterials can be obtained through the hierarchical integration of metal and semiconductor building blocks into higher order assemblies. For example, a nanoscale building block 'X' can be coupled using a spacer group with another nanoscale building block 'Y' to yield 'X-Y' type hybrid nanostructures (or with similar building block to yield an 'X-X' type nanostructure). Properties of these hybrid nanostructures can be tuned by varying the length of the spacer group. Design of hybrid nanomaterials with heterojunction can be achieved by bringing (i) metal and semiconductor nanoparticles, (ii) dissimilar metal nanoparticles and (iii) dissimilar semiconductor nanoparticles. Such hybrid nanomaterials possess properties that are fundamentally different from those of isolated components. Recent studies have shown that hybrid nanomaterials possess more promising functional properties than individual building blocks.

Organic-inorganic Hybrid Nanomaterials

Stable metal as well as semiconductor nanoparticles were prepared by capping with a monolayer of organic molecules. The ability of functional groups such as quaternary ammonium halides, amines, thiols, etc., to bind on to the surface of nanoparticles has been exploited for organizing organic molecules around them. The most widely used method for the functionalization of organic molecules on to the surface of metal nanoparticles adopts a two-phase synthesis reported by Brust et al and their modified procedures. Such monolayer-protected clusters (MPCs) can be visualized as three-dimensional assemblies possessing a core-shell structure (metal core covered with ligand shell; figures). Solubility of MPCs can be tuned from nonpolar to polar medium by varying nature of the stabilizer layer.

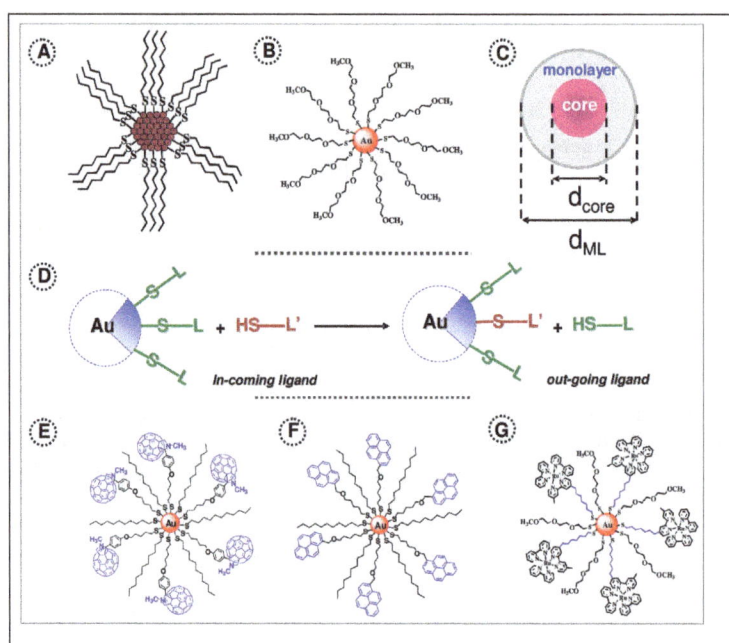

Monolayer-protected clusters (MPCs) having metal core covered with (A) alkyl thiol and (B) monothiol derivative of triethylene glycol as ligand shell; (C) schematic illustration of a core-shell hybrid nanostructure; (D) schematic representation of ligand exchange (place exchange) reaction for preparing mixed monolayers; (E–G) gold nanoparticles having mixed monolayers of (E) fullerenethiol and dodecanethiol; (F) pyrenethiol and dodecanethiol; (G) ruthenium trisbipyridine thiol and monothiol derivative of triethylene glycol.

Optoelectronic properties of MPCs can be tuned by linking photoactive molecules on to the surface of nanoparticles. Several methodologies have been developed for incorporating a desired number of chromophores on to the nanoparticle. For example, metal nanoparticles bearing mixed monolayers can be synthesized using ligand exchange (place exchange) reactions reported by Murray and coworkers. The addition of pre-synthesized nanoparticles possessing thiol bearing ligands to a solution of another thiolate molecule having diverse functionality results in the partial substitution of the protecting shell on the nanoparticle surface. This strategy enabled the functionalization of a desired number of photoactive or electroactive molecules around the metal nanoparticle.

Interesting physical processes arise when photoactive or electroactive molecules are linked on to metal nanoparticles. The most significant ones are electron/energy transfer processes, which can be further modulated by varying the size, shape and chemical constituents of nanomaterials. It is reported that Au nanoparticles in the size range of 2–5 nm can behave as conductor, semiconductor or insulator depending on their dimension due to quantum size effects. Optoelectronic properties of hybrid nanomaterials can be tuned by anchoring chromophores on to Au nanoparticles in this size range. Photoinduced electron transfer process from a chromophoric system to Au nanoparticle of ~2 nm was demonstrated by following transient spectroscopy. The effect of the transition behavior of metal nanoparticles was studied by Dulkeith et al by functionalizing lissamine molecules on gold nanoparticles of different sizes (1–30 nm) and isolated the resonant energy transfer rate from the decay rates of the excited dye molecules. The increase in lifetime with decrease in the nanoparticle size was indicative of the decreased efficiency of energy transfer. It is further reported that chromophore bound Au/Ag nanoparticles possess unique ability to store and shuttle electrons. A wide variety of chromophores have been functionalized on the surface of Au nanoparticles and proposed as active components in light harvesting systems. The significant photophysical events occurring in these hybrid nanomaterials are summarized in figure.

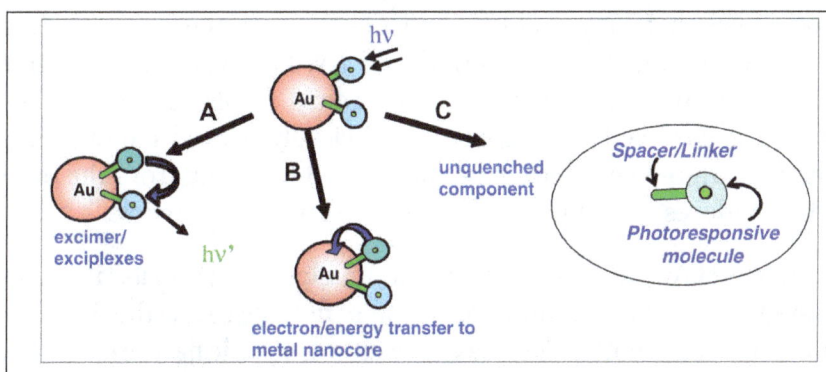

Excited-state deactivation processes in fluorophore-metal hybrid nanomaterials.

Photoactive molecules linked semiconductor QDs are widely used as hybrid nanomaterials for fluorescence resonance energy transfer (FRET) analysis. The use of QD based hybrid nanomaterials in the detection of biomolecular systems (proteins and DNA) have been summarized by Medintz and coworkers. Compared to organic fluorophores, QDs possess several unique photophysical properties, which make them attractive as biolabels: (i) broad absorption spectra and large Stokes shifts, (ii) molar extinction coefficients 10 to 100 times higher than organic dyes (iii) sizetunable photoluminescent emission with relatively high quantum yield and lifetime (iv) high resistance to photobleaching and chemical degradation. The native organic ligands used as a monolayer for protecting the QDs can be partially exchanged with a bioactive/photoactive molecule. The emission properties of QDs can be size-tuned to give better spectral overlap with a particular acceptor dye. It has been shown that the FRET efficiency can be increased by loading a central QD donor with multiple fluorophores. Unique photophysical properties of QDs opens up newer opportunities for the design of FRET-based biological assays, providing better distance resolution than traditional donor–acceptor FRET systems. However, the toxicity of QDs is a matter of concern, particularly for biomedical application, and newer generation QDs such as InP are now considered as alternatives.

Plasmon Coupling in Hybrid Nanomaterials

The transport of optical energy using materials that are considerably smaller than the wavelength of light is one of most challenging issues in the miniaturization of photonic components (due to problems associated with the diffraction limit of light). However, nanostructures can convert photons into surface plasmons that are not diffraction limited. Within the propagation length, the surface plasmon modes can be decoupled to light and this possibility offers tremendous opportunities in the design of nanoscale optical and photonic devices such as metal-nanoparticle based plasmon waveguides. Design of higher order hybrid nanomaterials (for example, one-dimensional arrays of noble metal nanoparticles with defined particle spacing) is an essential requirement for achieving this goal. Lithographic methods such as electron beam lithography are commonly used for the construction of higher order nanostructures and details are summarized in recent reviews. Maier et al have recently demonstrated the transport of electromagnetic energy over a distance of 0.5 µm in plasmon waveguides consisting of closely spaced silver rods. The waveguides were excited by the tip of a near-field scanning optical microscope and energy transport was probed by using fluorescent nanospheres.

Recent studies have shown that it is possible to fine tune the optical properties of metallic nanoparticles by their controlled organization into periodic arrays. Two types of interactions exist in organized metal nanoparticles: near-field coupling and far-field dipolar coupling. Near-field coupling (evanescent coupling) is observed in an ensemble of closely packed nanoparticles wherein they nearly touch each other. In the latter case, the dipole field resulting from the plasmon oscillation of metal nanoparticle induces an oscillation in a neighboring nanoparticle.

Closely packed 1D arrays of Au nanoparticles can, in principle, function as (i) guides of electromagnetic radiation (waveguides) allowing miniaturization of devices below the diffraction limit and (ii) interconnectors in optical and photonic devices. However, isotropic nature of spherical Au nanoparticles prevents the selective binding of molecules on surfaces which restricts the possibility of designing 1D array of nanomaterials by chemical functionalization methods. In contrast, the anisotropic features of Au nanorods allow their assembly in various orientations, and several attempts have been made for organizing Au nanorods using electrostatic/supramolecular/covalent approaches.

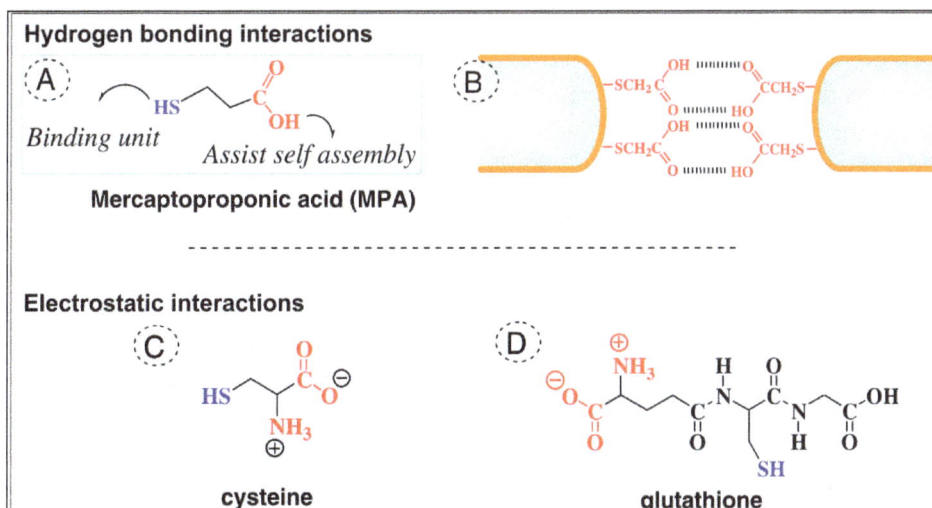

Molecules used for organizing Au nanorods using
hydrogen bonding and electrostatic approaches.

This includes the (i) linear organization of Au nanorods using biotin-streptavidin connectors and lateral organization through electrostatic interactions by varying the pH of the medium, (ii) longitudinal assembly through covalent functionalization by using dithiols, (iii) cooperative intermolecular hydrogen bonding by using 3-mercaptopropionic acid and electrostatic interaction by using cysteine and glutathione as shown in figure and (iv) end-to-end electrostatic assembly of Au nanorods on multiwall carbon nanotubes.

anotubes. On the basis of electron diffraction analysis and HRTEM studies, it is proposed that the end facets of Au nanorods are dominated by {111} planes and the side facets by {100} and {110} planes. It is reported that the thiol derivatives preferentially bind to the {111} planes of the Au nanorods and this specific interaction was further exploited for the organization of Au nanorods. The preferential functionalization at the edges of Au nanorods leads to the formation of 1D nanochains in the longitudinal direction. Based on detailed mechanistic investigations, it is concluded that the nanochain formation proceeds through an incubation step, followed by the dimerization and subsequent oligomerization of nanorods in a preferential end-to-end fashion. The clear isosbestic point observed in the time dependent absorption spectrum and dimers observed in the TEM micrographs confirms the involvement of the dimerization step in the chain formation process. Spectral changes were analyzed for a second-order kinetic process, and linearity in the initial period further supports the dimerization mechanism, which deviates with time due to the contribution of other complex processes (oligomerization).

More recently, studies are focused on the design of Au nanorod dimers and investigation of plasmon coupling in these systems as a function of their distance and orientation. The plasmon coupling in Au nanorod dimers linked through rigid molecules (e.g., 1,2-phenylenedimethanethiol) were found to be more pronounced due to effective dipolar overlap along their longitudinal axis compared to flexible aliphatic dithiol such a 1,6 hexanedithiol. These studies confirm that the nature of the linker group plays a critical role in plasmon coupling of metal nanoparticles. Such hybrid metal nanoparticles are promising components in the design of nanoscale devices, for example stable Au nanorod chains with effective dipolar overlapping can be used as elements in the design of plasmonic wave guides. Thus light waves can be channeled through hybrid nanomaterials depending on the assembly of Au nanorods. Rigid Au nanorod dimers are also promising as nanoscale interconnectors in future molecular electronics devices and the distance and orientation between the nanorods can be varied by choosing proper linker groups. A possible strategy for the design nanoscale molecular electronic devices using Au nanorod dimers is presented in figure.

We have recently developed a novel methodology for the preferential end functionalization of Au nanorods with nanoparticles by exploiting the electrostatic attractive interactions. The enhanced potential at the edges of Au nanorods preferentially attracts the positively charged Au nanoparticles, leading to their selective binding. Site specific binding results in a spontaneous bathochromic shift in the longitudinal plasmon band of Au nanorods which is dependent on the size of the nanoparticles. This concept can be further utilized for coating dissimilar metals/metal oxides on to the edges of nanorods and also for alloying specific domains of nanorods with other metals. Such hybrid materials are useful in surface enhanced spectroscopic studies. Another possibility is to selectively position molecules to specific domains of nanomaterials having enhanced electric field. This can lead to a large enhancement of various spectroscopic signals, finding applications in techniques like surface enhanced Raman spectroscopy (SERS) leading to single-molecule detection. Design of several 'metamaterials' with programmable physical and chemical properties have been reported by Shevchenko et al through the self-assembly of nanoparticles of two different materials into a binary nanoparticle superlattice.

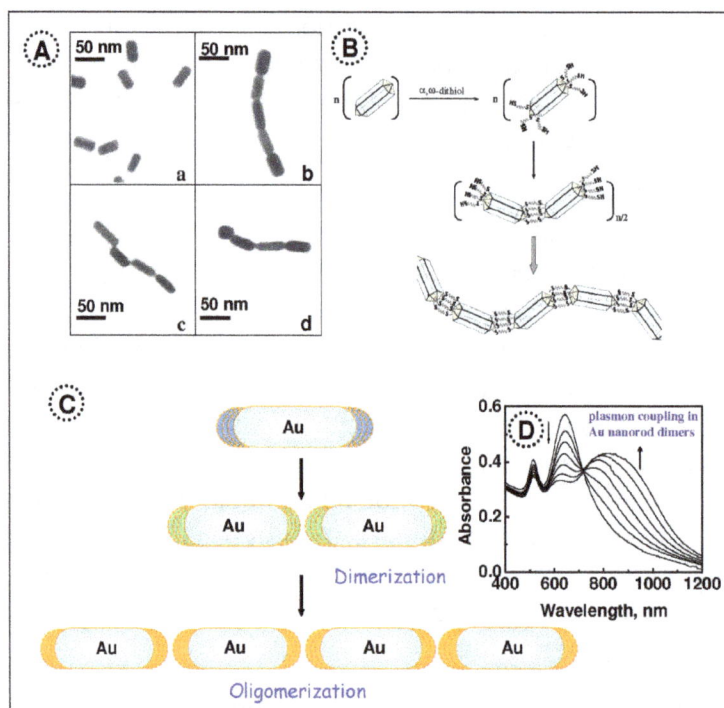

(A) TEM images of Au nanorods in the absence (a) and presence (b–d) of mercaptoproponic acid (MPA) illustrating the end to end assembly; (B) generalized scheme indicating the stepwise formation of nanochains through incubation, dimerization and oligomerization; (C) schematic representation of plasmon coupling in dimers and oligomers of Au nanorod and (D) Absorption spectra of gold nanorods in acetonitrile-water (4 : 1) recorded after the addition of MPA (0–8 μM). Decrease in the absorption of the longitudinal plasmon band, accompanied by the formation of coupled plasmon band of Au nanorod dimers, was observed, through an isosbestic point.

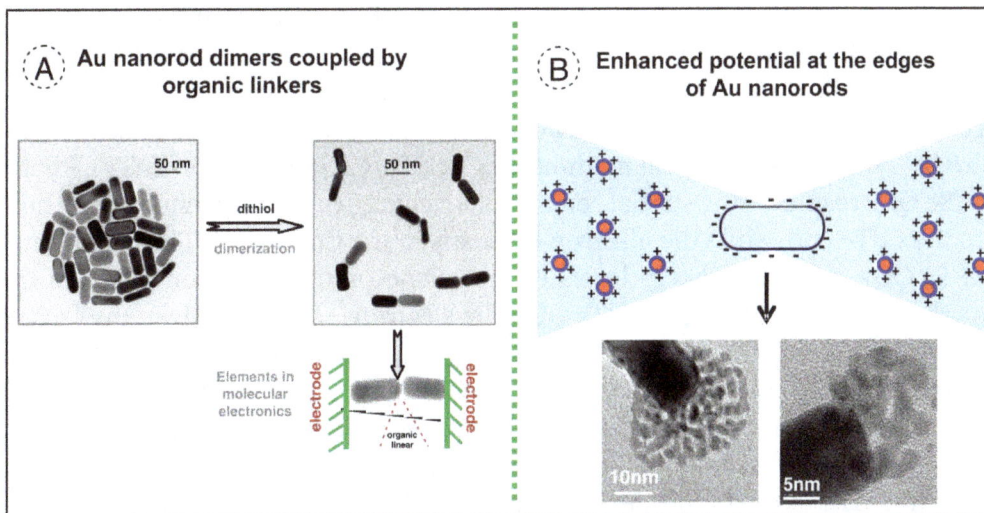

(A) Au dimers as elements in molecular electronics and
(B) enhanced potential at the edges of Au nanorods.

Heterojunctions in Hybrid Nanomaterials

Hybrid nanomaterials possessing heterojunctions are promising functional nanomaterials having potential application in optoelectronic devices. They possess novel properties that are fundamentally different from those of nanoscale components. Heterojunctions in hybrid nanomaterials can be obtained by bringing nanomaterials of (i) dissimilar metals and (ii) dissimilar semiconductors and (iii) semiconductor and metal.

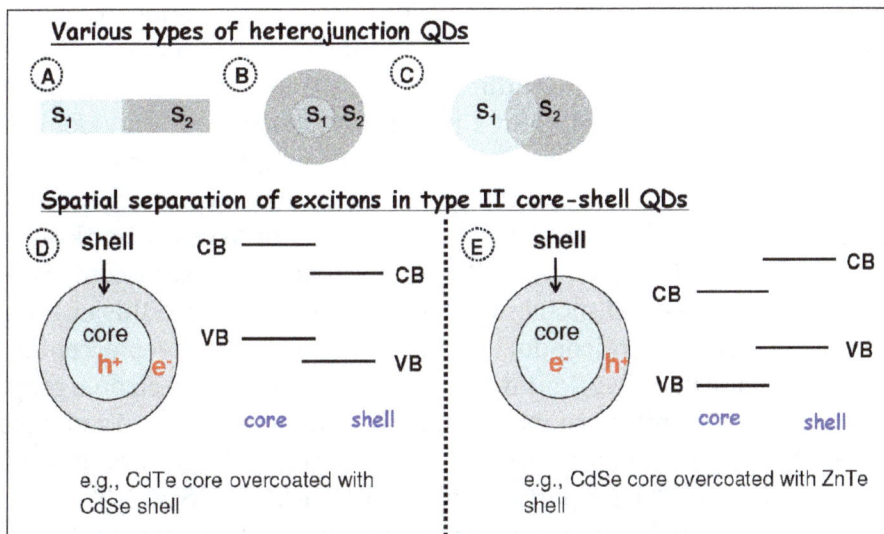

(A–C) Various types of heterojunction hybrid nanomaterials of dissimilar semiconductors and (D–E) schematic representation of the spatial separation of excitons in type II core-shell QDs and the corresponding energy levels of the valence and conduction bands of the core and shell.

Heterojunctions of dissimilar metal nanoparticles can form alloys having novel optical and electronic properties and these aspects are reviewed. Interestingly when two dissimilar semiconductor nanomaterials are in contact, the heterojunction created can assist the spatial separation of charge carriers (excitons) formed upon photoexcitation. This can be achieved by tuning the energy levels of the valence and conduction bands of semiconductors. If the valence and conduction band position of the semiconductor 'S$_1$' is lower (or higher) than semiconductor 'S$_2$', then one carrier is mostly confined in 'S$_1$', while the other one in 'S$_2$'. Core-shell nanostructures of this type are called type-II nanostructures. Such coreshell heterostructures can be synthesized by molecular beam epitaxy (e.g., GaSb/GaAs) as well as chemical methods (e.g., CdTe/CdSe and CdSe/ZnTe). Photoexcitation of CdTe/CdSe QDs results in the spatial separation of charge carriers (excitons); the hole is mostly confined to the CdTe core while the electron is mostly in the CdSe shell. This situation is reversed in CdSe/ZnTe QDs, since the energy levels of the valence and conduction bands of the shell are higher than that of the core. In CdSe/ZnTe QDs, the electron resides mostly in the CdSe core, while the hole is mostly in the ZnTe shell. Emission in type-II nanostructures originates from the radiative recombination of the electron-hole pair across the core-shell interface. Type-II QDs emit at energies that are smaller than the band gap of either material; the photoemission from CdTe/CdSe and CdSe/ZnTe QDs is observed at longer wavelengths than from the corresponding core and shell components. The emission properties of type-II QDs can be fine-tuned by changing the shell thickness and core size. For example, the emission spectra from CdTe/CdSe QDs can be tuned in the range from 700 nm to over 1000 nm by changing the core size and shell thickness. Type-II QDs are expected to have longer exciton decay times due to the spatial separation of charges; for e.g., the mean decay lifetimes of CdTe/CdSe heterojunction QDs is found to be much larger (57 ns) compared to CdTe QDs (9.6 ns). Heterojunction hybrid nanomaterials of dissimilar semiconductors are proposed as promising materials for photovoltaic applications due to the tunability in the emission band and longer lifetimes.

Bulk metals and semiconductors possess different electrochemical potentials, hence charge redistribution occurs at the contact junction so that the potentials are equilibrated (generally represented as band bending). When metals are doped on n-type semiconductor, charge transfer occurs to the metal resulting in the depletion of electrons at the semiconductor interface. For metal nanoparticle-semiconductor junction, it is difficult to apply traditional Schottky junction behavior due to issues related to depletion length and a fully accepted theoretical model is not yet developed. One of the theoretical models propose that the metal nanoparticles may create an interfacial 'Schottky-type' potential barriers on semiconducting substrates. However, recent experimental studies indicate that the Au nanoparticles form size-dependent 'nano-Schottky' potential barriers on semiconducting substrates that asymptotically approach the macroscopic Schottky barrier. Hence the semiconductor supported metal nanoparticle will experience this effect and create localized depletion regions on the semiconductor wall which act as the deep acceptor states. Other theoretical and experimental investigations dealing with the modified electronic properties of metal nanoparticle-semiconductor interface include metal nanoparticle-TiO$_2$, Ag nanoparticle-SbO$_2$, Mn nanoparticle-GaN nanowires and noble metal nanoparticleSWCNT systems. Methodologies for incorporating Ag, Au and Pt nanoparticles on to the surface of SWCNT and their potential applications as sensors and field emission transistors have also been demonstrated. It has been recently reported that the electron donation to the nanoparticle decorated SWCNT network of nanotube field emission transistor (NTFET), upon exposure to NO gas, is dependent on the work function of the metal. Based on these studies, it is concluded that the Schottky type potential barrier existing at the nanoparticle-SWCNT interface is intimately related to the work function of the metal. While SWCNT-metal nanoparticle systems are proposed for sensing applications, the possibility of utilizing these materials as components of light energy conversion systems has not been actively pursued.

Understanding Metals

Metals are substances which are good conductors of electricity and heat. They possess properties such as malleability, ductility, and lustre. This chapter delves into the aspects of thermal conductivity of metals, their physical and chemical properties, uses of metals and non-metals, and their advantages and disadvantages to provide in-depth knowledge of the subject.

All the things around us are made of 100 or so elements. These elements were classified by Lavoisier in to metals and non-metals by studying their properties. The metals and non-metals differ in their properties.

☐ *Main Group*
Al, Ga, In, Sn, Tl, Pb, Bi, Po.

☐ *alkali elements*
Li, Na, K, Rb, Cs, Fr

☐ *alkaline earth elements*
Be, Mg, Ca, Sr, Ba, Ra

General Physical Properties of the Metals

- The metals have a shiny appearance, they show a metallic luster. Due to their shiny appearance they can be used in jewellery and decorations. Particularly gold and silver are widely used for jewellery. In the old days, mirrors were made of shiny metals like silver. Silver is a very good reflector. It reflects about 90% of the light falling on it. All modern mirrors contain a thin coating of metals.

- Metals are mostly harder to cut. Their hardness varies from one metal to another. Some metals like sodium, potassium and magnesium are easy to cut.

- Metals on being hammered can be beaten into thinner sheets. This property is called Malleability. Most metals are malleable. Gold and Silver metals are the most malleable metals. They can be hammered into very fine sheets. Thin aluminium foils are widely used for safe wrapping of medicines, chocolates and food materials.

- Wires are made from copper, aluminium, iron and magnesium. This property of drawing the metal in to thin wires is called ductility. Most metals are ductile.

- Electric wires in our homes are made of aluminium and copper. They are good conductor of electricity. Electricity flows most easily through gold, silver, copper and aluminium. Gold and silver are used for fne electrical contacts in computers. Copper wires are used in electrical appliances while aluminium is cheaper is generally used for making electrical cables.

- Cooking utensils and water boilers are also made of iron, copper and aluminium, because they are good conductors of heat.

- Metals are general sonorous. That is they make a ringing sound when struck. Therefore, they are used for making bells. Metal wires are used in musical instruments.

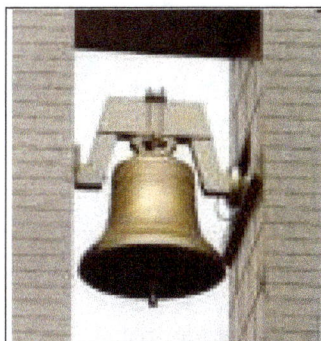

- Metals such as iron are very strong. Therefore, it is therefore, widely used in the construction of buildings, bridges, railway lines, carriages, vehicles and machinery.

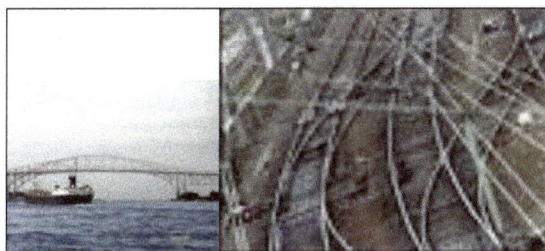

- All metals except Mercury, exist in the solid form at room temperature. Therefore, they retain their shapes under normal conditions.

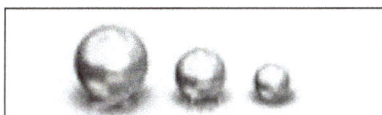

- Metals have high melting points.

- Metals have high tensile strength that is they can be stretched to some degree without breaking. Metals like tungsten have high tensile strength.

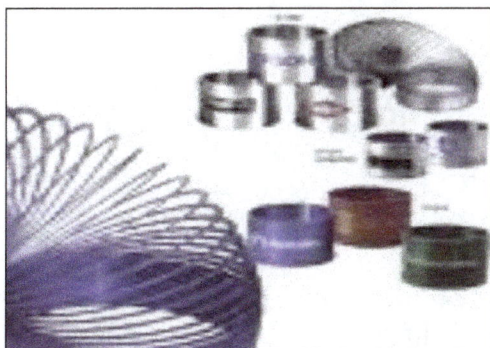

No two metals are absolutely identical. For example:

- Iron is magnetic and copper is not.

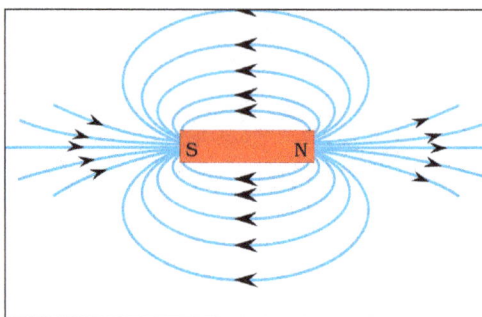

- Gold and platinum are malleable and ductile but do not react with water.

- Sodium is highly reactive and reacts vigorously with water to form a solution of sodium hydroxide.

General Chemical Properties

- Metals when burned in the presence of oxygen, they combine with oxygen to form metallic oxides which are basic in nature.

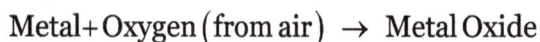

$$Metal + Oxygen\,(from\,air) \;\rightarrow\; Metal\,Oxide$$

For example:

$$2\,Mg + O_2 \;\rightarrow\; 2\,MgO\,(Magnesium\,Oxide)$$
$$MgO + H_2O \;\rightarrow Mg(OH)_2\,(Magnesium\,Hydroxide)$$

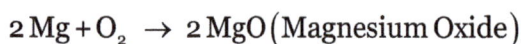

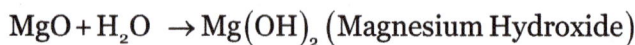

Metal hydroxide changes red litmus blue which shows its basic characteristics.

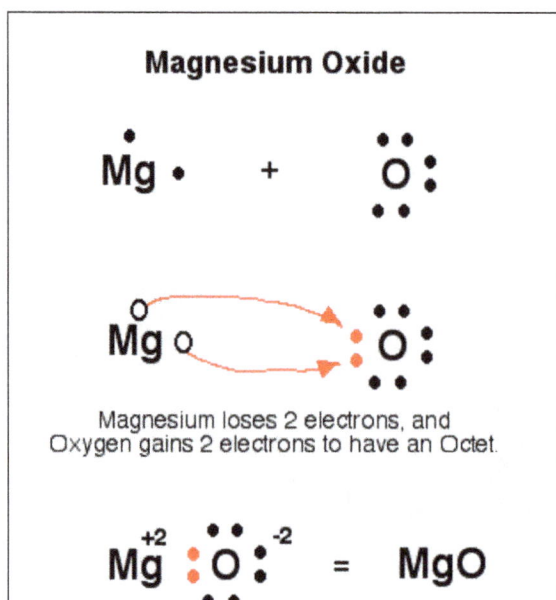

- Different metals react differently with water. Sodium reacts violently with water forming sodium hydroxide and hydrogran. Magnesium reacts mildly with water but vigorously with steam. Zinc and iron react mildly with steam. Copper, gold and silver do not react with water at all. Most metals, on reacting with water produce hydroxide.

$$2\,Na + H_2O \rightarrow 2\,NaOH + H_2$$

- Metals differ in their reactivity with acids. Most metals react with acids to produce salts and hydrogen.

Metal + acid $\rightarrow$ Salt + Hydrogen

$$Mg + 2\,HCl \qquad \rightarrow MgCl_2 + H_2$$

- Metals replace other metals. When an iron nail is placed in a test tube containing copper sulfate. The nail is coated with a layer of copper while the blue copper sulfate solution has turned greenish. The green solution is a solution of iron sulphate.

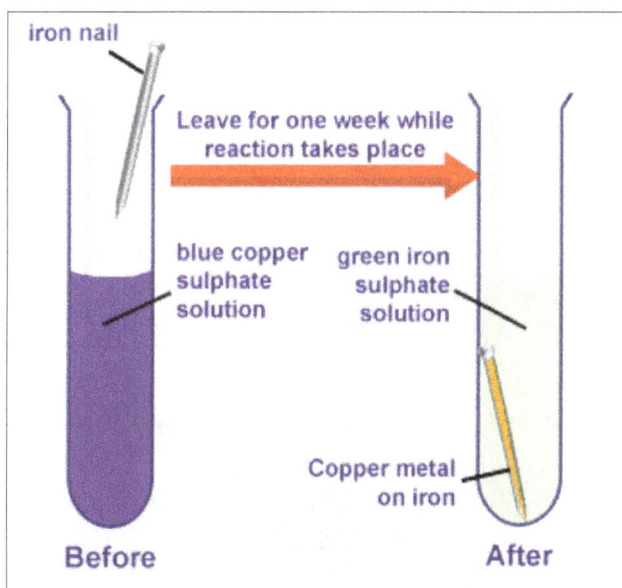

$$Fe + CuSO_4 \rightarrow Cu + FeSO_4$$

- Most metals corrode when they are exposed to atmosphere. For example, the iron gets rusty after sometime if it is not painted. Titanium is highly resistant to corrosion

Metallic Properties

In a metal, atoms readily lose electrons to form positive ions (cations). These ions are surrounded by delocalized electrons, which are responsible for conductivity. The solid produced is held together by electrostatic interactions between the ions and the electron cloud. These interactions are called metallic bonds. Metallic bonding accounts for many physical properties of metals, such as strength, malleability, ductility, thermal and electrical conductivity, opacity, and luster.

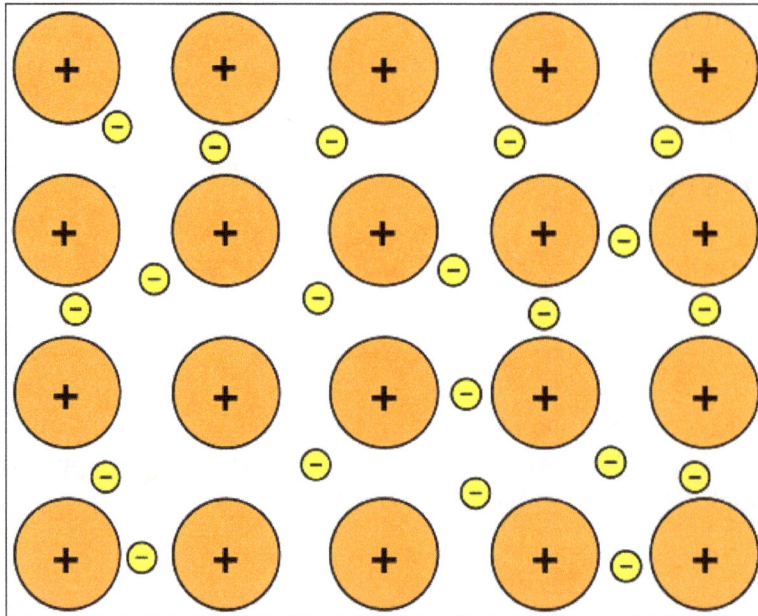

Metallic Bonding Loosely bound and mobile electrons surround the positive nuclei of metal atoms.

Understood as the sharing of "free" electrons among a lattice of positively charged ions (cations), metallic bonding is sometimes compared to the bonding of molten salts; however, this simplistic view holds true for very few metals. In a quantum-mechanical view, the conducting electrons spread their density equally over all atoms that function as neutral (non-charged) entities.

Atoms in metals are arranged like closely-packed spheres, and two packing patterns are particularly common: body-centered cubic, wherein each metal is surrounded by eight equivalent metals, and face-centered cubic, in which the metals are surrounded by six neighboring atoms. Several metals adopt both structures, depending on the temperature.

Metals in general have high electrical conductivity, high thermal conductivity, and high density. They typically are deformable (malleable) under stress, without cleaving. Some metals (the alkali and alkaline earth metals) have low density, low hardness, and low melting points. In terms of optical properties, metals are opaque, shiny, and lustrous.

Melting Point and Strength

The strength of a metal derives from the electrostatic attraction between the lattice of positive ions and the "sea" of valence electrons in which they are immersed. The larger the nuclear charge (atomic number) of the atomic nucleus, and the smaller the atom's size, the greater this attraction. In general, the transition metals with their valence-level d electrons are stronger and have higher melting points:

- Fe, 1539 °C

- Re, 3180 °C

- Os, 2727 °C

- W, 3380 °C.

The majority of metals have higher densities than the majority of nonmetals. Nonetheless, there is wide variation in the densities of metals. Lithium (Li) is the least dense solid element, and osmium (Os) is the densest. The metals of groups IA and IIA are referred to as the light metals because they are exceptions to this generalization. The high density of most metals is due to the tightly packed crystal lattice of the metallic structure.

Electrical Conductivity: Why are Metals Good Conductors?

In order for a substance to conduct electricity, it must contain charged particles (charge carriers) that are sufficiently mobile to move in response to an applied electric field. In the case of ionic compounds in water solutions, the ions themselves serve this function. The same thing holds true of ionic compounds when melted. Ionic solids contain the same charge carriers, but because they are fixed in place, these solids are insulators.

In metals, the charge carriers are the electrons, and because they move freely through the lattice, metals are highly conductive. The very low mass and inertia of the electrons allows them to conduct high-frequency alternating currents, something that electrolytic solutions cannot do.

Electrical conductivity, as well as the electrons' contribution to the heat capacity and heat conductivity of metals, can be calculated from the free electron model, which does not take the detailed structure of the ion lattice into account.

Mechanical Properties

Mechanical properties of metals include malleability and ductility, meaning the capacity for plastic deformation. Reversible elastic deformation in metals can be described by Hooke's Law for restoring forces, in which the stress is linearly proportional to the strain. Applied heat, or forces larger than the elastic limit, may cause an irreversible deformation of the object, known as plastic deformation or plasticity.

Metallic solids are known and valued for these qualities, which derive from the non-directional nature of the attractions between the atomic nuclei and the sea of electrons. The bonding within ionic or covalent solids may be stronger, but it is also directional, making these solids brittle and subject to fracture when struck with a hammer, for example. A metal, by contrast, is more likely to be simply deformed or dented.

Gold is a noble metal; it is resistant to corrosion and oxidation.

Although metals are black due to their ability to absorb all wavelengths equally, gold (Au) has a distinctive color. According to the theory of special relativity, increased mass of inner-shell electrons that have very high momentum causes orbitals to contract. Because outer electrons are less affected, blue-light absorption is increased, resulting in enhanced reflection of yellow and red light.

Stainless Steels

Nickel, molybdenum, niobium, and chromium enhance the corrosion resistance of stainless steel. It is the addition of a minimum of 12% chromium to the steel that makes it resist rust, or stain 'less' than other types of steel. The chromium in the steel combines with oxygen in the atmosphere to form a thin, invisible layer of chrome-containing oxide, called the passive film. The sizes of chromium atoms and their oxides are similar, so they pack neatly together on the surface of the metal, forming a stable layer only a few atoms thick. If the metal is cut or scratched and the passive film is disrupted, more oxide will quickly form and recover the exposed surface, protecting it from oxidative corrosion.

Iron, on the other hand, rusts quickly because atomic iron is much smaller than its oxide, so the oxide forms a loose rather than tightly-packed layer and flakes away. The passive film requires oxygen to self-repair, so stainless steels have poor corrosion resistance in low-oxygen and poor circulation environments. In seawater, chlorides from the salt will attack and destroy the passive film more quickly than it can be repaired in a low oxygen environment.

Types of Stainless Steel

The three main types of stainless steels are austenitic, ferritic, and martensitic. These three types of steels are identified by their microstructure or predominant crystal phase.

- Austenitic: Austenitic steels have austenite as their primary phase (face-centered cubic crystal). These are alloys containing chromium and nickel (sometimes manganese and nitrogen), structured around the Type 302 composition of iron, 18% chromium, and 8% nickel. Austenitic steels are not hardenable by heat treatment. The most familiar stainless steel is probably Type 304, sometimes called T304 or simply 304. Type 304 surgical stainless steel is austenitic steel containing 18-20% chromium and 8-10% nickel.

- Ferritic: Ferritic steels have ferrite (body-centered cubic crystal) as their main phase. These steels contain iron and chromium, based on the Type 430 composition of 17% chromium. Ferritic steel is less ductile than austenitic steel and is not hardenable by heat treatment.

- Martensitic: The characteristic orthorhombic martensite microstructure was first observed by German microscopist Adolf Martens around 1890. Martensitic steels are low carbon steels built around the Type 410 composition of iron, 12% chromium, and 0.12% carbon. They may be tempered and hardened. Martensite gives steel great hardness, but it also reduces its toughness and makes it brittle, so few steels are fully hardened.

There are also other grades of stainless steels, such as precipitation-hardened, duplex, and cast stainless steels. Stainless steel can be produced in a variety of finishes and textures and can be tinted over a broad spectrum of colors.

Passivation

There is some dispute over whether the corrosion resistance of stainless steel can be enhanced by the process of passivation. Essentially, passivation is the removal of free iron from the surface of the steel. This is performed by immersing the steel in an oxidant, such as nitric acid or citric acid solution. Since the top layer of iron is removed, passivation diminishes surface discoloration.

While passivation does not affect the thickness or effectiveness of the passive layer, it is useful in producing a clean surface for further treatment, such as plating or painting. On the other hand, if the oxidant is incompletely removed from the steel, as sometimes happens in pieces with tight joints or corners, then crevice corrosion may result. Most research indicates that diminishing surface particle corrosion does not reduce susceptibility to pitting corrosion.

Magnetism

The magnetic properties of materials were recognized by the ancient Greeks, Romans, and Chinese, who were familiar with lodestone, an iron oxide mineral that attracts iron objects. Although the attractive or repulsive forces that act between magnetic materials are manifestations of magnetism familiar to everybody, the origin of magnetism lies in the atomic structure of matter. Despite the fact that magnetism can be explained only by the quantum theory developed at the beginning of the twentieth century, qualitative predictions of magnetic properties can be made within the context of classical physics. Magnetic forces originate in the motion of charged particles, such as electrons. The electrons "spin" around their axis and move in orbits around the nucleus of the atom to which they belong. Both motions generate tiny electric currents in closed loops that in turn create magnetic dipole fields, just as the current in a coil does. When placed in a magnetic field, the tiny magnetic dipole fields tend to align with the external field.

According to their behavior in inhomogeneous magnetic fields, materials can be classified into three main categories: diamagnetic, paramagnetic, and ferromagnetic. Paramagnetic materials are attracted into a magnetic field. The main cause of this effect is the presence in the material of atoms that have a net magnetic moment composed of electron spin and orbital contributions.

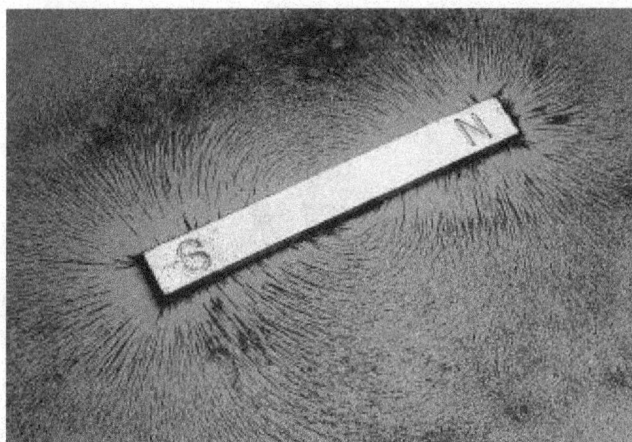

Iron filings in a circular pattern around a magnet, indicative of the field of force of the magnet.

When placed in a magnetic field, the magnetic moments of the atoms, which are otherwise randomly oriented, tend to align with the field and thus enhance the field. Paramagnetism is temperature dependent because increased thermal motion at higher temperatures impedes the alignment of the magnetic moments with the field. *Diamagnetic materials* are slightly repelled by a magnetic field. This effect occurs for materials that contain atoms in which the spin and orbital contributions to the magnetic moment cancel out. In this case, the interaction between the material and a magnetic field is caused by the occurrence of currents induced by the magnetic field in the atoms. The dipole fields corresponding to these currents are directed opposite to the applied magnetic field and cause expulsion of the material from the field. *Ferromagnetic materials* contain atoms that have magnetic moments that are aligned even in the absence of an applied magnetic field because of mutual interactions, creating a sizable net magnetic moment for domains of the material. The magnetic moments of domains can be randomly oriented unless a magnetic field is applied to the material.

Iron, cobalt, nickel, and their alloys are examples of ferromagnetic materials. These three elements are transition metals, and their atoms or ions have unpaired electrons in d orbitals. Rare-earth ions also have unpaired electrons situated in f orbitals. A detailed investigation of the properties of molecules that contain such metal ions in a magnetic field can provide significant information about how their electrons are distributed in orbitals. Typically, d orbitals of isolated atoms are degenerate. This situation changes when the metal ions are part of molecules in which they experience a nonspherically symmetric environment. Figures show the splitting of d orbitals for a transition metal ion that has six unpaired electrons and is situated in an environment of six atoms in an octahedral arrangement. Depending on the size of the splitting (the lighter shading in figure) and the interelectron repulsion, the metal ion may have four unpaired electrons or no unpaired electrons. This difference in electron distribution leads to significant differences in the magnetic properties of the molecules that contain such ions, with the former being paramagnetic and the latter being diamagnetic. When there are multiple metal sites in a molecule, the spins at different metal ions can be either ferro-(parallel) or antiferromagnetically (antiparallel) aligned to each other. Clever use of the magnetic properties for metal ions and of the interactions between spins manifested in molecular systems enables scientists to design and synthesize molecular systems with interesting properties, such as molecular magnets.

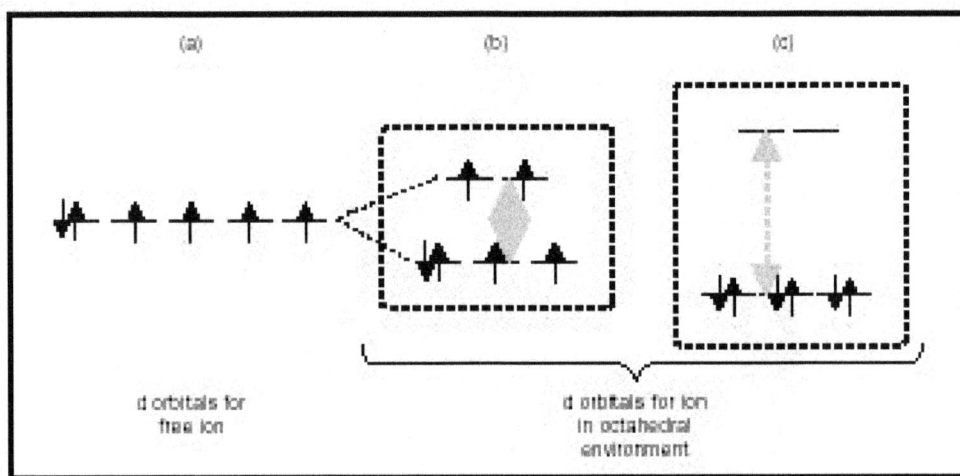

Splitting of d orbitals for a transition metal ion.

Magnetic materials are widely used for building technological devices and scientific tools. Classical examples are electromagnets that are used in motors, clutches, and breaking systems. The electromagnet makes use of an iron core situated in a solenoid through which electric current is passed. This current creates a magnetic field at the center of the solenoid that orients the magnetic moments in the domains of the iron core, which in turn results in a significant enhancement of the magnetic field at the core of the solenoid. Electromagnets can also be used to record information on magnetic tape, which has a ferromagnetic surface.

Finally, although atomic nuclei have significantly smaller magnetic moments than electrons, the study of their interaction with magnetic fields has many important applications. They enable the scientists in the biological and medical fields to elucidate the structure of biologically relevant molecules such as proteins and to diagnose diseases using magnetic resonance imaging.

Thermal Conductivity of Metals

Thermal Conductivity is a term analogous to electrical conductivity with a difference that it concerns with the flow of heat unlike current in the case of the latter. It points to the ability of a material to transport heat from one point to another without movement of the material as a whole, the more is the thermal conductivity the better it conducts the heat.

Let us consider a block of material with one end at temperature T_1 and other at T_2. For $T_1 > T_2$, heat

flows from T_1 end to T_2 end, and the heat flux(J) flowing across a unit area per unit time is given as:

$$\frac{\Delta Q}{A\Delta t} = J = -K\frac{dT}{dx}$$

Where, K is the thermal conductivity in Joule/meter-sec-K or Watts/meter-K.

Generally the heat transfer in solid has two components:

- Lattice conduction,
- Electronic conduction.

Both types of heat conduction occur in solids but one is dominant over the other depending upon the type of material.

In case of insulating materials, lattice conduction contributes to heat conduction. This is mainly due to the fact that in insulators the electrons are tightly held by their parent atoms and free electrons do not exist. Hence the heat is transferred from one end to another by vibration of atoms held in the lattice structure. Obviously insulators are bad heat conductors since they do not posses enough heat transfer capability due to lack of free electrons.

However in case of metals we have large number of free electrons, and hence heat conduction is primarily due to electronic conduction. The free electrons of the metals can freely move throughout the solid and transfer the thermal energy at a rate very high as compared to insulators. It is due to this that the metals posses high thermal conductivity. It is also observed that among the metals the best electrical conductors also exhibits best thermal conductivity. Since both electrical as well as thermal conductivity is dependent on the free electrons, factors such as alloying affect both the properties. Thermal conductivity of metals vary from 15 – 450 W/mK at 300K.

Wiedemann Franz Law

Wiedemann Franz law basically relates the two conductivities of metals, i.e. thermal and electrical conductivity with temperature. It states that the ratio of thermal conductivity K and electrical conductivity σ is proportional to the temperature of the specimen. G. Wiedemann and R. Franz in 1853 established on the basis of experimental data that the ratio:

$$\frac{K}{\sigma},$$

is constant at constant temperature.

In 1882 the Danish physicist L. Lorenz demonstrated that the relation,

$$\frac{K}{\sigma}$$

changes in direct proportion to the absolute temperature T.

$$\frac{K}{\sigma} = LT,$$

where, T = temperature.

$$L = 2.54 \times 10^{-8} W\Omega / K^2, \text{Lorentz number} \left(a \text{ constant} \right)$$

This law basically states that with increase in temperature the thermal conductivity of metals increases while the electrical conductivity decreases. We know that the two properties of metals are dependent on the free electrons. An increase in temperature increases the average velocity of the free electrons leading to increase in heat energy transfer. On the other hand increase in the velocity of electrons also increases the number of collisions of the free electrons with lattice ions, and hence contributes to increase in electrical resistivity or reduction in electrical conductivity. However this law has certain limitations. The proportionality does not holds true for all ranges of temperature. It is only found valid for very high temperatures and very low temperatures. Also certain metals such as beryllium, pure silver etc. do not follow this law.

Chemical Properties and uses of Metals and Non-metals

The materials that are present around us are grouped widely into metals and non-metals. They offer various uses in everyday life. Metal and Non-metals have some very interesting and useful chemical properties.

Chemical Properties of Metals

The chemical properties of metals are:

Oxidation

All metals except the noble metals, which is gold and silver react with the oxygen to form basic oxides. In the reaction, the Sodium reacts vigorously with oxygen and generates a lot of heat. For example:

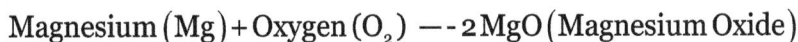

$$Magnesium\,(Mg) + Oxygen\,(O_2)\ --\text{-}\ 2\,MgO\,(Magnesium\,Oxide)$$

Rusting of Iron

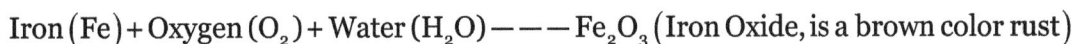

$$Iron\,(Fe) + Oxygen\,(O_2) + Water\,(H_2O) ---\!- Fe_2O_3\,(Iron\,Oxide, is\ a\ brown\ color\ rust)$$

Greenish Deposit on the Surface of Copper Vessels

The dull greenish material that is deposited on the surface of copper is a mixture of copper hydro oxide $[Cu(OH)_2]$ and copper carbonate $(CUCO_3)$ which takes place as:

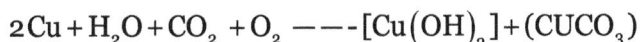

$$2\,Cu + H_2O + CO_2 + O_2\ --\text{-}\text{-}[Cu(OH)_2] + (CUCO_3)$$

Reaction of Metals with Water

The metallic oxides are basic in nature. Metals like sodium (Na) react vigorously with water at room temperature.

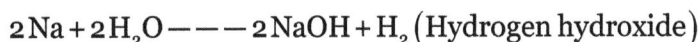

$$2\,Na + 2\,H_2O ---\!- 2\,NaOH + H_2\,(Hydrogen\ hydroxide)$$

The active metals like Potassium (K) and Calcium (Ca) reacts with water at room temperature. However, some metals do not react. For example – Iron reacts with water slowly.

Reaction with Acids

Acids react with metals to liberate hydrogen along with the corresponding salt of the metal:

$$Zinc + Hydrochloric\ acid\ --\ Zinc\ chloride + Hydrogen$$
$$Or,\ Zn + 2\,HCl\ --\ ZnCl_2 + H_2$$
$$Zinc + Sulphuric\ acid\ --\ Zinc\ sulfate + Hydrogen$$
$$Or,\ Zn + H_2SO_4\ --\ ZnSO_4 + H_2$$

When a burning matchstick is brought near Hydrogen, it burns with a pop sound.

Reaction with Bases

Metals react with sodium hydroxide to produce hydrogen.

Displacement Reactions

Some metals are capable of displacing other metals from their solutions. For example – Zinc (Zn) replaces copper from copper sulfate solutions.

Copper Sulphate + Zinc ——- Zinc Sulphate + Copper

$CuSO_4 + Zn —— ZnSO_4 + Cu$

More active metals displace less active metals from their solution, in general. In this case, Zinc is more reactive than Cu, so it replaces copper (Cu) from a copper sulfate solution. Implying the rule, a more reactive metal replaces a less reactive metal, however, a less reactive metal cannot replace a reactive metal.

Chemical Properties of Non – Metals

- Oxidation: Non-metals react with oxygen to form oxides which are acidic in nature.

- Reaction of Non-metals with Water: Non-metals do not react with waterthough they may be very reactive in the air. However, some non-metals react with air. For example – Phosphorous.

Uses of Metal

- We use metals are for various purposes like for making wires and sheets. For example – Copper and aluminum wires in electrical equipment's, especially for conduction of electricity.

- They are also extensively used in making automobiles, machinery, industrial gadgets, etc.

- The metals are used in making utensils and water boilers due to its property of being a good conductor of heat.

Uses of Non – Metal

- Non-metals like nitrogen and phosphorus are very useful in fertilizers for better plant growth.

- Non-metal like Sulphur is useful in crackers.

- We use a non-metal like chlorine in purple colored solution on the wound as the antiseptic.

- Many non-metals like chlorine, sulphur, iodine are very useful for medicinal purposes.

- Non-metal like oxygen is very essential for our life for respiration.

Advantages and Disadvantages

Aluminium

Advantages: The most abundant metal in the Earth's crust, Aluminium is relatively soft, durable, lightweight, ductile and malleable metal. It acts as a good thermal and electrical conductor and is also fairly corrosion resistant. Plus, it is theoretically 100% recyclable without any loss of its natural qualities and remarkably nontoxic.

Disadvantages: It's not particularly strong and is expensive compared to steel of the same strength.

Applications: Aluminium is almost always alloyed to improve its properties. It is commonly used in the transportation, construction and packaging industries.

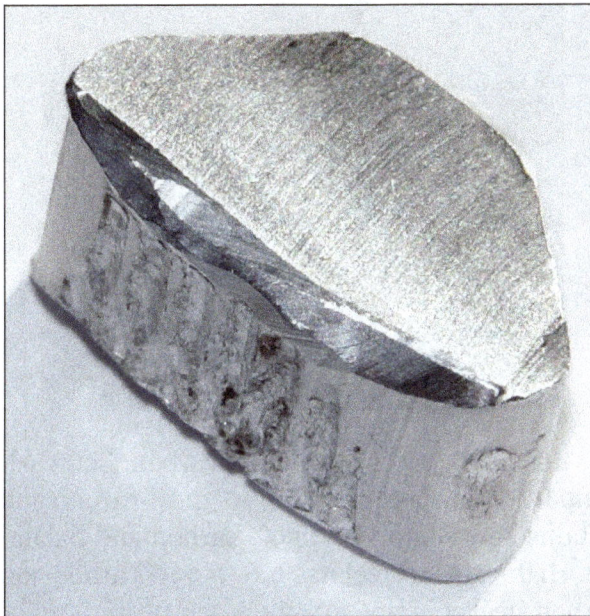

Bronze

Advantages: Bronze is an alloy consisting mainly of copper but the addition of other metals (usually tin) produces an alloy much harder than plain copper. Bronze resists corrosion and metal fatigue better, and conducts heat and electricity, better than most steels.

Disadvantages: Bronzes are generally softer, weaker and more expensive than steel.

Applications: Bronze is widely used for springs, bearings, bushings, automobile transmission pilot bearings and is particularly common in the bearings of small electric motors.

Carbon Steel

Advantages: Carbon steel's main alloying constituent is carbon. Low carbon steel is the most common and cost effective form. It contains around 0.05–0.320% carbon and is malleable and ductile. Medium carbon steel contains between 0.30–0.59% carbon and balances ductility and strength with good wear resistance. High-carbon steel has 0.6–0.99% carbon content and is exceptionally strong, while ultrahigh carbon steel contains 1.0–2.0% carbon and can be tempered to great hardness.

Disadvantages: Low-carbon steels suffer from yield-point runout and mild steel has a relatively low tensile strength.

Applications: Medium carbon is used for large parts, forging and automotive components. High-carbon steel is used for springs and high-strength wires. Ultra high carbon steel is used for special purposes like knives, axles or punches.

Nickel

Advantages: Nickel belongs to the transition metals. It is hard, ductile and considered corrosion-resistant because of its slow rate of oxidation at room temperature. It also boasts a high melting point and is magnetic at room temperature.

Disadvantages: Handling nickel can result in symptoms of dermatitis among sensitized individuals.

Applications: Nickel is valuable for the alloys it forms and roughly 60% of world production goes into nickel-steels. Specific uses include stainless steel, alnico magnets, coins, rechargeable batteries, electric guitar strings, microphone capsules, and special alloys. It is also used for plating and as a green tint in glass.

Titanium

Advantages: Titanium is corrosion resistant and has the highest strength-to-density ratio of any metallic element. Unalloyed it's as strong as some steels but less dense. Its relatively high melting point (more than 1,650 °C or 3,000 °F) makes it useful as a refractory metal. It is also paramagnetic and displays fairly low electrical and thermal conductivity. Disadvantages: Costly and laborious processes are needed to extract titanium from its various ores.

Applications: Titanium can be alloyed with iron, aluminium, vanadium, and molybdenum (among others) to produce strong, lightweight alloys. These are used in the aerospace, military, industrial process, automotive, agri-food, medical, and sporting industries to name but a few.

References

- Metallic-crystals, introchem-chapter: courses.lumenlearning.com, Retrieved 13 July, 2019

- Magnetism, Kr-Ma: chemistryexplained.com, Retrieved 18 June, 2019

- Thermal-conductivity-of-metals: electrical4u.com, Retrieved 27 August, 2019

- Chemical-properties-and-uses-of-metals-and-non-metals, chemistry-materials-metals-and-non-metals: toppr.com, Retrieved 18 March, 2019

- Advantages-disadvantages-metals-commonly-used-manufacturing: processindustryforum.com, Retrieved 12 February, 2019

Permissions

All chapters in this book are published with permission under the Creative Commons Attribution Share Alike License or equivalent. Every chapter published in this book has been scrutinized by our experts. Their significance has been extensively debated. The topics covered herein carry significant information for a comprehensive understanding. They may even be implemented as practical applications or may be referred to as a beginning point for further studies.

We would like to thank the editorial team for lending their expertise to make the book truly unique. They have played a crucial role in the development of this book. Without their invaluable contributions this book wouldn't have been possible. They have made vital efforts to compile up to date information on the varied aspects of this subject to make this book a valuable addition to the collection of many professionals and students.

This book was conceptualized with the vision of imparting up-to-date and integrated information in this field. To ensure the same, a matchless editorial board was set up. Every individual on the board went through rigorous rounds of assessment to prove their worth. After which they invested a large part of their time researching and compiling the most relevant data for our readers.

The editorial board has been involved in producing this book since its inception. They have spent rigorous hours researching and exploring the diverse topics which have resulted in the successful publishing of this book. They have passed on their knowledge of decades through this book. To expedite this challenging task, the publisher supported the team at every step. A small team of assistant editors was also appointed to further simplify the editing procedure and attain best results for the readers.

Apart from the editorial board, the designing team has also invested a significant amount of their time in understanding the subject and creating the most relevant covers. They scrutinized every image to scout for the most suitable representation of the subject and create an appropriate cover for the book.

The publishing team has been an ardent support to the editorial, designing and production team. Their endless efforts to recruit the best for this project, has resulted in the accomplishment of this book. They are a veteran in the field of academics and their pool of knowledge is as vast as their experience in printing. Their expertise and guidance has proved useful at every step. Their uncompromising quality standards have made this book an exceptional effort. Their encouragement from time to time has been an inspiration for everyone.

The publisher and the editorial board hope that this book will prove to be a valuable piece of knowledge for students, practitioners and scholars across the globe.

Index

www.ingramcontent.com/pod-product-compliance
Lightning Source LLC
Chambersburg PA
CBHW061318190326
41458CB00011B/3834